Geographic Information 1991

Geographic Information 1991

The Yearbook of the Association for Geographic Information

Edited by
James Cadoux-Hudson
Terraquest

and

Ian Heywood
University of Salford

Taylor & Francis
London • New York • Philadelphia

Miles Arnold

UK	Taylor & Francis Ltd., 4 John Street, London WC1N 2ET Miles Arnold, High Winds, Cassington, Oxford OX8 1DL
USA	Taylor & Francis Inc., 1900 Frost Road, Suite 101, Bristol, PA 19007

A catalogue record for this book is available from the British Library.
ISBN 0-74840-028-1

Typeset by Electronic Village Ltd., Richmond, Surrey

Printed in Great Britain by Burgess Science Press, Rankine Road, Basingstoke, Hampshire

Contents

Preface

Despite a new cover and a new title the structure of the AGI's *Yearbook* for 1991 has been altered little from previous volumes. Some minor changes have been made over the location of the directories which can now be found to the back of the book. In addition, a list of who's who in the AGI has also been incorporated.

The basic aim of the *Yearbook* is to provide a "snap shot" of what is happening in GIS/LIS in the UK for the benefit of the home and international communities. We see it as a taster and guide to what is happening; how, why and where, rather than a GIS *magnum opus*! Naturally, we have had to be selective and will inevitably have left something out. We therefore offer our apologies and ask that in future if you feel your activities are worthy of a mention please keep us informed.

Once again we are indebted to all those who have given up their time to make contributions in the form of articles, reports, reviews and so on. Behind the scenes a special thanks must go to Beverly Heyworth of Salford University Business Services Ltd., who co-ordinated the compilation of the digital manuscript; an achievement in itself.

Next year, with the move into Europe, there are already plans afoot to change the scope of the *Yearbook*. The aim, whilst maintaining UK focus, will be to enhance its appeal to a wider international audience. This in turn, we hope, will help increase the overseas profile of the AGI.

On the issue of change we would welcome any feedback on the structure, content, form and utility of the three *Yearbooks* so far. This should be sent to:

The Editors
The AGI Yearbook
The Association for Geographic Information
12 Great George Street
Parliament Square
London
SW1P 3AD

Finally, we hope that the contents of the 1991 *Yearbook* will be of value to everyone with a professional, academic or personal interest in GIS.

Ian Heywood
James Cadoux-Hudson
March 1991

PART I

The AGI in 1990/91

All enquiries regarding the AGI; matters of membership, publications, conferences, etc., should be directed to:

The Secretary,
The Association for Geographic Information,
12 Great George Street,
LONDON SW1P 3AD
Tel: 071 222 7000 Ext.226
Fax: 071 222 9430
Tlx: 915443 RICS G

1

AGI—A review of 1990

The last twelve months have been very important for GIS and the AGI. There have been many successes, all of them attributable to the active involvement of AGI members. This report details the activities of the individual committees since the last annual general meeting.

Council

The year began with its first Council election at the 1989 Annual General Meeting. The Council's objective has been to ensure that all aspects of the AGI's work are attended to by individual committees. Under the stewardship of the Council the level of activity has increased in all of the committees.

As part of its consideration of the AGI's future role the Council has drawn up a mission statement and a strategy paper for future activities.

- *To achieve the effective use of geographic information and its associated technologies.*
- *To represent the interests of the geographic information community in issues affecting or affected by geographic information systems (GIS).*

The Council has also represented the interests of the geographic information community in a number of issues. Advice has been given to the government's Tradeable Information Initiative regarding an audit of information of spatially related information held within central government departments. In the coming year the Council will be ensuring that a directory of "information on information" becomes available. Beyond that there is a need to expand the exercise to other areas, such as local government data.

Council has also been quick to point out the potential of geographic information and the danger of ignoring that potential to government. Council has voiced its approval for streetworks legislation whilst pointing out the need for recording more than the basic textural data. When proposals were made for changes to Land Registry (NI) Council was keen to ensure that these did not affect the value of LR(NI)'s geographic information. Council has also been developing links with PITCOM, the parliamentary IT committee.

The work carried out by the AGI is channelled through committees. Alongside Council sits the Management Committee which comprises the officers (Chairman, Senior Vice Chairman, Junior Vice Chairman, Honorary Treasurer and Honorary Secretary).

Management

The Management Committee has been actively involved in developing the detail in the Council's policy. It developed the business plan for the committees and has ensured that the AGI has been financially well throughout the year. More recently the committee has had opportunity to start developing the AGI's relations with the European Commission. One of the results of this is that CEC DGXIIIb2 used the 1990 AGI National Conference as the location for its opening workshop on the implications of GIS for the European information market.

Conference and future conferences

The conference committee has made a great success of the second national conference of the AGI. "AGI 90" was held in Brighton and attracted 600 conference delegates, 1300 exhibition visitors and 60 exhibitors. At the same time the future conferences committee has been working on a strategy for the next five years. This has resulted in the signing of a 5-year contract with conference organisers. AGI 91 will be held in Birmingham, November 20th—22nd.

Education, Training and Research

This committee has now run four very successful, regional seminars. The last of these focused on the AGI's Standards Committee, and attracted over 100 delegates. It has also sponsored a number of other seminars, the proceedings of which will be available in 1991.

Membership

The Membership Committee has been very active over the last year. Its aim has been to widen the membership base of the AGI and activities for members. The 1990 AGI Chairman, Mike Brand, gave a presentation about the AGI to a selected audience of major companies, after which we have had two sponsors join and have another in the process of joining. The Committee is also arranging an AGI presence at the different GIS software user groups (the first event attended was the Hoskyns "GGP" user group) and has reviewed benefits for members. As a result a site visits programme has been created. This will allow the membership access to a variety of organisations active in GIS, some of which would not ordinarily be open for visitors. Members will receive information about site visits by direct mail.

Membership Committee has also set the foundations for special interest and regional groups. These could be based on regions, industry sectors, or issues. The first such meetings were held at the 1990 National Conference. As a result two special interest groups have been set up, an environment group and a land and property group.

Promotions and Publicity

The Promotions Committee has set itself the task of raising the profile of GIS and the AGI in the public eye. It has placed several articles in journals and more recently has started a press release service. The latter activity seeks to respond to news items,in the press or on television and radio, with a comment on how GIS could have helped. It has taken some time but journalists are beginning to take note and are starting to seek us out for comment.

The committee has also redesigned and re-drafted the AGI brochure, of which copies are available from the Secretariat. In line with Membership Committee's priorities the AGI stand was taken to the Association for Metropolitan Authorities conference. The stand also appeared at the Property Computer Show in November 1990.

Commencing in 1991 the Promotions committee has also introduced four AGI awards: the Technology Award for applications in use in the UK; the Industry Award for implementations in use in the UK; the AGI Journalist of the Year, and; the AGI Student of the Year. The awards will be presented on May 23rd 1991.

Publications

Publications Committee has continued its good work in the publication of AGI News and the 1990 AGI Yearbook. The committee has also started the production of a "GIS Handbook". It is a large undertaking. When complete it will be a comprehensive and very practical reference book.

Standards

The Standards Committee has been steadily working on a number of different projects relating to data quality, spatial referencing, a directory of standards related to GIS, and a glossary of GIS and related terms. It has also been responsible for the development and maintenance of the National Transfer Format (NTF)—an essential standard for the geographic information community. It has commissioned work to allow NTF to become ISO 8211 compliant and to be adopted by the British Standards Institute. The committee has also been developing links within Europe in order that a European Transfer Format is developed and that the ETF is not imposed on individual nations as their national transfer format. As a result of this work AGI Council has approved the following policy on transfer standards:-

(1) An ISO 8211 compliant version of the National Transfer Format (NTF) to be developed as the standard for the transfer of spatially related digital data in the United Kingdom, and that the existing version should continue to be used and amended in the meantime;
(2) The development of a European Transfer Format (ETF) that allows for the separate development of individual nations formats such as NTF in (1) above;
(3) The development of an International Transfer Format (ITF) that allows for the separate development of the ETF in (2) above;
(4) The use of an ISO 8211 version of NTF to be developed as the basis of the ETF, as long as its use as an ETF does not conflict with (1), (2) or (3) above.

The Standards Committee has also overseen an investigation into the relational database model and its application in geographic information handling. This work is set to influence both vendors and the International Standards Organisation and is crucial to the future growth of GIS. All this work culminated in a very successful one day workshop organised by the Education, Training and Research Committee.

The committee has completed a survey of the AGI membership's priorities for GIS standards.

2

1991 plans

In retrospect 1990 was a very important year for the AGI with many "firsts" being achieved. Much work was carried out, the end results of which will not become available until later in 1991. The AGI will benefit in other ways from the 1990 achievements; the experience gained allows us to focus our attention more effectively, plan much more accurately and provide real direction for the GIS community.

The work will still be carried out throughout the AGI committees with the support of the Secretariat and under the guidance of the Council and Management Committee. However, the ever-increasing activity level will require closer involvement of Council. To this end the Council will focus on each committee's activities in turn at its 1991 meetings.

The statements which follow give an indication of the activity which will be organised by the 6 committees.

Conference Committee

The AGI now has two years of experience of producing the National Conference and Exhibition, and is capitalising on this by going into a five year planning cycle for conferences. The first in this series of conferences will be at the new Birmingham International Conference Centre (ICC) from 20th to 22nd November. Many locations are excellent for an exhibition or a conference but the ICC was purpose-built for a combined conference and exhibition. This will make it easy for the delegates to move from the main sessions to parallel sessions or the exhibition. Its location in the centre of Birmingham means excellent communication and a wide range of accommodation.

The committee are exploiting the ICC's conference facilities by inviting all organisations involved in GIS to organise sessions. This will give the delegates access not only to the general and technical papers, but also to papers of relevance to the applications of each and every delegate.

Education, Training and Research Committee

This committee will be producing the following review papers in 1991:

- Digital cartography
- Error and accuracy
- Spatial analysis
- GIS and healthcare
- GIS and routing
- Utilities and GIS

- Organisational issues
- Experience/best practice
- Coastal zones and GIS
- Remote sensing and GIS
- Methodology for cost-benefit analysis
- GIS for schools
- Ethics, privacy and confidentiality

In addition it will be sponsoring more seminars and developing further information for the GIS community.

Membership Committee

The plans formulated for the AGI membership activities in 1991 include Special Interest Groups (SIGs) and site visits. The SIGs planned for 1991 are both topic based; Land and Property and the Environment. Any member may initiate new SIGs whether based on a topic or on a sector of the GIS community.

There is a rolling programme of site visits, approximately one a month except for a break between June and August. The emphasis will be on those sites to which members would not ordinarily gain access and which play a key role in geographic information handling.

The AGI is also starting to visit different GIS software user groups. Any member wishing to have an AGI speaker at their user group meeting should contact the committee via the Secretariat.

The Membership Committee also organises the AGI's stand which is used to raise awareness at key locations. In 1991 it is due to go to the Association of County Councils, Association of Metropolitan Authorities and Association of District Councils conferences. These venues allow access to large numbers of local authority Chief Executives and elected members. The AGI stand will also be at the AGI 1991 National Conference where all members are invited to come and talk with members of the Committee.

Promotions and Publicity Committee

The Committee's task, as in 1991, is to raise the profile of geographic information systems and the Association. This will continue to be done in two ways; articles and press releases. Articles are placed in a wide variety of industry sector and computer journals. Press releases take two forms—statement of AGI policy in relation to specific issues, and immediate response to items seen in the news media.

The committee will also be running the first AGI awards. These will be presented on HMS Belfast on the 23rd May 1991. This year's awards are the Technology Award, Industry Award, Journalist of the Year, and Student of the Year.

Publications Committee

The Committee members are continuing their work with the *AGI News* and the *AGI Yearbook*. New ventures in 1991 will be the "GIS Handbook" and a series of small booklets. Both projects are intended to be practical guides. The booklet series will give essential facts and figures about a series of topics related to geographic information handling.

The Committee will also be developing a library of information which will reside at the AGI's office in London.

Standards Committee

The Committee has completed a review of the approach to GIS standards. The starting point for this was the 1990 survey of AGI members' priorities for GIS standards. The Committee has, as a result of the review, formulated a long range plan for GIS standards.

The expansion of the activities on standards is described by the Committee diagram in Figure 1.

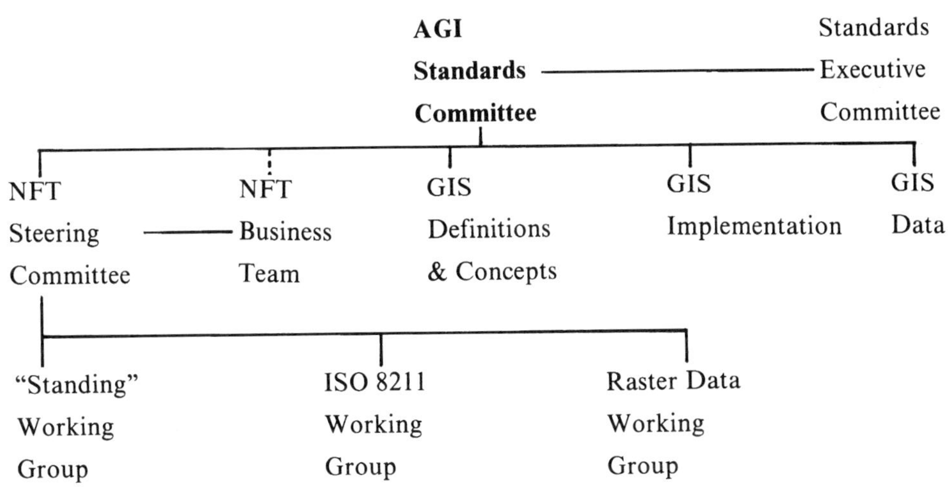

The plan includes work in 31 topics. In 1991 there are 6 objectives in 3 topic areas including work on SQL version 3, data quality benchmarks, quality assurance policy, quality assessment, and spatial data types.

The Committee will also be developing links with the British Standards Institute and the Merseyside Information Service.

3

AGI's long term plans

In 1990 the AGI Council devised a statement to encapsulate the Association's mission, recorded in Chapter 1, and all future activity and plans will relate to this statement.

The AGI's role as shown by the mission statement is to be an umbrella organisation for the whole geographic information community. It is not an exclusive club, but rather an all-embracing organisation. The organisation is the membership, but their work is to the benefit of all involved in geographic information handling.

The all-encompassing nature of the AGI's role and the enthusiasm it generates is reflected by the make-up of the Association's membership. On January 1, 1991, this stood at over 500, of which more than half were individuals. The membership, both individuals and organisations, is drawn from over 20 different industry sectors. These include: local government; central government; education; research; institutions and societies; financial services; property; emergency services; data capture and conversion; marketing, and data broking.

As previously stated the AGI's concerns are industry-wide and its activities benefit the whole geographic information community. These concerns include: standards; data availability; dissemination of information, and; best practice.

The AGI's first activities in standards have been well received both by the official standards organisations and by the vendor and user communities. The Standards Committee has spent a great deal of effort in 1991 on preparing a long term strategy for standards related to GIS and developing the AGI's role and future plans for that strategy. AGI is committed to furthering standards for the geographic information community through its own work and through collaboration with other bodies. AGI will use its own resources and where appropriate seek further funding from other organisations such as the European Commission. A document describing standards in GIS and how these standards affect you is available from the AGI Secretariat.

The AGI is also committed to increasing the amount of data available. This will be achieved through disseminating information about available information, encouraging the availability of information in digital form and addressing the problems of access to information. AGI has started this task by actively assisting the Tradeable Information Initiative. The AGI hopes to produce a database of government-held data which will be held at the AGI Secretariat and open to telephone enquirers. The next stage will be to expand the information for this service into other sectors.

The AGI is concerned to achieve the speedy dissemination of relevant information. This is not restricted to availability of data described above. It also includes information ranging from industry sector activity and general awareness through the latest research by academic institutions such as the Regional Research Laboratories.

Where there is an absence of existing information the Association plans to create its own publications. These will be created from the latest experience of the geographic information community, and will include guidance and best practice. This will be gathered through surveys and workshops and other committee-run activities. It will be disseminated through publications, workshops and the AGI's National Conference and exhibition.

The development of pools of experience and expertise within the AGI will be led by the "Special Interest Groups" (SIGs). These can be based upon issues, industry sectors or geographical region. The AGI 90 conference was the setting for two new SIGs; "Property and Asset Management" and "Environment". These groups will allow members to make contacts and share experience.

As the "UK umbrella organisation" the AGI has, and will continue to have, a major role in influencing central government. The AGI will lobby government primarily through its Council members. It will advocate the potential of GIS technology and try to influence the legislative frameworks to create?!opportunities for that potential to be realised. The Council has achieved some movement towards this objective by fitting this issue into the Parliamentary Information Technology Committee agenda for 1991.

The legal framework is a priority issue for the AGI and is likely to be the main focus of the new "Data Sharing Committee" being set up under the chairmanship of Professor Peter Dale. The legal framework affects all aspects of geographic information handling at all stages from capture through to use.

The Association is also keen to influence the Information Technology industry to meet the needs of the GIS community. This will be achieved through involvement in AGI's standards activities and demonstrating to the IT industry the benefit and the demand for GIS.

AGI is committed to education on GIS topics through the Education, Training and Research Committee's work and involvement with many parts of the education sector. The AGI will also pursue education on GIS topics for those outside the normal education sector by growing its successful conference and seminar programme. This will also be punctuated by member visits to sites of special interest.

The AGI's role in representing the interests of the GIS community also encompasses international relations. The Association plans to develop further its links with peer organisations abroad. Part of this international relations effort will be to secure resources from multinational sources for the UK GIS community.

In order that the Association can grow with the GIS industry the Council is laying firm plans to set the AGI on a more professional footing. This will necessitate the development of trading activities, particularly with publications. This, combined with securing long term funding, will allow the Association to pursue its more strategic activities such as standards.

The AGI will always be dependent on the generosity and the motivation of its members who supply liberal amounts of money, time and enthusiasm which the AGI requires to fulfil its mission. The AGI will continue to recognise this by continuing to develop an increasingly attractive and relevant package of membership benefits both core (free) and optional (at cost or subsidised).

4

Membership of Council and Committees

Council

Mr Peter Woodsford FRICS (**Chairman**), Laser-Scan Ltd, Milton Road, Cambridge CB4 4FY

Mr Frank Bennion, British Gas West Midlands, 5 Wharf Lane, Solihull, West Midlands B91 2JP

Mr Michael Blakemore ARICS, NOMIS, Mountjoy Research Centre, University of Durham, Durham DM1 3SW

Mr Michael Brand BA FRICS, Ordnance Survey NI, Colby House, Stranmillis Court, Belfast BT9 5BJ

Dr Guiseppe Cassettari, Kingston Polytechnic, Penrhyn Road, Kingston upon Thames, Surrey KT1 2EE

The Lord Chorley, 36 Queen Anne's Gate, London SW1H 9AS

Mr James Cogle, Registers of Scotland, Meadowbank House, 153 London Road, Edinburgh, Scotland EH8 7AU

Mr Martin Coulson FRICS, University College Swansea, Natural Science Building, Singleton Park, Swansea SA2 8PP

Mr Kevin Crawley FRICS, SEEBOARD plc, Electric House, Wellesly Road, Croydon, Surrey CR9 2AR

Mr John Custance, Department of the Environment, 2 Marsham Street, London SW1 3EB

Prof Peter Dale MA FRICS, Polytechnic of East London, Longbridge Road, Dagenham, Essex RM8 2AS

Mr Richard Emmens, Hampshire County Council, The Castle, Winchester, Hampshire SO23 8UD

Dr Sara Finch, Logica Communications and Electronic Systems Ltd, 64 Newman Street, London W1A 4SE

Mr Ian Gilfoyle BSc MRTPI, Cheshire County Council, Commerce House, Hunter Street, Chester CH1 2QP

Dr Roy Haines-Young ARICS, University of Nottingham, University Park, Nottingham NG7 2RD

Mr Philip Jeanes, P A Consulting Group, Cambridge Laboratory, Melbourn, Royston, Hertfordshire SG8 6DP

Mr John Leonard FRICS, Ordnance Survey, Maybush, Southampton S09 4DH

Mr Alastair MacDonald FRICS, Ordnance Survey, Romsey Road, Maybush, Southampton SO9 4DH

Ms Cathrine McAllister, Glasgow District Council, 231 George Street, Glasgow G1 1RX

Mr Michael Nicholson BSc FRICS, Property Intelligence Ltd, Ingram House, 13–15 John Adam Street, London WC2N 67D

Dr David Parker, University of Newcastle upon Tyne, Newcastle upon Tyne, NE1 7RY

Mr Nick Pearce FRICS, Coopers & Lybrand Deloitte, Plumtree Court, London EC4 4HT

Mr John Rowley, Geobase Consultants Ltd, 28 Church Road, Epsom, Surrey KT17 4DX

Mr Peter Shand FRICS, Miles Arnold, High Winds, Cassington, Oxford OX8 1DL

Mr Nigel Sheath FRICS, ICL (UK) Ltd, 127 Hagley Road, Birmingham B16 8LD

Mr Brian Shephard FRICS, Taywood Data Graphics, 309 Ruislip Road East, Greenford, Middlesex UB6 9BQ

Mr Gurmukh Singh FRICS, Pinpoint Analysis, Mercury House, 117 Waterloo Road, London SE1 8UL

Mr Petcr Smith, H M Land Registry, Lincoln's Inn Fields, London WC2A 3PH

Mr Peter Todd FRICS, CMG Information Services, Telford House, Tothill Street, London SW1H 2DX

Mr Thomas Waugh FRICS, University of Edinburgh, Drummond Street, Edinburgh EH3 9EU

Management

Mr Peter Woodsford FRICS (**Chairman**), Laser-Scan Ltd, Milton Road, Cambridge CB4 4FY

Mr Kevin Crawley FRICS, SEEBOARD plc, Electric House, Wellesley Road, Croydon, Surrey CR9 2AR

Prof Peter Dale MA FRICS, Polytechnic of East London, Longbridge Road, Dagenham, Essex RM8 2AS

Mr Alastair MacDonald FRICS, Ordnance Survey, Romsey Road, Maybush, Southampton SO9 4DH

Mr Gurmukh Singh FRICS, Pinpoint Analysis, Mercury House, 117 Waterloo Road, London SE1 8UL

Mr Peter Todd FRICS, CMG Information Services, Telford House, Tothill Street, London SW1H 2DX

Conference

Mr Nick Pearce FRICS (**Chairman**), Coopers & Lybrand Deloitte, Plumtree Court, London EC4 4HT

Mr Derek Hilder, DWH Associates Ltd, 20 Rother Street, Stratford-upon-Avon, Warwickshire CV37 6NE

Mr Gurmukh Singh FRICS, Pinpoint Analysis, Mercury House, 117 Waterloo Road, London SE1 8UL

Mr Alistair Keddie, Northumbrian Water Group, Northumbria House, Regent Centre, Gosforth, Newcastle-upon-Tyne NE3 3PX

Mr Peter Clegg, Cliffe House Cottage, High Bradfield, Shelfield S6 6LJ

Education, Training & Research

Mr Michael Blakemore ARICS (**Chairman**), NOMIS, Mountjoy Research Centre, University of Durham, Durham DM1 3SW

Mr Derek Hilder, DWH Associates Ltd, 20 Rother Street, Stratford-upon-Avon, Warwickshire CV37 6NE

Mr Jeffrey Willis, Merseyside Information Service, Suite 301, Royal Liver Building, Pierhead, Liverpool L3 1JH

Mr Gurmukh Singh FRICS, Pinpoint Analysis, Mercury House, 117 Waterloo Road, London SE1 8UL

Miss Vanessa Lawrence, Longman Group UK Ltd, Longman House, Burnt Mill, Harlow, Essex CM20 2JE

Mr Martin Coulson FRICS, University College Swansea, Natural Science Building, Singleton Park, Swansea SA2 8PP

Ms Cathrine McAllister, Glasgow District Council, 231 George Street, Glasgow G1 1RX

Membership

Mr Philip Jeanes (**Chairman**), P A Consulting Group, Cambridge Laboratory, Melbourne, Royston, Hertfordshire SG8 6DP

Mr Brown, British Railways Board, British Rail (SR), 13L Southern House, Wellesley Grove, Croydon CR9 1DY

Dr David Parker, University of Newcastle-upon-Tyne, Newcastle, NE1 7RY

Dr Brian Kelk, British Geological Survey, Keyworth, Nottinghamshire NG12 5GG

Dr Jonathan Budd, Nature Conservancy Council, Northminster House, Northminster, Peterborough PE1 1VA

Mr David Hendley, DH Associates, 10 Challoners Way, East Molesey, Surrey KT8 0DW

Ms Dorothy Pugh, Cambridgeshire County Council, Shire Hall, Castle Hill, Cambridge CB3 0AP

Promotion & Publicity

Mr Michael Nicholson BSc FRICS (**Chairman**), Property Intelligence Ltd, Ingram House, 13-15 John Adam Street, London WC2N 67D

Mr Per Dirdal FRICS, Alper Systems Ltd, Milton Road, Cambridge CB4 4FQ

Mr Gurmukh Singh FRICS, Pinpoint Analysis, Mercury House, 117 Waterloo Road, London SE1 8UL

Mr Roland Cunningham FRICS, ICL Ltd, Observatory House, Windsor Road, Slough, Berkshire SL1 2EY

Mr Alastair MacDonald FRICS, Ordnance Survey, Romsey Road, Maybush, Southampton SO9 4DH

Publications

Mr Nigel Sheath FRICS (**Chairman**), ICL (UK) Ltd, 127 Hagley Road, Birmingham, B16 8LD

Dr Michael Isherwood, London Fire & Civil Defence, 8 Albert Embankment, London SE1 7SD

Ms Julia Blackmore, National Joint Utilities Group, 30 Millbank, London SW1P 4RD

Dr Roy Haines-Young ARICS, University of Nottingham, University Park, Nottingham NG7 2RD

Mr Roger Moore, Institute of Hydrology, Maclean Building, Crowmarsh Gifford, Wallingford, Oxfordshire OX1 8BB

Dr David Maguire, University of Leicester, University Road, Leicester LE1 7RH

Mr Peter Shand FRICS, Miles Arnold, High Winds, Cassington, Oxford OX8 1DL

Mr James Cadoux-Hudson, Mayfield, Buckland Road, Childswickham, Broadway WR12 7HH

Dr Ian Heywood, Department of Geography, University of Salford, Salford M5 4WT

Standards

Mr John Rowley (**Chairman**), Geobase Consultants Ltd, 28 Church Road, Epsom, Surrey KT17 4DX

Mr Chris Gower, Berkshire County Council, Shire Hall, Shinfield Park, Reading RGZ 9XG

Ms Mandy Lane, Institute of Terrestrial Ecology, Merlewood Research Station, Grangeover Sands, Cumbria LA11 6JU

Mr. Nick Green, Digital Equipment Co. Ltd, Hampshire House, Wade Road, Basingstoke RG24 OPL

Mrs Kathryn De Groot, De Groot Carto Consultancy Ltd, 1 25 Victoria Road, New Barnet, Hertfordshire EN4 9PE

Mr John Turner, Ordnance Survey, Romsey Road, Maybush, Southampton S09 4DH

PART II

Current GIS issues

As activities in the GIS field are advancing rapidly virtually all of the contributions contained in this book could be considered to be about current GIS issues. In this section we have tried to identify those issues which are perhaps more pertinent to the development of GIS activities at the present time. Naturally, this is a very personal selection. We do, however, consider that the issues discussed in this section will be those which have a marked effect on the development of GIS well into the current decade. For example, A. Elders raises the frightening question "Do we, as users, really understand the limitations of spatial data?" F. Bennion mentions that despite new versions of software being redeemed, all too frequently, there is little hope of it ever catching up with developments in hardware; which leaves even less hope for the rest of us to keep pace! The remainder of the contributions in this section examine other issues, including access to education, and set the scene for the rest of the book.

5

Trends in hardware and software

Frank Bennion
British Gas plc

Introduction

Many of the current developments in the Information Technology (IT) field are of direct interest to the GIS community. There is a move away from traditional data processing to integrated systems which include information processing and presentation. This trend will continue under the influence of user pressure.

Hardware

Hardware will be dealt with under four main headings: computers, storage, peripheral devices and communications.

Computers

The limits of effective single processor computers are being reached and it is becoming increasingly rare to find such designs in high performance machines. To increase performance a number of options are being pursued. The first is the parallel computer which has potential for GIS applications. The major drawback at present is the lack of compilers which will automatically convert sequential code into parallel code efficiently. The second is the use of Reduced Instruction Set Computer (RISC) chips. These are increasingly used in high performance engineering workstations. Finally the increased availability of Application Specific Integrated Circuit (ASIC) chips makes the option of embedding software increasingly attractive. Such devices will always run more quickly than the equivalent software code. Work is also proceeding on optical computers and colour portable microcomputers, allied to optical discs, open up possibilities for out-of-office use of GIS data.

Storage

The amount of data generated by GIS applications calls for effective storage media. There is pressure from Operations Sections for unattended back-up. These needs are being addressed by suppliers in the form of optical discs, now available in re-writable form, and improvements to the technology of both 8mm video tape and 4mm Digital Audio Tape (DAT). Each type now has the possibility of juke-box type multiple units with storage capacities up to 50 gigabytes. As user pressure for standards bears fruit it is likely that optical disc units will be able to read the three main types of disc by detecting the type automatically.

Magnetic media still have a place for lower volume use and multi-disc packs with storage capacities approaching 1 gigabyte are now available. Solid state storage is still slow to become available in sufficient capacity and may not survive the pressures from disc and tape media.

Peripheral devices

There is a wide variety of devices including plotters, printers and Computer Output to Microfilm (COM) devices.

The need to use vector plotters where high quality output is needed is now removed as raster plotters can reach similar standards. The only drawback is the cost but as the quantities required increase so costs will fall. Improvements are continuing with raster plotters which will further improve quality and production speeds. Printers still continue to improve but for graphical type output they have limited application. The most promising is the thermal printer which is now available up to A3 size. Laser printers are monochrome at present but again colour could be an option if sufficient demand is there. Where some form of out-of-office/hours facilities are required these are provided mainly by film technology in the form of fiche or aperture cards. Equipment is available to produce master frames automatically but fiche still requires manual intervention to fit frames in a jacket. The longer term needs are most likely to be met by computing equipment.

Communications are included as they are often the platform on which a GIS depends for data collection. Local Area Networks (LAN) are currently available at speeds up to 10 Mbits/sec but improvements will quickly increase this speed by a factor of about 5. Wide Area Networks (WAN) are usually slower because of the restraints imposed by available bandwidth. Even so the British Telecom (BT) Megastream service offers a 2Mbits/sec rate. The Integrated Digital Services Network (ISDN) is also being put in place by BT. Several private network suppliers have suitable facilities. Satellite links have certain restricted use particularly for long distances. Finally the rapid development of facsimile has opened a useful channel for transmitting small sized graphics speedily and has the advantage that it uses an international standard.

Software

Software will be considered under five main headings: operating systems, databases, parallel languages, artificial intelligence (AI), development methodologies and user interfaces.

Operating systems

As part of the move to open systems, pressure is coming on suppliers to isolate applications from the computer operating system and any associated proprietary software. One choice gaining considerable support is the move to standardising UNIX. At present there are two main contenders but the European Community have said that only one standard will be acceptable for future government and authorities contracts. Because many GIS applications run on workstations, which use UNIX already, this is an important factor in users' choice of platform.

For use on microcomputers MS-DOS still holds the major share of the market. OS-2 has been very slow to penetrate probably because of the lack of package software and its need for the more powerful micro computers.

Databases

The debate as to the suitability of a particular database type for GIS continues. The final choice may well rest on the outcome of the AGI SQL Working Party's deliberations and any resulting standards. There seems to be a need for some form of object-oriented facility regardless of which database finally emerges. Object-oriented databases (strictly object-oriented systems use an environment which extends beyond the database) are being developed and may yet become the predominant database in use. It is already clear that even relational databases cannot meet all the needs of GIS and particularly those associated with version management. Distributed databases are now becoming a reality although not all the problems have been resolved.

Parallel languages

Many applications in computing are suited to parallel processing. This in itself is sufficient reason for development work to proceed. The main benefit will be that, in conjunction with multi-processor hardware, computations speeds will increase considerably. GIS has many potential uses for parallel processing, for example searching multiple databases concurrently.

The only true parallel language at present is OCCAM from Inmos but that start is being built on. A real need is for a compiler that will automatically convert suitable sequential code to parallel code. Work is under way with such a compiler for FORTRAN, the language often used for GIS.

Artificial intelligence

Expert systems and artificial intelligence systems have been available for many years but have been slow to come into general use. As with GIS the difficulty has been the processing power required to drive any significantly sized system. Their main use would seem to be in removing many of the operations which have to be done manually at present and perhaps a more significant use will be in the sorting and analysis of large volumes of imprecise or incomplete data.

Development methodologies

In an attempt to control software development in a formal and logical way considerable effort has been put into the development of tools to assist the process. This technique is known as Computer Assisted Software Engineering (CASE) for which there are a number of tools available. For major government software contracts the use of SSADM (Structured System Analysis and Design Methodology) is mandatory. Unfortunately there is no tool yet available which can model spatial data let alone deal with uncertainty in data availability. They will probably have to wait for GIS data definitions to be made available. This does not preclude limited use at the initial stages of a project.

User interfaces

The way information is presented to the user is often the difference between the success and failure of a system. The general move is now to the use of menu-driven options allied to a multi-window option. The two main contenders are MS-Windows (usually found on microcomputers) and X-Windows which has found favour in the workstation market. It is likely that for GIS an interface based on X-Windows will be the chosen option.

Conclusions

There is no sign of software development catching up with the rate of hardware developments. This does mean that useful software for the GIS market will be slow in coming, particularly until suppliers see an increasing and more stable market. In this respect the work of the AGI Standards Committee is important in that stable standards encourage software development done in the knowledge that efforts will not be wasted by fundamental changes.

6

Copyright in GIS databases

Nicholas Higham
S.J. Berwin & Co

Is there copyright in a database?

Judges do not know everything; in a recent celebrated case, a judge was heard to ask counsel whether "Gazza", the hero of England's world cup team, was well known in England. That same judge would need also to ask what a geographic information system might be, and he would be surprised, but I hope impressed, by the extraordinary advances that have taken place in this field.

The circumstance in which this judge would most likely encounter GIS would be a case involving infringement of copyright. And once our judge had been educated in GIS, he would be quick to appreciate that a GIS database enjoyed copyright protection and that, in addition to any remedies that the owner of the database might have in contract against his customer, the owner could also enforce his copyright against any person who copied or used the database without permission.

Let us take this back a stage. In the UK, under Section 1 of the *Copyright, Designs and Patents Act 1988* ("the Act"), copyright is described as a property right which subsists in original literary and artistic works (among other works). In Section 3 of the Act, a literary work expressly includes "a table or compilation" and, under Section 4, an artistic work includes any drawing, diagram, map, chart, or plan or photograph, irrespective of artistic quality.

Some data on a GIS database will consist of names of places or information which categorises them; that will take the form of a digitised record of maps and drawings. These are artistic works, and remain so despite being converted into binary form. In addition the digitised maps may also form part of the compilation of data which enjoys separate copyright as a literary work.

To enjoy copyright protection, a database must be *original*. Very often the person compiling a database will extract his material from one or a number of other publications and re-format these. Seldom will he compile his database only from truly original research.

Even in this case, however, the court will readily find that the compiler has exercised sufficient skill and labour in the selection and recording of the database for it to be "original" and entitled to a separate copyright under UK Copyright Law.

Who owns the copyright, and for how long?

The first owner of copyright in a literary or artistic work will be the individual who created it and, under Section 12 of the Act, the copyright will continue for 50 years from the end of the calendar year in which that individual dies. In many cases, of course, the individual will be an employee and, in that case, but not necessarily where he is an independent consultant, the copyright will automatically vest in his employer.

What is copyright?

Under Section 16, the owner of the copyright in a literary or artistic work has the exclusive right (again among other things) to do the following acts in the United Kingdom:

(1) to copy the work;
(2) to issue copies of the work to the public; and
(3) to make an adaptation of the work, or to copy or issue copies of that adaptation.

If any other person does any of these acts in relation to the whole, or any substantial part, of a database without the consent of the copyright owner, he will infringe the copyright.

While neither computers nor software were expressly referred to in earlier copyright legislation, certain of the special factors which apply to computer-related products are taken into account under the Act. Thus, a literary work is defined as including a computer program, which will therefore enjoy copyright protection on the same basis as any other literary work and to "copy" is defined as including "the making of copies which are transient or incidental to some other use of the work". This last provision was included at a request of the computer lobby to make it clear beyond doubt that use of a computer program, or accessing a computer database, would still constitute infringement even if the only copy made was within the internal memory of a computer or on a screen, and was transient or incidental to some larger purpose.

Rarely, of course, will any unauthorised user reproduce the whole of a database. It is much more likely that he will access some part in the course of a particular project. This raises the question of whether what he is copying or using is sufficiently substantial to constitute an infringement. While this issue is, perhaps, at its most complex in the context of a database, it is not altogether new. In a case last century, the proprietor of a local newspaper was held to infringe copyright by regularly extracting from another journal, which provided national livestock prices, certain details of the agricultural prices achieved for products in its area of circulation. It was held that, although the volume of material taken on any one occasion was very small, the fact that the newspaper regularly extracted information of this type was sufficient to amount to an infringement.

The owner of the copyright in a database may therefore bring proceedings in the English Courts against any person who:

(1) makes a copy of any substantial part, or regularly makes copies of small parts, of a database; or

(2) regularly uses a workstation or terminal in the UK to access parts of a database (wherever the database is located),

unless, in either case, he does so with the consent of the copyright owner, and acts strictly within the constraints of his licence.

The only possible exception to these rules is the exemption under the Act for "fair dealing with a literary work for the purposes of research", but, although there is no case on the point, this exemption is unlikely to apply to a commercial company which accesses a database in the course of its business and with a view to making the information available to its clients for profit.

The danger faced by the database compiler

Earlier, in the discussion on originality, the case of the newspaper and agricultural prices was mentioned. That case clearly shows that any person who, in compiling a new database, draws on earlier copyright works must consider whether or not he needs a licence to do so from the owner of the copyright in the earlier work; he very often will and cannot generally compel the owner to grant the licence he needs. Any licence must be so worded as to authorise whatever the database owner intends to do and anything that his customers, and even perhaps their customers, may need.

The relationship between the database owner and his client

Ordinarily, there will be a contract between the owner of a GIS database and the person who uses it. The user may perhaps be provided with a copy of the database on disk, or the database may be stored on a particular computer system and made available for access remotely by a terminal. In either case, the contract should specify cxactly what the user may do, and therefore what he may not do. For example:

- Is user limited to a single processor?
- Can the user use the database only for his own internal research, or for wider purposes?
- Does the database owner enjoy sufficient rights from the owners of the copyright in any source material to grant the intended licence to the end-user?

Particular problems arise where the user is himself an intermediary, in the sense of using the database as a tool in the provision of services to his own clients. Is the user then entitled to reproduce extracts from the database and provide them in any (and if so which) form to his own clients?

If the user is given a simple general authority to use, copy and publish the database, he could even go so far as to set up a service in competition with that provided by the database owner. The licence must therefore be limited but, in that case, the user must ensure that the limitations are not so severe as to deny him the right to use the database in the way he intends.

Users should periodically remind themselves of the terms of any licences which authorise use of any GIS database, or indeed any computer program. Owners of software and databases, particularly PC-based software, are rightly becoming more active in enforcing their rights against companies which take a limited licence to use a product on a single PC, and then allow the product to be replicated and used throughout the company.

The international position

As the United Kingdom is a signatory to the Berne Convention, any person who compiles a database in the UK will automatically acquire, without the need for any formality, the same rights of copyright in each other country which is a member of the Berne Convention as are available to its own citizens. He does not obtain *UK* copyright in, say, France or Germany; he obtains *French* or, as the case may be, *German* copyright in those countries, just as a French citizen would, in the reverse case, obtain UK copyright protection here. While in general the leading industrial, and many other, countries have similar laws of copyright, as indeed they are required to do under the Berne Convention, there can be some differences in the extent or nature of protection available.

7

The nature of mappable data

Alex Elder
Bartholomew

Introduction

Any discussion on the topic of the accuracy of spatial data can only be of benefit if we firstly attempt to understand the nature of spatial data. We are, after all attempting to record the natural and man-made phenomena which are a part of our daily lives and which too often we assume are quite unambiguous in their positional and qualitative characteristics. Cartographers, who have traditionally dealt with the uncertain nature of spatial data have developed an understanding and knowledge of its provenance and fitness for use.

In order to map anything we must be able to answer two questions:

- Where is it? and
- What is it?

Any feature which is not definable in its position is not mappable as we cannot know where to put it, and any feature which is uncertainly defined as to its nature is not mappable because we are unable to define its characteristic.

Looking at this situation more closely we can see that there are degrees of uncertainty in all information and it is often extremely difficult to decide whether a feature is mappable or not. In this situation we must consider the nature of the data itself and the methods by which this information was collected and classified.

Data collection

The easiest information to collect is that which is clearly defined and has a fixed position. The science of geodesy and the mathematics of map projection have built up over the centuries an accurate framework for the collection of information on features existing on the Earth's surface.

Knowledge about the Earth's shape and size is essential if we are to collect information and make maps of its surface. It is necessary to know about its size in order to make maps to scale, and shape is equally important as it influences the kind of mathematical transformation required to map it on a two-dimensional surface. For historical and political reasons, as much as for scientific merit, approximately fifteen different measurements for the shape and size of the Earth (or spheroid) are in current use. The need to establish the spheroid used in calculating or measuring the position of features is particularly important at large scale mapping as these maps are based on original surveyed information. At smaller scales any errors in calculation of measurement are negated to a large extent by the ability of a human to draw or digitise accurately.

The techniques used in land survey and photogrammetry are designed to measure and plot, very accurately, many features of the Earth's surface. For example, relief (contours), and man-made features such as administrative boundaries and buildings, can be drawn or plotted accurately, within the limits of scale and generalisation. However, suppose the position of a feature is not clearly defined, or fixed. Some water features illustrate this clearly; in marshy areas it is often extremely difficult to establish the boundary between dry land and inundated areas. Many features may not be constant in their position through time. The best example of this is perhaps coastline where the edge of the land has to be defined by some average measurement such as "high-water mark" or "mean-spring tide". Where variations with time are irregular it may only be possible to define an approximate shape or alignment to intermittent lakes and rivers.

The spatial features discussed so far have all been visible, but non-visible features are often mapped. For example, isohyets (lines of equal rainfall) representing mean rainfall over a period of time are defined by actual measurement while other categories of isolines are measured indirectly, for example population density isolines. As density cannot exist along a line a subjective decision must be made as to where the line should be.

We can see that even when we think we have all the information necessary to record and map spatial features accurately there are many cases where the nature of the data or its provenance are such that there cannot be an absolute definition of its precise position or alignment. Taking this discussion further there is a whole branch of information which is collected in an even less definite way. Statistical data, for example census information, can only be mapped accurately in terms of the smallest areal unit at which the data was collected. This potential problem can be further compounded by a process of aggregation prior to the publication of the information, and a lack of understanding of the sampling methodology adopted during the original survey.

Data manipulation

There are an increasing number of GIS users who have little experience with or appreciation of the data that they find relatively easy to acquire and use to justify or assist the decision-making processes within private or public sector organisations. We are now witnessing the introduction of low cost, personal computer-based GIS packages, offering a variety of spatial analysis tools. These packages are often advertised as easy to learn and use and this may be true, but can the user really be expected to appreciate the proper application and more crucially the correct interpretation of the results of manipulating spatial data?

Organisations collect or acquire information to meet their needs. To be successful a GIS must be able to manipulate this information and produce a model or map of a part of the Earth's surface in a manner supporting the organisation's identified requirements. The GIS will typically store spatial information composed of a set of locational data (position, geometry and topology) and a set of related semantic or attribute data (name, classification or description). To construct a meaningful model or map the GIS will process and manipulate this information in order to position and generalise features in such a way that the user is able to interpret in a clear and unambiguous manner their characteristics and relationship.

Positioning is the translation on the plane of the map of the geographical location of the feature. The defined location of a point feature can be considered as the intersection of two coordinate axes, either geometrical (cartesian) or geographical (latitude, longitude). Similarly a string of coordinates can define a linear feature or a polygon edge.

The defined location of a feature is regarded by many as an item of information which is unalterable, yet traditionally the cartographer must adapt the positioning of that feature according to the scale of the map by means of generalisation in order to keep the map clear and readable. Generalisation is the selection and simplification of the representation of features appearing on the map. It is carried out to aid the legibility of the map and should always take place with proper reference to the scale and purpose of the map. There are two main factors which influence generalisation; scale and the intended purpose of the map or model.

If we examine maps of the same area at different scales it is immediately apparent that there is a progressive simplification of the detail as scale decreases. Obviously then scale has a profound effect on the degree of simplification necessary. The smaller the scale the greater the degrees of generalisation. The reasons for this are self-evident. It is impossible to represent all features at small scales with the same degree of detail or positional accuracy with which they can be shown at larger scales. The representation must be simplified in order to maintain legibility.

There are many different types of maps and the features represented on any particular map should always be selected to fulfil the purpose of that map. Hence, the purpose of the map primarily determines the features which have to be shown on the map and in doing so it also determines the degree of simplification necessary in their representation. Scale and purpose are closely related; the scale must be chosen to suit the particular or intended purpose of the map. If the degree of generalisation is such that the features to be shown cannot be adequately distinguished then the scale is too small to suit the map purpose and should be enlarged.

One of the most difficult aspects with regard to generalisation is that it is a very subjective operation. Professor Imhof gives a very apposite description of generalisation when he writes:

> "The objective of generalisation is the highest accuracy possible in accordance with the map scale, good geometric informative power, good characterisation of the elements and forms, the greatest possible similarity to nature in the forms, clarity and good legibility, simplicity and explicitness of the graphical expression and coordination of the different elements."
> From 'Kartographische Gelandedarstellung'

Spatial information representing natural and man-made features cannot be truly and accurately represented, within acceptable levels, at all map scales. In 1:10,000 and larger it may be possible to maintain the planimetric accuracy of the original survey. As soon as one begins to represent features by symbols, for example double line roads, planimetric accuracy is no longer sustainable.

Attempting in this short paper to describe some of the peculiar qualities of spatial data and the cartographic principles and techniques in common use, the intention has been to persuade GIS users to examine carefully and critically all spatial data in terms of their requirements. The majority of spatial data available to existing and future GIS users is likely to be digitised from existing mapping. It is therefore important that the basis on which the information was collected and the processes through which the information passed prior to and during the construction of the map or model are understood by those who digitise and manipulate this information in a GIS.

8

Tradeable information—A review of 1990

Peter Dale
Polytechnic of East London

1990 has seen some movement in the development of national policies on the exchange of spatial data. The Central Computer and Telecommunications Agency (CCTA) published a guide on "Managing Information as a Resource" and the Department of Trade and Industry (DTI) produced a revised version of their Tradeable Information Guidelines. Following on from the VANGUARD initiative (1988) which sought to promote the increased use of Value-Added and Data Services (VADS) in the UK there would seem to be a framework for the greater sharing of spatial information.

None of the above reports addresses the specific problems of handling spatial data and hence the Association for Geographic Information (AGI) has been exploring what role it can play in creating more open access to relevant data sets. In particular, it has approached Government with a view to the more speedy and effective implementation of the Tradeable Information Initiative (TII) and has held discussions with the Inter-Departmental Working Group on the TII (TIIWG).

The TII has its roots in 1983 when the Information Technology Advisory Panel presented a report on "Making a Business of Information". In its analysis, it argued that as a matter of policy:

> "We would give prominence first to the stimulation of the UK Industry through the release of Government-held information, and secondly to the protection of the commercial value of information."

In response, the Government gave the responsibility for developing and implementing information policy to the Department of Trade and Industry. This in turn set up the TIIWG. In 1986 the Tradeable Information Guidelines were first published. These have now been revised version by the DTI, taking into account such developments as the *Copyright, Designs and Patents Act of 1988* and the Commission of the European Communities publication "Guidelines for improving the synergy between the public and sectors in the information market".

The Tradeable Information Guidelines have, amongst others, the stated aims:

(a) to promote growth in the UK information services market on a commercial basis
(b) to promote efficiency and thus improve competitiveness in the economy at large, through the use of commercial, electronic information services; and
(c) to make as much Government-held information as possible available for the information sector to turn into electronic information services.

The Guidelines do not define precisely what constitutes Tradeable Information. They recognise that information is a valuable commodity to which further value can be added through a variety of means, including accessibility, agglomeration, structure, transaction and re-utilisation. They identify three types of trading activity, namely:

(i) information produced by Government for one purpose and sold on to the private sector for the same purpose;
(ii) information collected by Government but processed by the private sector for use by Government; and
(iii) information collected by Government for one purpose but resold to the private sector for some totally different purposes.

The Guidelines advise Departments to carry out an audit of their information holdings to see if any are "tradeable". A series of detailed recommendations offer suggestions on who should take the lead in negotiating with the private sector, how the information should be priced and what terms need to be built into any contract with the information industry. Finally, the Guidelines offer a set of model contract clauses, amongst which is a recommendation that "No warranty is given by the Data Provider or Controller as to the accuracy or comprehensiveness of the Data".

The AGI has encouraged more open access to Government-held data and commented to the TIIWG on the draft of a questionnaire that sought to identify spatially referenced datasets held by Government departments. The questionnaire asked, amongst other matters, for the following information:

(1) Title of datasets.
(2) The extent to which the data can or have been made available under the TII.
(3) What items of information are contained in the datasets.
(4) Their time span and frequency of update.
(5) The purpose for which they were collected, the methods of collection including sampling techniques and from whom they were collected.
(6) Any limitations on their use.
(7) The type and extent of locational referencing.
(8) The area of cover, the smallest area for which the data were collected and the smallest area for which they could be released.
(9) The medium on which the data are held, including software.
(10) Whether the datasets contain any data relating to individuals.
(11) When and under what conditions the data will be made available.

Details of some 400 datasets were received. The intention of the TIIWG has been that the "information about information" should be as widely disseminated as possible through the creation of a database, preferably on-line, that is accessible to outside bodies. In September 1990, an invitation to submit proposals on "The Dissemination of Information about Spatially Referenced Data Sets Held by Government Departments" was issued. Proposals had to indicate how the information would be disseminated, how publicity would be given to the service and what financial arrangements were proposed.

A number of problems, such as the update of the information and the manner of charging for it, have to be worked out. The AGI offered to provide an appropriate service but was only one of several contenders to do so. At the time of going to press, the successful offer is not known. What however is clear is that by

1991, it should be possible to find out what datasets are held by Government to which the general public can obtain access.

References

Central Computer and Telecommunications Agency, (1990). *Managing Information as a Resource*. London: HMSO.

Department of Trade and Industry, (1990). *Government Held Tradeable Information—Guidelines for Government Departments in dealing with the Private Sector*.

VANGUARD, (1988). *Value Added and Data Services Interworking Study*. London: HMSO.

9

GIS in local goverment

Richard Buxton
Principal Associate, Coopers & Lybrand Deloitte

Local government is once again in a state of anticipated change. One of the first moves of Michael Heseltine after his return to the Department of Environment at the end of November 1990 was to announce a wide-ranging review of local government structure and finance. The abolition of the community charge, directly elected mayors and the possible introduction of a unitary system of local government are just three of the items that are open for evaluation. At this stage it is impossible to predict the outcome; no options are excluded from the debate and uncertainty is the only thing we can be sure of.

Information is always a valuable resource, for without information decision-making can at best be an intuitive hunch and at worst a blunder into the dark. In times of change, the value of that information resource and of the tools to analyse that information rises still further. Those local authorities who have invested in information, and in appropriate information technology, will find themselves more readily able to cope, whatever the nature of the eventual changes.

Change in local government is not only restricted to the Heseltine review. During the course of the past few years almost every service in local government has been subject to some form of externally imposed change. The current agenda includes environmental matters and increased government powers to intervene in cases of perceived under-utilisation of public sector land holdings.

Much of the information required to respond to this change is of a geographic nature, with geographic information lying at the heart of most local authority services. The importance of geographic information and of the technology to analyse this information was recognised by the four national local authority associations in the formation of the Geographic Information Steering Group (GISG) in May 1988. GISG was concerned with all aspects of geographic information and geographic information systems (GIS) in relation to local government. The primary aim of GISG was:

> "to advise the local authority Associations on policy; interests in geographic information and through the Associations influence Government and other relevant national agencies."

GISG was responsible for commissioning, through LAMSAC, two major GIS needs studies at both County Council and District Council levels and in addition supported a number of other GIS initiatives, including product profiles. With the demise of LAMSAC in October 1989, GISG decided to carry out a review of its role, structure and work programme and this review was carried out in the period from May to September 1990.

In the two years from the formation of GISG in May 1988 until the start of the review in May 1990, many changes had taken place in the field of geographic information and GIS. Those two years had seen the formation of the AGI, continuing progress on the Ordnance Survey digitising programme and active moves by a number of local authorities to implement GIS. Yet despite this progress, major uncertainties continued to exist. The rate of change in the GIS market place was faster than ever before, whilst the assessment of costs and benefits remained as difficult as ever. Some work on standards was progressing, but to what extent would this take into account the specific requirements of local government?

In carrying out the review of its role, structure and work programme, GISG was concerned to find some means of ensuring that the most appropriate support and advice could be provided to local government on geographic information issues. At the same time, GISG recognised the importance of working, wherever appropriate, with other organisations active in the GIS field, both to coordinate resources and to ensure that the local government voice was fully represented.

The local authorities themselves confirmed the importance of GIS to their own operations. Out of the total of 247 authorities who responded to a questionnaire sent out during the course of the GISG review, over 85 per cent identified as "essential" some form of external support on GIS and geographic information issues. Of the 247 respondents, 46 stated that they already had some form of GIS, whilst only 29 indicated that they did not intend to invest in GIS over the next five years.

High up the list of priorities identified by the local authorities were the preparation of a standard functional specification for GIS in local government and the development of a standard cost/benefit methodology. In addition, the authorities who responded considered it important to ensure that local government was well represented on the range of technical and standards issues, including close liaison with the utilities, the Ordnance Survey and the AGI.

The nature of the work programme and the importance of the representational and liaison functions led to the conclusion that meeting the needs of local government would, at least in part, require the appointment of a full time GIS Advisor. It was, however, also apparent that many of the other components of the work programme could be best carried out by making full use of a range of resources, including the local authorities themselves, the Regional Research Laboratories and specialist consultancies. Other projects were identified as suitable for joint ventures in collaboration with the AGI, the Ordnance Survey or even private sector companies.

The appointment of a full time GIS Advisor to local government implied some form of organisational structure, particularly to ensure direction and accountability. The formation of the Local Government Management Board (LGMB) from April provided the opportunity for an appropriate organisational framework to be established.

The Local Government Management Board will have wide ranging responsibilities for supporting local authorities on the full range of management issues. Information management in its widest sense is one such issue. Within the context of information management in local government today, geographic information has particular importance. Geographic information and GIS are therefore inseparable from the fundamental objectives of the new Local Government Management Board.

LGMB, as it has begun to emerge into daylight in advance of its formal launch, has been positive in its recognition of the importance of providing local authorities with support on geographic information and GIS issues. The new GIS Advisor to local government will be appointed as a key member of staff, with responsibilities for all GIS matters. In addition, the GIS Advisor will be commissioning and supervising a range of projects to meet specific needs. Information dissemination will also be a high-priority task, to ensure that all local authorities are able to take advantage of the positive programmes that are being carried out in their name.

Formal accountability for the new GIS Advisor will be through the Chief Executive and Board of LGMB, to local authority associations and hence to the local authorities themselves. In practice, however, the GIS Advisor will be supported by an Advisory Group of local authority representatives, ensuring close links between the GIS Advisor and the community he or she will serve. The present Geographic Information User Panel will also continue to have an important role, providing a forum for sharing experience and for carrying out specific project tasks commissioned by the GIS Advisor.

The Geographic Information Steering Group has discharged is primary function by putting into place a mechanism that will provide a secure platform from which to meet the needs of local authorities in relation to geographic information and GIS. Through the Local Government Management Boards, the new GIS Advisor to local government will be able to represent local government interests with a single voice and will ensure that a local government input can be provided into the important deliberations currently underway on GIS standards.

With the continued decrease in cost of computing power, the rapidly increasing potential for storing large volumes of information, the moves by many commercial GIS vendors away from proprietary operating systems and the range of new functionality offered by the latest generation of GIS, it is clear that powerful tools now exist to assist local authorities in their critical information management tasks in an age of change. Through the GIS Advisor, it should be possible to coordinate resources, commission projects, disseminate information and provide clear guidance to assist local authorities to take full advantage of the potential offered by the technology of GIS and, by doing so, provide more efficient and effective services to their communities.

10

GIS in the curriculum

David Unwin
University of Leicester

Introduction

Despite the relative recency of the introduction of GIS into the education system, there is already evidence of a concerted attempt to provide education and training both about GIS and, less commonly, with GIS. Education about GIS involves students learning the underlying theory and the practical application of GIS to help solve real world problems. Education with GIS involves using them to help learning over a wide range of other courses (Thompson, 1990). Amongst the various educational initiatives have been the publication of a series of suggested course syllabuses and curricula, the authoring of low cost GIS software designed for educational use (see Fisher, 1989), the collation of computer aided instructional software such as GIST and DEMOGIS (Raper and Green, 1989; Maguire, 1989a), and a series of conferences around the general theme of education for GIS (for summaries of this activity see Unwin, 1991; Unwin and Dale, 1990).

A concern central to all this activity has been the definition and delivery of curricula for GIS and there have already been a number of published suggestions. These range from reports of individual attempts at specific institutions (for example Nyerges and Chrisman, 1989; Green and McEwen, 1990), through collected designs for an entire syllabus (for example Unwin *et al.*, 1990) to the collation of a complete curriculum by the US National Center for Geographic Information and Analysis (NCGIA, see Goodchild and Kemp, 1990).

The curriculum in education

Defining and delivering a curriculum specifically for GIS is no easy matter. Unlike the schools where such debate is commonplace, in higher education curricula are seldom discussed or opened for scrutiny. In fact the GIS curricula already published are pretty unusual. In my ten years of association with the major international journal concerned for teaching geography in higher education (The *Journal of Geography in Higher Education*), I can only remember it publishing, or even considering for publication, a handful of complete curricula over the entire content of geography and I am not aware of anything that has ever been published that is anywhere near as wide-ranging and ambitious as the NCGIA Core Curriculum in GIS. The evident concern for the curriculum shown by GIS educators is thus to be welcomed, but the central thesis of this note is that such concern must be informed by a more careful attention to basic design issues.

An initial problem is to appreciate exactly what is meant by the term itself. The bland dictionary definition 'a course of study' gives little away and educational theorists invariably give a much wider definition that includes explicit statements of ideology, general long-term aims, specific, testable, short-term objectives, assessment procedures, and so on. By this definition a curriculum is not simply a syllabus of topics to be covered, yet this is often the assumption made in higher education. Viewed from this educational perspective, of the published 'curricula' for GIS, only the NCGIA attempt even remotely approaches completeness. It cannot be emphasised too strongly that there is abundant evidence to show that syllabus content is not what most influences student learning. It is the extra components of the curriculum, such as attitudes to study, assessment tasks, and so on, that most define what students ultimately remember and use in later life.

Secondly, curricula in education should be designed and there are a number of formal models of the design process (see Gold *et al.*, 1990) that are surprisingly similar to those proposed in software engineering. Amongst the models that have been suggested are:

(1) 'Top-down' approaches that start from a clear statement of broad educational aims, refine these into a series of explicit and testable objectives, and then devise teaching strategies, content and assessment methods to meet these aims and objectives. As with software, most of the relevant literature tends to favour this approach.
(2) 'Bottom-up' approaches that begin by defining the subject matter to be taught and then fit teaching methods and assessment around this. The literature tends to caution against the use of this approach since it usually leads to bad design.
(3) Various 'middle-out' strategies that design by refinement of a central core of materials that most probably already exist. In practice, this is probably the most common approach.
(4) Student-centred design that begins by an examination of individual student needs and attempts to provide course materials to meet them. The problem with this model is of course that only seldom do educators 'listen to the learner' and, even if they did, it is by no means clear that students would have a correct perception of the field.

Software engineers will recognise the various dilemmas and difficulties attendant on the application of any of these models in designing curricula for GIS, but it is clear that all the published efforts have used the content-driven, 'bottom-up' approach of (2) and are arguably the worse for it.

GIS characteristics and the curriculum

It seems to me that, in common with many new technologies, GIS have a number of characteristics that make formal curriculum design difficult. First, by definition, GIS integrate ideas from many of the traditional disciplines in education. This means that responsibility for the design and provision of appropriate curricula does not, and should not, rest with any one discipline. In practice the 'Geographical' in GIS has meant that academic geography has led in the provision of GIS courses and I have argued elsewhere that this has been both a strength and a weakness (Unwin, 1991). Although some of the direct antecedents of GIS (computer cartography, remote sensing, spatial analysis) were firmly located in geography departments, many of its technical underpinnings (such as database management and design, geometry and algorithms) were not. Similarly, from an applications viewpoint, GIS should not be the sole preserve of geographers since potential benefits exist for virtually any science that has a spatial content. It is also the case that the generally empirical, data-driven ethos of GIS research was very much out of fashion in the academic geography of the 1980s. The scathing attack by Taylor (1990) in which he refers to GIS as 'positivist geography's great revenge' and 'an empiricism that is anti-geography' would be echoed in many geography department common-rooms. From a curriculum design point of view, it is doubtful if anyone from a purely geographical background is able adequately to balance the material that goes into the curriculum or to specify educational aims and objectives that fully address what a complete education in GIS should provide. An examination of curricula such as that devised by the AUTOCARTO Education Trust Group (Unwin *et al.*, 1990) will make it plain that issues to do with management are not well covered; the viewpoint is clearly rooted in the spatial analytical tradition of the academic geography of the 1960s to which has been added a flavouring from computer science.

Secondly, GIS are a technology that can be approached from at least two differing perspectives which in turn lead to very different educational aims. On the one hand, it is necessary for students to be educated in the principles and concepts that underlie GIS but, on the other, they also need to be trained in how to drive existing GIS to produce useful results. Of course academics, such as the designers of the NCGIA and AUTOCARTO curricula, have tended to favour education in basic principles but it is by no means clear what either the industry or even the students might prefer and surveys such as that reported by Willis and Nutter (1990) do not seem to help much in making the choice. This 'education vs training' debate permeates almost all the curriculum. It clearly must influence the overall aims and objectives, but it also affects the modes of delivery and the content that is offered. A basic dilemma concerns the conflicting demands of depth and breadth. The curriculum must have sufficient depth to enable students to handle the technical issues that arise in the implementation of any GIS (for example, issues of georeferencing, database design, errors and error analysis) but at the same time needs to provide students with a good understanding of the wider implications of what they are doing in relation to the problem being addressed and the various management issues that will inevitably arise.

This 'breadth vs depth' dilemma can be understood by reference to a standard taxonomy of educational objectives due to Bloom (1956; see also King, 1976). Bloom's taxonomy has six major categories from knowledge, through comprehension, application, analysis and synthesis to evaluation. The ordering of these categories is intended to be broadly hierarchical, each measuring a more complex behaviour than its predecessor and also subsuming it. Reference to this hierarchy makes it clear that although we are asking for all to be addressed, in practice the published GIS curricula tend to address only knowledge, comprehension and, possibly, application. Analysis, synthesis and evaluation are harder to teach and may well require a variety of teaching strategies.

Thirdly, the entire field of GIS has evolved very rapidly relative to the speed at which developments can possibly be incorporated into curriculum structure. This has had a number of consequences. The recency of most GIS development has meant that there is a critical shortage of staff able to teach about them and a related shortage of resources to support this teaching. Normally, there is a reasonable supply of qualified educators willing to enter into careers in higher education. These teachers are able to draw on models of curriculum practice based on their own experiences or have a background in research and applications that leads to a pretty clear idea of what should make up a curriculum. None of these conditions is met in GIS education.

My impression is that most of the early educational developments in GIS were undertaken by academic staff with a background in the 'quantitative revolution' that is supposed to have occurred in geography during the 1960s and 1970s or, somewhat more rarely, their research students rooted in the same tradition of spatial analysis. This pool of labour has dried up with remarkable rapidity and the well known 'defections' to the private sector have further diminished the available supply.

Similarly, the existence of a curriculum also implies that supporting materials in the form of textbooks, library journals, software, teaching datasets and so on are readily available. The speed of development of the GIS industry has meant that, until very recently there was no choice in the selection of textbook, and journals such as the *International Journal of Geographical Information Systems*, *GIS World* and *Mapping Awareness* have been hard to acquire at a time of deep cuts in funding in higher education. Although aficionados are well-aware of the existence of some very good public domain GIS software for teaching use and a modest number of carefully constructed datasets to illustrate GIS functions (Fisher, 1989), there remains a perception throughout the education system that education in GIS must involve students using one or other of the expensive proprietary production systems. Whatever the model of curriculum development chosen, and whatever the result, shortages of qualified staff and supporting resources have been major constraints on developments.

The National Curriculum: The future for GIS?

Thus far, the emphasis in this note has been towards higher education but there is no reason why education with and about GIS should be confined to that sector. A GIS full of data describing a specific country, such as the BBC Domesday IV disk (see Maguire, 1989b), is a marvellous resource for teaching. Similarly, the technology and its associated concepts form one plank in an emerging *spatial information science* of relevance to all disciplines that have a spatial content. The assumption made by Taylor (1990) that such a statement or ones like it constitute 'the imperialism of the new geography' is simply untrue. Spatial information science takes a part of geography but adds to it concepts and methods from a wide range of other disciplines.

In the long term, one of the most important curriculum developments of relevance to GIS has not taken place in higher education but lies in the recently established National Curriculum for Geography (DES and the Welsh Office, 1990). *Geography for Ages 5 to 16* has been criticised by professional educators (Lambert, 1990; Graves *et al.*, 1990) for a perceived concentration on facts at the expense of concepts but it does form a major statement of the importance of many traditional geographical concerns. In particular, it reasserts the importance of students having knowledge and understanding about real places, the physical environment for its own sake (as well as in relation to human activities), and, quite critically for the future of GIS, of also acquiring skills in handling geographical information. Of the seven listed attainment targets, one for skills (AT-1) forms a continuing theme and these skills are interwoven into the specification of the other six targets. By the age of 7, for example, children should understand basic coordinate systems, mapping and field recording. At age 14 this understanding will have been widened to include plotting relationships, understanding most types of thematic map, and the use of maps 'as an aid to making locational decisions'. Level 10 of this skills attainment target, to be reached at the age of 16, has as its statement of attainment that pupils should 'construct a composite map for a specified purpose by overlaying separate distributions of thematic data and evaluate its effectiveness as a Geographical Information System' (DES and Welsh Office, 1990, page 20, emphasis added). Although it is under the 'skills' attainment target that one finds most reference to mapping and geographic information, small print elsewhere in the document makes it quite clear that maps and mapping are once more on the geographical educator's agenda. This can only be good news for those of us who teach GIS in higher education and for those of you who work in the industry.

References

Bloom, B.S. (1956). *Taxonomy of Educational Objectives. Handbook 1: Cognitive Domain*, New York: Mackay.

Department of Education and Science and Welsh Office (1990). *Geography for Ages 5 to 16*, York: National Curriculum Council, 120 pp.

Fisher, P. (1989). Geographical information system software for university education and research. *Journal of Geography in Higher Education*, **13**, 69–78.

Gold, J.R. *et al.* (1990) *Teaching Geography in Higher Education: a manual of good practice*, Oxford: Blackwells, (especially pp. 193–227).

Goodchild, M.F. and Kemp, K. (1990). Developing a curriculum in GIS: the NCGIA core curriculum project. In "GIS Education and Training, Collected Papers", University of Leicester, 20–21 March 1990

Graves, N., Kent, A., Lambert, D., Naish, M. and Slater, F. (1990). First impressions: a discussion of the Interim Report of the Geography Working Group for the National Curriculum from one departmental team. *Teaching Geography*, **15**, 2–5.

Green, D.R. and McEwen L.J. (1990). GIS as a component of information technology options in higher education geography courses. In Foster, M. and Shand, P. J. (editors) *AGI Yearbook 1990*, London: Taylor & Francis and Miles Arnold, 287–294.

King, R.L. (1976). Assessment in geography: approaches to formulating objectives. *Studies in Higher Education*, **1**, 223–232.

Lambert, D. (1990). The National Curriculum is coming. *Journal of Geography in Higher Education*, **14**, 92–97.

Maguire, D.J. (1989a) DEMOGIS Mark1 : an ERDAS based GIS tutor. *In Proceedings of AUTOCARTO 9*. Falls Church, Va: ASPRS/ACSM, 620–630.

Maguire, D.J. (1989b) The Domesday interactive video system in geography teaching. *Journal of Geography in Higher Education*, **13**, 55–68.

Nyerges, T.L. and Chrisman, N.R. (1989). A framework for model curricula development in cartography and geographical information systems. *Professional Geographer*, **41**, 283–293.

Raper, J.F. and Green N.P.A. (1989). Development of a hypertext based tutor for geographical information systems. *British Journal of Educational Technology*, **9**, 3–23.

Taylor, P.J. (1990). Editorial comment: GKS. *Political Geography Quarterly*, **9**, 211–212.

Thompson, D. (1990). G.I.S.—a view from the other (dark?) side: the perspective of an instructor of introductory geography courses at University level. In "GIS Education and Training, Collected Papers", University of Leicester, 20–21 March 1990.

Unwin, D.J. (1991). The academic setting of GIS. In Maguire, D.J., Goodchild, M.F. and Rhind, D.W. (editors) *A Manual of GIS*, London: Longman, in the press.

Unwin, D.J. and Dale, P.F. (1990) An educationalist's view of GIS. In Foster, M. and Shand, P.J. (editors) *AGI Yearbook 1990*, London: Taylor & Francis and Miles Arnold, 304–312.

Unwin, D.J. and eight others (1990). A syllabus for teaching geographical information systems. *International Journal of Geographical Information Systems*, *4*, 457–466.

Willis, J. and Nutter, R.D. (1990). A survey of skills needs for GIS. In Foster, M. and Shand, P.J. (editors) *AGI Yearbook* 1990, London: Taylor & Francis and Miles Arnold, 295–303.

PART III

1991 and beyond

Census information is an important source of data for many GIS applications. By the time this volume is published the 1991 Census will have been and gone. The contributions in this section examine the value of census data as a resource and look at a range of markets for its potential application and use. These perspectives are drawn from both the data provider and potential public and private sector users. In particular, the growth in value-added census products is discussed by J. Beaumont in some detail.

1992 is a much heralded date. It will produce no dramatic change but it does signal the start of more international cooperation. That is not to say that cooperation is not already taking place—it is; notable examples are CORINE and some of the military programmes. In these early projects it has become apparent that problems exist in knowing "what data is where" and what quality it is. A further problem occurs when exchange of data is attempted—the linkages are such that in some instances it would seem easier to redigitise, which is clearly a waste of effort.

In looking ahead it is hoped that two areas be addressed as a priority by the EC member states:

- easing access to data sources and pedigree;
- standards for international data exchange.

11

Digital boundaries for the 1991 Census of Population in Scotland

Frank Thomas
General Register Office for Scotland

Introduction

The General Register Office for Scotland, GRO(S), is currently digitising the boundaries of unit postcodes in Scotland. This work is part of the preparation of output areas for Small Area Statistics from the 1991 Census of Population. Unlike 1981, and unlike England and Wales in 1991, Output Areas will not be the same as the areas used for enumeration.

Postcode maps and directory

Since 1973, GRO(S) has mapped and maintained the boundaries of unit postcodes. The Post Office assign addresses to a postcode; GRO(S) draws a boundary around these addresses so that the entire lawn surface of Scotland is covered. Once its boundary has been drawn, a postcode is assigned to an electoral ward, to a civil parish and to about 15 other area types.

The codes for these area types are stored on a *postcode directory* which is used within and outside the Office to assign postcoded data (eg. births, deaths and marriages) to larger areas. Data for a subset of these area types is extracted for addition to a GB postcode directory maintained by OPCS.

1981 Census in Scotland

The postcode was used in the conduct of the 1981 Census in two ways. In the planning of workloads for enumerators, estimates of the number of households in each postcode were prepared. Postcodes were thereby assembled into enumeration districts (EDs). The second use was in editing Census data; because an ED contained only a given set of postcodes, the postcode on a household form in an ED had to be one of that set.

Small Area Statistics (SAS) were produced for each ED comprising some 40 tables of aggregated counts. The user could thus link Census data to postcoded data from other sources. Further, the 1981 ED was added to the area types coded on the postcode directory.

All postcodes introduced since the 1981 Census have been assigned a 1981 ED code.

1971 Census in Scotland

Records from the 1971 Census were postcoded and thus aggregated to 1981 EDs and, where possible, to 1981 SAS tables layouts. The resulting re-formatted SAS (RSAS) were made available at about the time of the 1981 Census.

ED planning for the 1991 Census in Scotland

In 1981, EDs in Scotland, with an average of about 120 households, were about two-thirds the size of those in England and Wales. As part of a wider policy to reduce unnecessary differences in the Census north and south of the border, GRO(S) decided to plan 1991 EDs from scratch to bring workloads up to levels in England and Wales. (In practice, for geographic reasons, many rural EDs would remain unchanged.) A second reason for increasing workloads was that savings in pay by having fewer enumerators were expected to exceed the cost of replanning. The methodology of ED planning in Scotland was the same as in 1981 (ie. based on postcodes). The decision to replan led to 1991 EDs being abandoned as output area because they would not give continuity with 1971 and 1981.

Consultation with users on output areas

GRO(S) consulted users on their requirements for output areas. Various options were identified during consultation. GRO(S) decided to create an output area (OA) to meet the main requirements which were:

(1) continuity with previous Censuses; and
(2) greater flexibility than in 1981 of aggregation to *ad hoc* areas.

The creation of OAs would be governed by two constraints. First, for confidentiality, no OA should contain fewer than 16 households and 50 residents. These thresholds, which also apply to England and Wales and which are double their equivalents in 1981, were announced when the 1991 Census was debated in Parliament at the end of 1989.

The second constraint was that some of the work could only be done once the actual counts of households and residents were available for comparison with the confidentiality thresholds. This phase of the work must be contrived to cause the minimum delay to output. The first, pre-Census, phase would create *provisional* output areas (POAs) based on the estimated household counts used in ED planning. When actual household and resident counts are known then each POA will *either* pass the confidentiality checks and be confirmed as an OA *or* fail and be merged with one or more neighbouring POAs so that an above-threshold OA is formed.

The provisional output area

The POA is constructed using the estimated household counts of the postcodes currently assigned to a 1981 ED. Either these postcodes can be divided into two or more clumps each with at least an estimated 40 households or they cannot. If they can then each clump becomes a POA. If not, the whole 1981 ED (ie. the postcodes assigned to it) becomes a POA.

Each POA is also assigned a "best neighbour" so that later, if mergers are necessary when creating OAs, they may be carried out automatically.

Topography: the identification of neighbouring postcodes

In order to select postcodes to form clumps and to select "best neighbours", it is necessary to know which postcodes have common boundaries. A number of methods were considered:

(1) Plan POAs in the same way as EDs were being planned. This was to inspect the current postcode maps, draw 1981 ED boundaries, mark each postcode on the map with its estimated household count, split 1981 EDs into clumps if possible, record the allocation of postcodes to POAs in a database held on computer, and maintain this database when postcodes or estimated household counts are amended.

(2) Construct a "neighbouring postcode database" by inspecting maps. Each pair of postcodes with a common boundary would be recorded in the database held on computer. Postcodes would be aggregated into POAs automatically with software to be written.

(3) As (2), but instead of inspecting maps and recording neighbouring postcodes in a database, we would digitise postcode boundaries and create the database automatically.

After taking into account the costs of staff and equipment required, it was concluded that option (3) would be the cheapest. We also felt sure that this option would offer Census users and others considerable benefits. Digitised boundaries of postcodes, OAs and other areas (eg. the 1981 ED) could be displayed and otherwise manipulated by computer. Consultation with users confirmed this view.

Implementation

A team has been set up in Output Processing Branch of GRO(S) Census Division and equipped with four digitising tablets. The task of capturing the boundaries of about 130,000 postcodes from about 5,500 maps is, at the time of writing, about 84 per cent complete.

When the postcodes for a local government district have been digitised, the data are checked against the postcode directory and against the postcodes used by the ED planners. Any mismatches are corrected clerically before the neighbouring database is created. Various other checks allow for errors in mapping and digitising to be identified and corrected. Postcodes are then assigned to POAs and POAs are given best neighbours. This assignment will be made available to users before the Census for them to plan how best to incorporate SAS for OAs into their information systems. Also, they will be able to comment on whether the OAs likely to result from the POAs coincide with the areas they use.

Apart from the programs to carry out the above tasks, GRO(S) are also writing software to check and, if need be, merge POAs into OAs once actual household and resident counts are known.

1991 Small Area Statistics

Consultations on the content of the SAS are now complete. OPCS have started work on behalf of both Census Offices on the systems to produce about 80 specified tables containing about 9,000 cells. The smallest areas for which SAS will be produced are OAs in Scotland and EDs in England and Wales. Users will need to know the boundaries of these areas.

Boundary Products

GRO(S) has published its Boundary Products Prospectus (obtainable from contact points given below) listing for sale some 12 different products. The choice of product by users depends on, among other things, the hardware and software they have access to. Some possibilities are described below.

Users who wish to handle 1991 SAS in a similar way to the way they handled 1981 output will have to buy a product, the "OA-postcode index file", and obtain copies of the maps used by enumerators. These maps show postcode boundaries. (As in 1981, one copy may be offered to local authorities on long-term loan.) The user can thereby identify the postcodes belonging to an OA and draw its boundary.

Other users who wish to use hard copy may choose to buy OA boundaries on clear film. This product can be used to overlay the user's own information mapped at the same scale.

Users with the requisite hardware and software will be able to use OA boundaries in digital form. SAS data for OAs can then be manipulated and displayed with OA boundaries. Some users may wish to go further and obtain digitised unit postcode boundaries and the "postcode-OA index file". They would then have boundaries to match postcoded data from non-Census sources, and also be able to create OA boundaries themselves to aggregation.

A price of £1 per postcode boundary on magnetic medium or 50p on clear film (excluding media costs) should enable GRO(S) to cover the costs of digitisation. Consortium arrangements will be considered.

Further work

GRO(S) took the step of digitising initially for the specific purpose of assembling postcodes into OAs. Decisions still to be taken include:

- How to maintain digitised boundaries after work for the 1991 Census is complete;
- How best to exploit the data in producing mapped Census output or in meeting requests for Census output not simply expressed as groups of areas coded on the postcode directory.

Contact points in GRO(S)

For further information, please contact Frank Thomas (031–314 4217 or GTN 7166 217) or Denis Owens (031–314 4283 or GTN 7166 283) at GRO(S), Ladywell House, Ladywell Road, Edinburgh, EH12 7TF.

Further reading

Clark, A.M. and Thomas, F.G. (1990). *The Geography of the 1991 Census*, Population Trends No. 60, London: HMSO.
Boundary Products Prospectus, October 1990, GRO(S).

12

The 1991 Census: A personal perspective from within local government

Graeme Horn
Greater Manchester Research and Information Planning Unit

When data became available from the 1981 Census there were two tiers of local government in England and Wales. The county and district authorities differed not only in scale of operation but also in their functions. Counties had a wide, strategic level of geographic interest while districts tended to take a more local view of the world. Since 1986 there has been only one tier of local government in Greater London and the metropolitan counties.

Local government was one of the largest purchasers and is a principal user of Census data. However the perspective on the Census from within local government varies not only from one authority to another but also between the two tiers of government. County authorities' immediate interests are in their standing relative to other counties and in comparisons between their constituent districts. District authorities share the latter interest in seeing how they compare with their neighbours but are most interested in information at the more local, i.e. sub-district, level. Census data is used extensively for these forms of comparative research.

While such research will no doubt form the core of Census based research within local government a further need has emerged since the last Census. There is now a great call for locally based information for demand forecasting, planning and review of local government services. The introduction of competitive tendering has generated new demands for data to inform the targeting and costing of services. The 1991 Census should prove invaluable in meeting these information needs.

Most counties joined with their districts to form consortia for the joint purchase of 1981 Census data coverage of their administrative areas. The Census product most widely used in local government is the Small Area Statistics (SAS). Complete sets of county, district, ward and enumeration district level data were usually purchased. Special Workplace and Migration Statistics (SWS and SMS) were also bought under similar arrangements. This avoided duplicating purchases and kept costs down.

The two tiers of local government also worked closely over the ways they used Census data. Many of the smaller district authorities relied on their county authorities to process the Census data for them. However even the smallest district authorities are likely to hold a complete set of SAS for their areas—though it might only be in the form of paper copy or microfiche in a central library or planning department. Local authorities have played an important role in making Census data widely available to the general public.

The 1981 Census, in particular, led to the publication of abstracts and digests of comparative Census statistics by most county and many district authorities.

Local government also took the lead in developing computer software for handling data from the 1981 Census. Contributing authorities provided the seed-corn funding and ongoing finance for the development of the SASPAC and MATPAC software packages. Similar arrangements have been made to fund the development of software capable of handling 1991 Census local base statistics and SAS. SASPAC is the most common software package in use at local government mainframe computing installations. MATPAC being more specialised is less widely used.

Given how widely available both the raw data and SASPAC are it is perhaps surprising that more use has not been made of the Census within local authorities. This is, of course, a generalisation and is not meant to belittle the research that has been carried out by many authorities. However there can be few, if any, local authorities who can say that they have extracted every useful piece of information from the last Census.

There are many reasons why, particularly at district level, fuller use of the Census has not been made by local government. Firstly there is the matter of human resources. Census research is time-consuming and requires expertise. The majority of local authorities, which includes most districts, just do not have the resources to devote to the acquisition of anything more than basic expertise in the handling of Census data. Most authorities do though employ at least one person who has some familiarity with the Census; expertise has been developed most fully in county authorities and especially in county planning departments. The abolition of the Greater London and metropolitan county councils broke up a well-respected group of expert Census users in the English conurbations.

A second factor which limited opportunities for Census analysis was the way computing resources were managed in most local authorities. In the early 1980s, when Census data came on-stream, most local authorities had some mainframe computing capability. Now, as then, most computer resources are under the control of central support agencies. It is very much the exception within local government for researchers in service delivery agencies to have control over their own information technology. Most Census-based research is probably still carried out through the submission of *ad hoc* requests for non-interactive, batch processing at remote installations.

Somewhat perversely, SASPAC, which helped make Census data widely accessible, does not make the researcher's life any simpler. SASPAC is very good at producing standard format pages, tables and listings of Census counts. It cannot however be regarded as an analysis package. Technically its success owes much to its portability between mainframe environments. On the downside, the SASPAC procedures for aggregating data to customised zones which are not based on standard administrative areas (wards, districts,etc.) and for selecting specific cells for output are onerous to say the least. Any analyses at, for example, county level which require reaggregation of ED-level data to a nonstandard geography are major undertakings. The inflexible nature of the software has surely killed off many potential research projects which call for flexibility and ease in spatial manipulation of data. Thankfully GIS software is now available to facilitate such research.

Since the last Census, local government has experienced the widespread introduction of new information technology. The potential for putting this investment to work in analysing data from the 1991 Census would appear to be great. However the design brief, mainly set by local authorities, for the development of software to 'analyse' 1991 Census data is set almost in the dark ages as far as information technology is concerned. A survey of local government user requirements identified an overriding need for another mainframe-based product. It also showed sufficient, but secondary, demand for the development of a microcomputer-based product. The data processing capabilities of the software, requested by local authorities, will again be similar to SASPAC, i.e. generation of listed output for standard Census areas and user-defined zones. A cynic's view of what local government wants might be of a repackaging of old software, designed for the lowest common denominator machine environment, and capable of generating lots of paper with numbers on it. The lack of innovation is most disturbing.

The next Census should open up whole new areas for research. Much of the higher level Census geography used in 1991 will be the same as in 1981. The scope for comparative analysis of 1981 and 1971 census data was greatly hindered by the reorganisation of most local government boundaries in 1974. Since then most administrative boundaries have remained unchanged so there exists greater scope for spatial analysis of inter-Census change. Unfortunately, the development brief for SASPAC91 specifically excludes the ability to interrogate both datasets at the same time. Another development that is taking place is the combining of Census data with other sources, e.g. digital boundary data, in GIS. Add to the above the significant advances that have been made in the computer-aided presentation of data and there should be an explosion in the amount of information coming out of the Census. Local government is well placed to take the lead in this—resources permitting.

The expertise is already available in Greater Manchester. The first full implementation, by Salford University Computing Department, of SASPAC on a microcomputer happened at Greater Manchester Research in 1988. Everything that can be done with SASPAC on a mainframe or minicomputer can be done on a PC—only quicker, interactively and in real time. Add to this a menu-driven command writer and even what is a most user-unfriendly software can be used by the non-expert. This might sound like a significant advance. That might be true technologically but the software is still the same and can do no more in one environment than it can in another. The real advance is in how the SASPAC output is further processed and presented.

1981 Census data allowed advances to be made in the mapping of statistics. Within local government much of this mapping was done using GIMMS after preprocessing by statistical packages like SPSS. The import and export of data between such packages, even in the same computing environment, caused many problems. In spite of these difficulties a lot of good work was done—even if the end result was often only a choropleth map. This might sound dismissive of what were, for the time, real achievements. Although the situation has changed, a lot of local government users are still going to be locked into mainframe computing. The problem is that most of the technical developments have taken place outside this realm.

As a benchmark, at Greater Manchester Research it is possible to (1) produce a SASPAC case-wise listing of selected 1981 Census variables for the 214 wards in the County, (2) reprocess this output in a spreadsheet for export to (3) a mapping package and present the data for wards classified in quintiles—in under 10 minutes from start to finish. Technically this is elegant but its real beauty lies in the immediate impact of a map that a printout of counts alongside their Census geography labels certainly does not have.

Another advance has been the interrogation of Census data using GIS software.

Attaching Census-based attributes to ED centroids within a GIS has opened up many avenues for research. For example, pseudo-ED polygons can be generated for spatial classification of Census variables; contour mapping techniques can be applied to stratify Census data spatially; and point-in-polygon techniques can be used to extract Census profiles for user-defined areas.

At the time of writing, data from the 1991 Census should become available within twelve months. From my perspective, the scene seems set for local government, in Greater Manchester at least, to turn that data rapidly into information through the application of new techniques which are already being applied in practice.

13

The 1991 Census of Population: A marketing perspective

John Beaumont
University of Bath

Background

For many of the readers of this *1991 AGI Yearbook*, by the time you read this paper, you will have "participated" in Britain's nineteenth decennial Census of Population in a series which dates back to 1801. The large-scale datasets available from the Census are of fundamental significance for the operation of central and local government. Similarly, as has been demonstrated by the recent growth in so-called "market (or geodemographic) analysis", Census datasets are also extremely important for private sector organisations; applications areas include marketing information systems, branch location analysis, direct marketing and credit referencing (see Beaumont, 1990 for more details).

Since CACI's provision of 1971 Census data and geodemographic analyses based on ACORN (A Classification Of Residential Areas) in 1977, a number of other organisations have entered this growing market (see the Trade directories in this book and note that a fundamental objective of ESRC's Regional Research Laboratory (RRL) Initiative is that the RRLs will become "self-financing" through the products and services that they provide to external clients).

With a cost for the 1991 Census of slightly less than £135 million (reported at the end of December 1989 to the House of Commons Fourth Standing Committee on Statutory Instruments), it is pertinent in the context of "tradeable information" to highlight the various markets for this dataset. Attention in this paper is concerned exclusively with the private sector, although there are many parallels in the public sector (see also the local government perspective in this section).

1991 Census of Population

Marketing is no longer focused on products and services, rather it is or should be concerned with customers and satisfying their requirements. As with any new or updated product, for the exploitation of the 1991 Census, it will be necessary to understand customers' (or potential customers') needs. Importantly, owing to a range of development factors including the greater significance given by organisations to an external marketing orientation and the segmentation of their customers, the requirements to be able to profile customers and small neighbourhood areas will continue to be important. The Census is a comprehensive dataset that describes the social, economic and demographic mosaic of the country. Its secondary use in market analysis will be given a major impetus by the availability of 1991 Census data (although, somewhat surprisingly, many organisations still seem satisfied with 1981 data!).

In addition, the new nature and type of Census products and services should facilitate further market analysis applications. For instance, the "core" Census dataset, the Small Area Statistics (SAS), will be available by the output areal units, Enumeration Districts and postcode-based pseudo-Enumeration Districts. Neither standard Enumeration Districts nor postcode-based units will be provided with digitised boundaries. For England and Wales, the Office of Population Censuses and Surveys will produce a Postcode-Enumeration District directory to enable spatial linkages; for Scotland, the digitisation of postcode boundaries is being undertaken and will be available. ESRC is considering the financing of a pilot project to test the feasibility and costs of recognising and digitising 1991 Census output boundaries in England and Wales.

From SAS, new geodemographic discriminators, describing similar and dissimilar neighbourhoods, can be envisaged, although many companies will have to think carefully about direct linkages with their earlier 1981 versions (such as CACIs' ACORN, CCN Systems' MOSAIC, CDMS's SUPER PROFILE, and Pinpoint's PiN).

An exciting and very significant new dataset from the 1991 Census will be a Sample of Anonymised Records (SAR). To date, micro-level Census-based analyses in Britain have been precluded. Many past analyses based on SAS have suffered from the so-called "ecological fallacy", which means it is very difficult to make reliable inferences about relationships. There have been formidable methodological and interpretative problems. Significant new analyses, however, will be possible at this important decision-making unit level. ESRC has led both the academic consultation on the Census (see Wrigley, 1990 for more details) and the specific initiative for the agreement to release this data (see Marsh *et al.*, 1990 for more details). The SAR will comprise two samples, 2 and 1 per cent, from the 1991 Census:

- the 2 per cent sample—a rectangular file of individuals, with fairly detailed administrative geographic (local districts amalgamated to areas of 120,000 minimum population size);
- the 1 per cent sample—a hierarchical file of households and all individuals in them, with only the standard region identified.

The majority of basic Census variables will be on the SAR dataset. Special precautions will need to be taken in providing SAR to avoid accidental identification of any individual. More detailed classification systems than existing geodemographic discriminators should be feasible, focusing on household structures which have direct impacts on consumers' behaviour and consumption patterns.

As well as new data, it can be envisaged that data distribution and support will be extended and enhanced. During the 1980s, on-line access to different datasets became increasingly important, as did the availability of microcomputer-based systems (such as CACI's InSite System). For the 1991 Census, CD ROM-based products with supporting manuals and training can be expected. OPCS and ESRC funded a project at Birkbeck College and the University of Manchester to develop an on-line system for 1991 Census data. The project has demonstrated that, technically, such a system can work very well and be relatively cost-effective even for large-scale national datasets; the constraint, however, remains the perceived confidentiality problems. An update of Rhind's (1983) *Census User's Handbook* is essential; there may also be scope to design and develop a "hypertext"—type overview of the 1991 Census as stand-alone introductory material for active learning by trainees on personal computers.

Discussion

The markets and their requirements for the 1991 Census should be much larger and more diverse than for the 1981 Census. On the other hand, it must be stressed that the continued growth and development of the market analysis industry is likely to exhibit a reduced overall reliance on Census data *per se*. Many organisations are recognising the power of establishing and developing their own customer databases, and some companies have gone into the business of collecting and maintaining "lifestyle" databases. Important longitudinal, rather than cross-sectional, analyses become feasible. (For example, note ESRC's Interdisciplinary Research Centre on "Micro-social Change in Britain" at the University of Essex which intends to build up a panel dataset of 5000 households monitored in annual waves).

In terms of strategic positioning of the "market analysis" suppliers, it is interesting to note that, to varying degrees and with different specific emphases, the evolutionary stages of product development (and repackaging) and service provision are from data provision through statistical/analytical consultancy to Decision Support Systems. As with many (service) industries, "niche" players will come to characterise the industry as they attempt to provide "added value" to customers. This competitive and (industry–academia) collaborative environment will receive major stimuli from the availability of 1991 Census data.

To conclude, looking beyond 1991 in the context of the Single European Market (with its 320 million plus consumers or its 850 million plus consumers, depending on your definition of Europe), it is a strong personal belief that there is a market for the derivation of a pan-European geodemographic discriminator, which, in part, would be based on Census datasets.

The future is exciting, although unknown. There is an important role for education and training to ensure the Census datasets are exploited fully. It will be important to be professional marketeers, at least in the sense of proving a quality customer service.

References

Beaumont, J R (1990). *An Introduction to Market Analysis*, Norwich: Geo Books.

Marsh, C., Arber, S., Hobcraft, J., Lievesely, D., Openshaw, S., Penhale, B., Skinner, C., and Walford, N. (1990). The Case for a Sample of Anonymised Records from the 1991 Census. *Journal of the Royal Statistical Society*, Series A, in the press.

Rhind, D. (1983). *A Census User's Handbook*, London: Methuen.

Wrigley, N. (1990). ESRC and the 1991 Census, *Environment and Planning, A*, **22**, 573–582.

14

Geographic information systems—A European initiative

Peter Dale
Polytechnic of East London

This meeting was held "behind closed doors" at the Metropole Hotel, Brighton, on the day prior to the AGI National Conference, 1990. It was attended by 36 speakers and delegates from 10 EC countries representing a variety of organisations—governmental, quasi-governmental, commercial and academic. Data and systems producers were in a majority though there was some voice from users. The AGI was represented by its Chairman, Mike Brand and the Chairman of the Publications Committee, Peter Dale.

The meeting was opened by Martin Littlejohn of the CEC DGXIII-B2 who outlined the current thoughts of the CEC on IMPACT (Information Market Policies and ACTions) where an initiative into the developments in geographic information systems is under review. The Commission has identified a number of barriers to the development of pan-European activity, including:

- a lack of standardisation in data capture;
- inadequate map coverage especially in less developed areas of the Community;
- inconsistent locational referencing and projection systems amongst existing datasets;
- problems with copyright, royalties and data protection;
- data transfer formats to facilitate inter-country data exchange;
- inconsistent feature coding and classification systems;
- a lack of internationally accepted data structures;
- limited availability of common datasets;
- a lack of coordinated training of professional GIS users;
- discrepancies in the utilisation of GIS between countries of the EC; and
- the poor transfer of skills and technology to Less Favoured Regions (LFRs).

The morning session was taken up by presentations that focused on the shortage of information about what is happening in the different European countries and the extent to which they embrace GIS concepts and practices. Indeed much of the day was spent in data exchange, that is in informing delegates of what is going on. The need for a European Transfer Format (ETF)—a logical extension of our own approach in the UK to the National Transfer format (NTF)—was extensively discussed. There would be need for a central body to maintain such a standard and neither CERCO (Comité Européen des Responsables de la Cartographie Officielle—a body consisting of the heads of European national mapping agencies) nor AM/FM Europe were considered suited to such a function. The CEC might have a role as facilitator in creating a coordinating body but did not see their role as executor. Standards are, however, matters that it might address and these formed the basis for most of the discussion during the morning. A number of delegates felt that better data modelling was a prerequisite to the development of a European set of standards. The NTF was however developed without such a model which was subsequently discovered to be embedded within it. A distinction was made between the needs of the AM/FM and the environmental agencies but these were not explored and the debate in the end was inconclusive.

The afternoon session began with technical presentations on the work of one vendor and on the needs for data for in-car navigation. This was followed by a presentation from Portugal on some of the problems of LFRs (Less Favoured Regions) and finally there was a progress report on CORINE. There was little of controversy, the emphasis being on the development of awareness in delegates of some of the issues. The subsequent debate largely ignored the earlier presentations. It reiterated the problems of standards, explored the value of cost-benefit analysis and the possibility of learning from the experiences of others, and touched on (though failed to debate) issues relating to training and education.

The organisers, MVA Systematica, then summed up the lessons which differed little from those specified before the beginning of the workshop. The aim had been:

- to identify likely future developments;
- to establish the requirements of both users and providers of services;
- to identify key issues restricting the development of GIS applications within the EC;
- to examine the need for Commission initiatives to overcome barriers; and
- to advise on the feasibility and likely effectiveness of such approaches.

It was difficult to see immediate tangible results from the deliberations, other than a clear statement of a need for pan-European standards, especially for data exchange. There is clearly value in the exchange of ideas on an EC country basis and for that it was not unreasonable to restrict severely the number of those attending to ensure some degree of balance. Those unable to attend missed little. There was an invitation from the Project Officer for IMPACT II (a five year project for which funds have not yet been agreed, so no initiative may be forthcoming) to anyone interested to submit evidence to the Commission. One had the impression that the CEC was genuinely keen to receive such representations. The background information for the workshop had been fairly widely circulated, especially by the AGI in the UK. MVA Systematica have asked that views should be expressed through them for onward transmission to the Commission. There is need for a wider range of issues to be identified—together with the benefits to the EC of any further IMPACT investment.

Reprinted from AGI News

15

Eurodata—Myth or reality?

Helen Mounsey
Coopers & Lybrand Deloitte

Introduction

Listen to a conversation between a group of GIS users, and sooner or later their talk is likely to turn to the question of spatial data—the problems of establishing spatial databases and the difficulties in data collection from paper documents. For some users, these are quite literally the million dollar questions, as an increasing body of experience of the true costs of map digitising reinforces this figure as reasonable for a large GIS user.

A growing number of organisations are seeking to defray the high costs of data collection by marketing the information they have collected. As Blakemore (1990) has noted, information as a saleable resource is big business. There is, however, a catch to this. For information is only valuable on the open market if that market knows of its existence. Knowledge of and access to the dataset are critical to the commercial realisation of value. Although many datasets distributed by commercial data-brokers are well known, knowledge of and access to many datasets originating from government sources is still hazy. The efforts of groups working to alleviate this problem (for example the AGI committee working with the Department of the Environment as part of the Tradeable Information Initiative) represent a welcome contribution.

Now expand the horizons beyond Great Britain, extend the requirements to European data, and the difficulties noted by Blakemore multiply 25-fold (give or take a country or two). Many UK-based users of GIS are still too busy with the problems of their own national data to have any use for its European counterpart. In anticipation of a post-1992 upsurge of interest in this topic, however, this paper seeks to provide pointers to sources of European data, and to highlight some of the difficulties which may arise in the pursuit of data from Continental sources.

Topographic data

Basic topographic coverage of an area is likely to form the starting point for any digital database. This might include altitude, hydrography and man-made features (settlements, road, railways, etc). Such data may be used simply as a "backcloth" or locational framework for the portrayal of other spatial datasets. Alternatively, it may represent a source of data for spatial analysis and modelling.

Since 1986, CERCO (Comité Européen des Responsables de la Cartographie Officielle) has been working to establish a European Territorial Database (ETDB). Work Group V of this group is charged with defining the minimum information content for ETDB to meet the needs of a diverse range of users. These include (at least potentially) other cross-European multi-topic database builders, and those who require consistent, high quality information for specific end products such as in-car navigation systems. CERCO argues that responsibility for development of component national parts of ETDB should rest with survey and other official organisations at a national level. Thus its efforts to date have been directed towards the development of an international standard for data collection and exchange (discussed further below) to underpin the consistent development of ETDB.

In the meantime, the user wanting to assemble a topographic database for Europe, even in paper form, is likely to face some difficulty. Parts of Europe enjoy some of the highest standards of topographic mapping anywhere in the world. However, international comparisons of cartographic products are very difficult given the differences in scale and content. The largest scale at which complete topographic coverage for the 12 member states of the European Community is available is 1:100,000. Venture outside the EC and the coverage does not improve. Information about the sales and coverage of national topographic mapping is often difficult to obtain, but Parry and Perkins (1987) provide a valuable reference document in this area.

A second problem relates to the information content and accuracy of topographic maps. Surprisingly little effort has been directed towards consideration of map accuracy (Rhind and Clark, 1988). Cassettari (1989) has highlighted the differences between topographic maps for six countries in Western Europe. These include content, style of presentation, currency and frequency of update. These will require reconciliation before a consistent topographic database for the continent is produced.

The current availability of national topographic data in digital form is difficult to establish and no single document is readily available to provide such information. Many national survey organisations are known to have initiated digitising programmes, although the scales, content and spatial coverage of the resulting data are not well known internationally. Users who require data for any particular country are best advised to approach the national survey organisation, as listed in Parry and Perkins (1987).

NATO represents a major player on the future field of digital topographic data. Specifications for the raster scanning of the (military) 1:50,000 series in each member country are to be released in 1991. These will allow for the development of a consistent raster product for use as a "backcloth" for display of other datasets. The specifications do not attempt to resolve national differences in content, style of presentation or accuracy, however, and there are no plans to vectorise the resultant raster information. Longer term plans include the development of specifications for a number of vector-based products at 1:50,000 scale. These include products for terrain analysis, road networking and air navigation.

Many users may not have a requirement for medium or large scale topographic data across Europe even if it were available, if only because of the likely volumes of data involved. The area of the European Community alone is 2.25 million km^2 (the equivalent of at least 1600 Ordnance Survey 1:50,000 sheets, depending on layout). Data volumes which might result from a digitising programme are impossible to calculate as they depend on database content and structure, but are likely to be measured in gigabytes. Few GIS users have practical experience in handling such large databases. However, the experiences of researchers working with large datasets derived from satellite imagery suggest that problems of data storage and rapid access are non-trivial.

The most consistent European digital topographic databases currently available are derived from maps at 1:1 million. Several commercial organisations offer such datasets (for example Petroconsultants, Bartholomew and the Automobile Association). Plotted at source scale, 1mm on paper represents 1km on the ground. Whilst useful for small scale data display they are low resolution, highly generalised representations of reality, with limited applications for detailed spatial data analysis and modelling. It seems likely that a database derived from products within the range of scales 1:100,000 to 1:500,000 will have the widest range of applications in the foreseeable future. A database derived from satellite imagery is technically possible, and would be up-to-date, consistent and unaffected by variations in national data collection practices. Without the addition of "human" features (principally place names and administrative boundaries), however, it would find limited applications. A composite of both sources seems to represent an appropriate solution.

Socio-economic data

Topographic data by itself may be of limited use to many customers. Additional information in the form of population and other socio-economic indicators (employment, education, health, industrial production, etc) is often required for detailed applications. Such data is often difficult to access on a national basis, a problem compounded in an international framework.

The Statistical Office of the European Community (SOEC) has as its remit a requirement to assemble socio-economic statistics for the 12 member states and to validate the resultant datasets, which are included in the EUROSTAT programme. Much of the data that it holds is at national level only, and thus of limited use in the detailed study of regional variation across Europe. However, there is a well defined hierarchical subdivision of countries into the so-called NUTS regions. These are based on the national administrative subdivisions in any member state (for example standard regions, counties and districts in the UK). The use of existing administrative units for data collection lessens the problems of data availability. However, the great variation in size between regions across Europe presents a continuing challenge to users (for example the standard international definition of industrial sectors, so that true patterns of unemployment by sector might be assessed).

SOEC undertakes this role, and as a result is developing valuable and consistent time series of data for the member states of the European Community. It also ensures the availability of adequate supporting documentation. The requirement for consistency and standardised definition of attributes imposes a limitation on the geographical resolution of available data. Internationally consistent data is often not available at a detailed level. In contrast, assembling what local sources are available into regional and national datasets can present a considerable statistical and bureaucratic challenge. As a result, much of the Eurostat data is available across Europe to NUTS level 3 only (counties in the UK). Nevertheless it represents a valuable starting point for the comparative regional studies in Europe. Although now outdated, Hudson *et al.* (1984) show how extensive use can be made of EUROSTAT data for inter-regional comparison.

Statistical data by administrative region are of greatest use in a GIS only if digitised boundary information is also available. In the UK, digitised ward boundaries (NUTS level 5) are available for the whole country, and some other member states are in a similar position. However, the practicabilities of reconciling digital boundaries at this level are immense. Many have been digitised to variable levels of accuracy, on national projections and with differing co-ordinate systems. The CORINE programme (see below) has assembled the detailed boundaries for the Benelux countries into a single database, but assimilation of the remaining digital data (where available) represents a considerable challenge. Level 3 boundaries in digital form, based on a source map at 1:1m, are available from this programme however.

Environmental data

The requirement for sound, objective and reliable information on the state of the European environment was recognised by the European Community in the early 1980s. The CORINE programme (Co-ordinated Information on the European Environment) was established in 1984 to supply information to underpin the development of Community environmental policy. The database developed since the formation of the programme represents a valuable source of consistent information on aspects of the environment. It will continue to be maintained by the European Environment Agency, once this is established.

As with the EUROSTAT programme, the CORINE programme aims to assemble a consistent and well documented database of information relating to the whole of the European Community. Environmental data has proved harder to obtain than socio-economic statistics, but significant progress has been made and the database now includes information on climate, soils, vegetation, water quality and air pollution amongst other topics. CORINE (1990) catalogues the availability of data from this source.

Challenges to data collection

Although programmes such as EUROSTAT and CORINE have made significant contributions to the cause of cross-European data availability, there is still much to be done. The principal challenge still remains that of data availability, consistency and quality. As Briggs and Mounsey (1989) have noted, inconsistencies arise even within datasets due to differences in timeliness, measurement methods, spatial coverage and density of sampling. Discrepancies may be hidden behind inconsistent use of terminology, imprecise definitions and poor documentation. Until there is definition of European standards for data collection, the responsibility lies with the user to ensure that the data is fit for purpose and the results of analysis sound and applicable.

A European Transfer Format

Members of CERCO and many other users of European spatial data have long recognised a requirement for a single transfer format for data exchange, where data is available. Several national data transfer standards are now in place, and users who wish to access data which spans an international boundary will need to develop interfaces between often incompatible formats. As Sowton (1990) has shown, in the worst case a country such as France with its own transfer format requires an interface to seven adjoining countries. The development of a European Transfer Format (ETF), and of interfaces between this and existing or proposed national formats, would reduce the requirement for interfacing to one (the national format to the ETF), in contrast to the seven required at present.

Some progress has been made towards the development of an ETF. National standards are already widely used in Denmark, Finland, Germany, Holland and the UK, and are under development in France, Norway and Sweden. In addition, ISO 8211 forms an international standard for data exchange, and a military exchange standard, DIGEST, has also been proposed.

CERCO have defined the requirements for an ETF. These include a need:

- to make possible the physical exchange of data;
- to enable the geographic content of the data to be understood; and
- to be compatible with ISO standards.

Currently, the UK National Transfer Format (NTF) meets the first two requirements, and an ISO compatible version is under development. Although NTF could form the basis of an ETF, any decision is dependent on a number of factors:

- an evaluation of the new ISO 8211 compatible version of NTF is required;
- verification of its compatibility with the data model and structure of the European Topographic Database is also required;
- rigorous testing is required, to cover the transfer of topographic data, and also a wide range of other datasets including typical utilities information, digital boundary information and associated socio-economic and environmental attributes, etc; and
- it is not without competitors for this role (for example, DIGEST).

The development of an ETF is not a panacea to all problems of data availability and compatibility. Any development must tread a fine line between simplicity for ease of understanding and flexibility for maximum application on the one side, and sufficient structure and rigour to ensure consistency on the other. As an example, the coding system used by Ordnance Survey (Northern Ireland) includes codes for utility data, whilst responsibility for the equivalent data for Britain rests with the National Joint Utilities Group and is not defined within NTF. Whilst this ensures the simplicity and flexibility of NTF, user specification of codes allows considerable scope for variation. The answers to such problems are neither clear nor well defined, and responsibility for consistency rests with the user in the long term.

Conclusion

This paper has attempted to address the question of availability for Europe—what exists where and in what form? Although many international organisations and programmes have made significant progress in addressing the issues behind these questions, a large scale Europe-wide database of consistent topographic data and supporting attributes is still some way off. The obstacles along the path to the goal are not primarily technical ones, but those of cost. As Blakemore (1990) has noted, the cost of gathering and validating data rises disproportionately as greater statistical precision is demanded. The development of a sound case for investment in a European database is still a difficult one.

References

Blakemore, M. (1990). Sharing data—whose, why, how? Proceedings, *AGI National Conference*, Brighton, 21–23 October 1990.

Briggs, D.J. and Mounsey, H.M. (1989). Integrating land resource data into a European geographical information system: practicalities and prospects. *Applied Geography* **9**(1), 5–21.

Cassettari, S. (1989). Scaling the heights. *Geographical Magazine*, August 1989.

CORINE (1990). *Examples of the use of the results of the Programme 1985–1990*. Brussels, Directorate General XI (Environment, Consumer Protection and Nuclear Safety).

Hudson, R., Rhind, D.W. and Mounsey, H.M. (1984). *An Atlas of EEC Affairs*. London: Methuen.

Parry, R.B. and Perkins, C.R. (1987). *World Mapping Today*. London: Butterworths.

Rhind, D.W. and Clark, P.K. (1988). Cartographic data input to global databases, in Mounsey, H.M. (editor), *Building Databases for Global Science*. London: Taylor & Francis.

Sowton, M. (1990). Towards a European Transfer Format. In *Towards a Common International AM/FM Transfer Format*, Proceedings, AM/FM Workshop, Montreux, 1–2 October 1990.

16

Spatial data overview: UK vs the Rest of the World

Robin McLaren
Know Edge Ltd, Edinburgh, Scotland

and John Tulip
Prime (Canada) Ltd, Vancouver, Canada

Due to the relatively small extent of the UK and the strategic and fiscal significance of mapping imposed by successive governments, the UK has always supported an excellent mapping programme. This eminence was sustained through the 1970's by the UK and Northern Ireland's Ordnance Surveys' initiatives to push the UK into the forefront of digital mapping worldwide. Although faltering in the early 1980's, the recommendations of the Chorley Commission (DoE, 1987) have now set new priorities to promote the growth of the UK GIS market in the 1990's. However, are there sufficient data sources to support the demands of this expanding market? Are we a data-rich nation or have we been relegated to the realms of the data-developing nations? This article explores the current status of data resources in the UK and compares this against other examples from around the world to establish our "league" position and to identify trends that could influence the UK situation; round the world in a thousand words.

In the UK it is currently difficult to assess the availability of GIS-compatible data sources since no definitive inventory is currently maintained. Although Appendix 3 of the Chorley Report (DoE, 1987) lists the major datasets of spatially referenced information held by central Government and local authorities, it is out of date and gives no indication of the availability and quality of the datasets. However, two current metadata (data about data) investigations are leading the way for the creation of a national spatially referenced information inventory. The Department of the Environment has just recently (December 1990) compiled an inventory of spatial information held by central Government Departments as part of a wider Tradeable Information Initiative (Garnsworthy, 1990). This database of metadata is due to be published in the public domain by a third party in 1991. The Royal Institution of Chartered Surveyors (RICS) is currently embarking on creating a similar inventory of environmental datasets available within the UK and this should be available in a similar timeframe. The availability and cost of these identified datasets must still be negotiated with each of the corresponding Departments. These initiatives will certainly raise the GIS user's awareness of the existence of the information, but may also cause equal frustration due to a potential lack of resources, mandate and incentive for co-operation by the data owners. Further work is required to determine the market value of this tradeable information, in its raw or added-value form, and to resolve the associated institutional and legal issues before the information will be released.

The problems of dataset inventories and the legal problems associated with their wider use are still major issues to be resolved worldwide. No countries currently maintain a definitive inventory of available spatial data. Due to this deficiency some unofficial ones have been created. In the USA, for example, the *GIS Source Book* (*GIS World*, 1989) quotes some 300 sources of digital, spatial data, including:

Base map	130
Land use	60
Environmental	55
Demographic	20
Utility and transport	20
Land & property	10
Gazetteer	10

62 per cent of these datasets are managed by Federal agencies, 13 per cent by State agencies, 7 per cent by Municipal authorities and 17 per cent by the private sector.

The disclaimer is still the temporary solution to potential copyright and litigation controversies and market value issues are still being formulated. However, the USA has a distinct advantage over the UK since its Constitution has a Freedom of Information Act allowing the public right of access to Government-created information including software. This, in conjunction with an advanced communications network infrastructure, has encouraged the growth of a considerable information services industry where 75 per cent have been implemented by the private sector, as opposed to Europe where 70 per cent have been established by the public sector (EC, 1990).

Of the spatial information available within the public domain in the UK, the greatest national assets are the Ordnance Survey's large scale, digital, topographic mapping and the Land Registry information. The 1/1,250 component of the urban digital mapping is due for completion in the UK in 1992. In other countries most large scale, urban, topographic mapping is not provided by central government organisations, but is supplied on an ad hoc basis by municipal authorities or is even created by private companies for their internal use. Therefore, the UK has the distinct advantage of encouraging integrated urban GIS applications by having a complete, consistent level of detailed urban digital mapping.

However, at smaller scales the UK is less well endowed. The rural areas covered at 1/10,000 and smaller have no digital topographic mapping despite several pilot projects and much debate. This unfortunately encourages discrete data capture projects that never reach the public domain and discourages the use of GIS within resource management and environmental impact assessment applications. This is certainly not the case in Canada, for example, where most of the provinces are undertaking major programmes to create digital information at 1/20,000 for the entire province directly from original surveys. This is being primarily driven by the need for better resource management over their vast territories. Similar strategies can be found in the USA and Australia.

A major step forward in the availability of spatial data in England and Wales was achieved in December 1990 with the partial "opening" of the Land Registry information, to bring it in line with most land registry systems worldwide, including the Land Registry in Scotland. Hopefully, this a first step towards more extensive openness and ease of access as is evidenced in many countries in Europe, North America and Australia where an open cadastral system has historically operated. The corresponding large scale, cadastral mapping, defining the land parcels and information associated with ownerships, has become the basis for integrated, multipurpose, land information systems. Considerable success has been achieved in Sweden with the much publicised Swedish Land Data Bank. This computerised system of property and land registration interacts with many other information services and acts as a land information system in the broadest sense; applications include planning, natural resource management, banking, and real estate brokerage. Users have access through Videotex or directly through remote terminal connections.Although annual costs of the computerised approach are twice as high as the former manual system, the revenues collected enable substantial and increasing profits to be realised. Similar systems have been successfully implemented in Austria, the Maritime Provinces in Canada, New Zealand, and several states in Australia. This integrated approach must be a high priority development for the UK, extending even further the progress made in Northern Ireland, for example.

There has been considerable criticism of the UK GRO's decision not to base the 1991 England & Wales Census on enumeration districts based on unit postcodes. Their non-standard approach is contrary to equivalent activities in the USA. As a referencing framework for the 1990 USA census, the Federal Government created a topologically correct street centreline network with street block level information where warranted (acronym TIGER). This is having a major impact on the extent of locational and demographic analysis being performed in the country and the way spatial information is disseminated.

With many of the UK Government Departments delaying the release of datasets into the public domain and growing market pressure for the availability of datasets, it is inevitable that private enterprises will create and market alternate information. A good example of this development is the gazetteer information marketed by Pinpoint Analysis. This private dataset market could well expand through information services, as it has in the USA.

This brief comparison of spatial dataset availability in the UK with the rest of the world has identified several key areas where the UK has fallen behind and where it could benefit from the experiences of these other countries. These areas include spatial referencing of census data, provision of medium scale digital topographic data, freedom of access to Government-held information, and the availability of land parcel and ownership information as the basis of a National Land Information System. However, it should be remembered that the key issue is not just the quantity and variety of information available, but the ability to integrate this information, to add value, and to educate the users in the potential applications of the data. This is the ultimate objective to be pursued.

References

Department of the Environment, (1987). *Handling Geographic Information*, Report of the Committee of Enquiry Chaired by Lord Chorley, London: HMSO.

European Community (1990). EC Commission Communication on Establishing an Information Services Market. *Communications of the ACM*, *33* (4) (Special

Issue on the EC '92), 426–432.
Garnsworthy, J. (1990). The Tradeable Information Initiative. In Foster, P. and Shand, P.J. (Eds), *AGI Year Book 1990*, pp. 106–108.
GIS World (1989). *GIS Sourcebook*.

17

Data supply for geographic information systems: Challenges for the nineties

Hans Mesker and Jan Pongers
AM/FM European Division

The nineties

The theme of this paper "Data supply for geographic information systems" is very current and fitting in our timeframe. Looking back at the past 10 years we see that this period has shown many developments in the field of AM/FM and GIS. We expect these turbulent developments to continue into the nineties. It is our challenge to cope with these developments; we will try to contribute to that.

AM/FM, GIS and Society

If we talk about AM/FM/GIS in the nineties it may be a good thing to ask ourselves: what makes GIS so important and why should there be international strategies? The answer can be short. GIS is important because it is part of the basis of our society. The importance of good, reliable and cost-effective GIS is for many people today, but especially tomorrow, "the basis of existence". Legal, economic and managerial aspects play an important role in developing socially responsible geographic information. Beyond that there is a complex relationship between geographical information and society which justifies the "interference" of government with geographical information.

Co-ordination

Apart from technological renewal and managerial developments a major tenet of a good geographical information system is cost control. Co-ordination can contribute to that, and in this sense it differs from centralisation. Co-ordination means that standards and agreements have to take place on an (inter)national level.

A good example of this would be the development of the Cables and Pipes Information Centres in Holland. For example, nowadays everybody in Holland who has to do excavation work can be informed in advance about the position and routing of cables and pipes. A lot of damage is thus prevented.

To achieve this a government policy must be guiding and conditioning, which means that there is a lot of room for co-operation. In a general sense the preference is that such co-operation will be spontaneous and not obligatory. The purpose must be a good, economical, geographical information structure that continuously answers the social needs of reliable and actual data. For that today, we are in large part dependent on technological developments.

We do not know yet all the possibilities of "1992". One thing is sure, if we want to co-operate within Europe we must have international standards for data exchange, networking, etc. The E.E.C. is one of the international bodies who can help to achieve this.

Information technology

In the field of AM/FM/GIS we see a growing use of information technology that can contribute to a more effective and more efficient distribution system, involving management and governmental behaviour. Managerial and organisational choices are needed. From an organisational point of view we believe that information technology must be fully integrated into management structure and culture. Top management must pay full attention to information technology projects and overall management.

The use of information technology requires considerable investment in people and capital. Therefore we must strive for co-ordination of tasks now done by different organisations (nationwide and international) or by parts of organisations at company level.

We must strive to prevent data being collected more than once, avoid repeated updating, and take care that the collected data satisfies certain quality conditions. Duplication is socially wasteful when the availability of exchangeable geographical information is not only of economical but also vital importance. A responsible investment in information technology also means seeking better structural communications within a company and between companies, governments and societies.

Goals for the nineties

Cost control of AM/FM/GIS as a whole

An important step in a good direction is to prevent "island-automation". When every country, organisation (or part of an organisation) or institute builds its own system without looking around at others, this will not usually result in much improvement of efficiency of information. Island-automation frustrates information exchange and hinders co-operation.

Improvement of the consistency of data

Internationally and nationally we are working very hard on activities in the field of normalisation and standardisation. In particular we can mention the development of a standard exchange format for cartographical information.

In future we have to work within local government on a so-called Municipal Function Design structure. MFD's will support the internal automation activities of individual municipalities. We must make clear which information is to be exchanged between municipal companies and the (international) government. The importance of Municipal Functional Designs is especially the fact that they can contribute to the selection of data that in future can be exchanged automatically.

Achieving a higher level of service for the customer

This has to do with the quality of services. Clear, transparent information and a good accessibility to data are necessary. An international advisory group is needed to look at how geographical information, in a wide sense, can be made easily accessible to as many users in as many places as possible.

Achieving a more effective communication in the field of geographical information within government

For several years there has been a heavily automated information exchange between governmental organisations in the field of geographical information. The next step will be to set up data communication networks between those organisations, both national and international. As far as users are concerned, from 1991 municipalities should have a "helpdesk" function where everybody can access municipal information.

Source registers

An important contribution to achieving the four mentioned goals for the nineties are the so called "source registers". In the GIS field the policy should be to have the Large Scale Base Map widely available. Among other things this aims to offer a basis for the exchange of cartographical information and through that spread the cost of making and updating base maps. For example in Holland 40 per cent of the territory is now covered by the Large Scale Base Map, but the production of the map is stagnating. This stagnation is caused mainly by the different wishes and financial conditions of cadastre, municipalities and utilities. A governmental advisory group has been ordered to producc alternative solutions.

Conclusion

We have focused on some international strategies considering AM/FM/GIS. Governments will play an important role in creating conditions to make it possible to develop successful AM/FM/GIS applications. Cost-control, technological development and exchange of geographical information are the major challenges for the nineties. We really hope and trust that our effort within AM/FM/GIS European Division will contribute to coping with these challenges.

Biographical note

Hans Mesker is Managing Director of Synergie Consultancy B.V. Jan Pongers is Managing Director of INGECON B.V. Both companies and their directors are pioneers in the field of GIS services in Holland as well as in Europe.

PART IV

GIS and the environment

At the present time green issues have a high priority on the agendas of both public and private organisations. The application of GIS technology to seek green solutions is not new. In fact, as the paper by I. Heywood in this section illustrates, concern about our natural environment fuelled initial developments in GIS over three decades ago.

Over the last few years, however, there has been a noticeable increase in the number and range of environmental problems seeking a GIS solution. The volume and diversity of subject matter covered by the authors in this section serves to illustrate the point. C. Bird, for example, looks at the miles behind the use of GIS as a tool to assist in land use monitoring whilst the pressure group Friends of the Earth examine the persuasive power of GIS for environmental education.

18

Managing our future environment: Some GIS issues

Ian Heywood
Department of Geography, University of Salford

Introduction

In the UK green issues and the need to comply with new environmental legislation are being used to justify funding for a wide range of GIS projects in both the private and public sectors. Numerous GIS vendors and prospective GIS users have seized upon this opportunity to promote GIS technology as a new solution to collating and managing environmental data. There is, however, nothing really new in using GIS technology to collate information about the environment. In fact, as several authors (Tomlinson, 1989; Peuquet and Marble, 1989) note, GIS emerged as a response to complex environmental issues that were first highlighted by the environmentalist movement of the 1960's.

The first, and still predominantly the major, role for GIS in environmental management is strategic, concerned with resource inventory. Finding out where things are and what state they are in is still one of the major reasons used to justify expenditure on GIS. In this respect, the current uses to which GIS are being put have developed little in the last 30 years. A second important role is in prioritising the status of environmental resources and the need to determine the impact associated with their usage. The analytical use of GIS to help with these tasks, however, has predominantly been limited to the calculation of areas statistics, simple map overlay analysis and limited spatial queries. In fact much of the methodology for performing these types of tasks was developed during the construction and implementation of the world's first GIS, The Canadian Geographic Information System (CGIS). In this article some of the major issues which are likely to influence the development of GIS as an effective tool in environmental planning and decision-making are discussed.

Environmental GIS issues

If GIS technology is to find wider application in environmental management than resource inventory and simple impact assessment there are a number of issues which users and developers need to address. These include: the choice of basic spatial units for environmental reporting; the volume and availability of spatially referenced environmental data; the quality of this data; whether or not the analytical functionality and modelling capabilities of present-day systems are sufficiently sophisticated to cope with the needs of environmental managers, policy makers and scientists; and finally, to what extent GIS user interfaces and visualisation methods satisfy the needs of environmental users.

Anyone working in GIS will recognise that this list of issues mirrors a research agenda for the development of GIS technology as a whole (Maguire, 1990). However, each aspect of the list can be looked at from an environmental perspective.

Basic spatial units for environmental science

The principal problem with a large amount of environmental information is the lack of consistency in the use of a basic spatial referencing unit. Currently there are only two nationally agreed guidelines for standardisation on spatial referencing in the UK—the Ordnance Survey (OS) National Grid and the Postcode (DoE, 1987). The use of the postcode is generally only relevant where environmental surveys relate to households or other properties. The use of grid referencing satisfies the requirements for the spatial referencing of point-specific monitoring programmes. However, neither of these systems provides a solution for the large amount of environmental data which are not point-specific. These data require some form of zonal referencing system, but at the present time there is no nationally or internationally agreed standard for the zoning of geographical data.

The problem might be solved by the adoption of existing geographical zones such as the census districts or by the definition of new environmentally based zones. There is a need for environmental agencies to agree on some UK standard to ensure future compatibility of environmental information and to avoid the current dilemma faced by those attempting to integrate social and economic information which is aggregated into over fifty different spatial units (Openshaw *et al.*, 1987).

Data availability and access

Environmental data is collected by a wide range of organisations, collated in many forms and stored on numerous computer systems. There is currently no single agency which keeps track of digital environmental data or any data directories which could guide prospective users to relevant data sources. Despite the attempts to create central repositories for social and economic data at both national (for example the Economic and Social Research Council Data Archive) and regional level (for example the Midlands Regional Research Laboratory data directory) there have been few equivalent environmental initiatives. The Environmental Information Centre (EIC) at the Institute of Terrestrial Ecology, Monks Wood, Cambridge (set up in 1989) is the only really comparable project (see the paper by Brown in this section).

That is not to say that there has been a lack of activity in building environmental databases in the UK; rather that they have concentrated on being application-specific rather than for generic use. Examples of such databases would include the UK National Parks Database (Bird and Taylor, 1990); the Walter Database (the Wales Terrestrial Database Project) (Bracken and Higgs, 1990); and the database being constructed for the Nature Conservancy Council. The only project to date which has focused on collating environmental data in a spatial framework for more broad-based use is the Europe-wide CORINE project (see the paper by Briggs in this section).

Perhaps the quest for a central repository or set of directories is somewhat Utopian since even if such a system could be implemented the current problems of data standards, pricing and access might well limit its usefulness. The problem of standards may well sort itself out as the GIS marketplace stabilises. The real issues lie in ensuring that standards are dictated by the requirements of environmental managers and are not imposed by system developers.

Price and access, on the other hand, are perhaps the two major data issues which are most likely to constrain GIS technology from developing beyond an inventory technology. In order to develop new techniques of environmental spatial analysis it is essential that researchers in both public and private organisations be allowed to 'play' with data. Unfortunately, with the growth in the use of digital technology the value attached to data has increased. Currently, we are in a period where we are unsure how to price data. Should we make it available cheaply so it becomes widely used or should we price it highly to restrict its use and ensure the greatest profit is made by the collector?

Unless we can resolve issues of pricing and distribution of environmental data then the application of GIS technology will remain restricted. It would be somewhat ironic if global environmental applications of GIS were constrained not by the limits of the technology but by the cost of data.

Data volume and quality

Data volume and data quality are perhaps two of the most frequently underestimated parameters in the development of any GIS strategy. The problems of data volume are multiplied in environmental GIS applications because many users need to use digital data from remotely sensed systems or 'real time' digital monitoring programmes. Most environmental research also has a temporal component which requires multiple datasets to be analysed for the same geographical area, compounding the data volume problem.

Much of the current data volume problem is due to less-than-efficient storage and data compaction methods. Despite advances in computer technology and ever-increasing disk space there is a need to develop intelligent spatial filtering methods which can be used to reduce data volumes by pre-processing before storage. Artificial intelligence, knowledge-based systems, automated generalisation procedures and spatial interpolation routines all offer solutions for reducing data redundancy and duplicity. However, few, if any, of the GIS available offer these features as part of their spatial data handling toolkit. The development and implementation of effective data compaction and intelligent filtering techniques will be essential during the 1990's if our ability to analyse data is to keep pace with our ability to collect it.

Data quality problems are inherent in most environmental analysis. In GIS these problems are compounded because of the inaccuracies associated with locational referencing and the procedure of combining datasets containing different levels of error. Most GIS users are aware that their source data contain error. In a recent survey (Coward and Heywood, 1990) all respondents stated that they knew their source data contained errors. More worrying was the fact that 30 per cent of the respondents stated that that this error was ignored when performing analysis. To date one of the primary justifications for ignoring the error effect in processing of spatial data is that an answer (even of dubious quality) is 'better than nothing'. In the UK at least this justification is holding up because we are still experiencing a honeymoon effect associated with the use of GIS technology for most applications. The consequences of bad environmental policy formulated through the use of GIS technology are still to be tested. What, for example, will be the consequences for the environment or the legal implications of a bad environmental policy?

It is, therefore, absolutely essential that over the next few years research focusing on the handling of error in a GIS framework (see the article by Carver and Openshaw in this volume) is translated into real solutions for gauging the confidence of environmental decisions.

Analytical functionality and modelling

Parallelling the development in GIS technology over the last two decades has been an equally important growth in the use of computers to help model a range of environmental processes. These models range in scale from those designed to predict the dispersal of pollutants in individual salt marshes through to modelling the potential effects of global warming. In all cases, the purpose of these models has been to help scientists understand and predict the behaviour of the environment under a given set of circumstances. One of the exciting challenges that face environmental researchers during the 1990's is the convergence between computer model building and GIS. In moving towards this goal, three key areas of research can be identified, all of which are, in part, already receiving attention. These areas include spatial model calibration and testing, data presentation and spatial model building.

Spatial model calibration and testing

One problem readily acknowledged by the developers of many environmental computer models is that their geographical application has been restricted by the problems associated with finding enough data about the character of a location for calibration. The value of linking existing, spatially illiterate, models with a GIS approach is that the GIS can be used as a 'data gateway' through which existing complex models can gain access to large sources of geographical data. This database can then be used to help calibrate and test the application of the model for different geographical regions.

Data presentation

The presentation of information from complex computer models has traditionally relied on statistical, tabular or graphical output. In many cases this provides a confusing picture. If the model is designed to investigate an environmental process which has a geographical perspective then the logical step is to take the results of the model back into the GIS and use it to present information in the form of maps, 3D images or even computer movies.

Spatial model building

Primarily because of the problems of access to digital spatial data, few computer models have explored the geographical dimensions of model building. Few, if any, of the existing environmental models have rules built in to handle issues such as adjacency, containment, spatial consistency, etc. The exciting challenge over the next few years will be the modification of existing models to draw on the strengths of GIS for handling and modelling the spatial relations of environmental phenomena.

The user interface

In common with other professionals, environmental scientists and managers do not have enough time to become proficient in using the majority of GIS products available. Currently, few GIS have been designed with the application in mind; most offer a generic set of tools which, with the right amount of skill and knowledge, can be customised. As Medyckyj-Scott (in this volume) states, the design of the user interface has been neglected by most developers of GIS software. The fact is that environmental applications of GIS will not become widespread until the technology becomes more user friendly. System developers with eyes on the environmental market would therefore do well to look at the design and packaging of their GIS with the prospective client in mind.

Visualisation

The degree to which information about the state of our environment is disseminated to the public is an issue of debate with, on the one hand, the environmental pressure groups calling for complete and open access to information, whilst others, both in government and industry, considering that only limited information should be made available due to the complex nature of most environmental issues.

Since dissemination of information must start with the education of the recipients of that information, the visual nature of GIS technology can be a powerful tool. As Openshaw *et al.* (1989) indicate, the use of a GIS at a public inquiry into the siting of a nuclear waste dump could go a significant way in de-mystifying siting procedures by showing visually the stages which the planners have gone through to determine the preferred site. Such a visual breakdown of the environmental planning process, though by no means justifying a decision, could be used to break through the written jargon that often accompanies most descriptions surrounding environmental planning. In fact such a technique has been used effectively in the province of Manitoba, Canada to explain the siting criteria used to select the location for a toxic waste dump. In addition, the GIS was also used to illustrate to the general public what the siting options would be under a range of different criteria. Closer to home, ITCERES (International Information Technology Centre For Education and Research In Environmental Science) established under the European COMETT (Community Action Programme in Education and Training For Technology) at Southampton University is linking GIS technology with new developments in audio-visual software to produce interactive computer videos to help in environmental education.

Conclusion

As the recent article on the countryside survey being carried out by the Institute of Terrestrial Ecology (ITE) (Barr, 1990) illustrates, the environmental application of

GIS technology is still typified, at least in the UK, by environmental inventory programmes and simple impact assessment techniques. Research is being carried out to broaden the use of GIS through the development of analytical and modelling capabilities. However, perhaps the most fundamental problems which are likely to limit the use of GIS into the 1990's are those associated with data quality, data access, data volume, data compatibility and visualisation techniques. The *ad hoc* funding of research in public and private sector institutes may well provide a solution in the long term. In the short term, however, co-ordination is vital at both national and international levels if GIS is to be used efficiently as a tool to aid environmental policy-making. The problem at national level is who will be prepared to take the lead?

References

Barr, C. (1990). Countryside Survey 1990, *NERC News*, October, 4–6.

Bird, A.C. and Taylor, J.C. (1990). GIS for landscape monitoring: An operational system for National Parks in England and Wales. *EGIS'90, Proceedings of the first European Conference on Geographical Information Systems*, Amsterdam, April 10–13, pp. 68–77.

Bracken, I. and Higgs, G. (1990). The role of GIS in data integration for rural environments. *Mapping Awareness*, **4** (8).

Coward, P. and Heywood, D.I. (1991). Aspects of error modelling in spatial decision making. *Proceedings of the 3rd National Canadian Conference on GIS, Ottawa, March, 1991.*

Department of the Environment (DoE) (1987). *Handling Geographic Information.* Report of the Committee of Enquiry chaired by Lord Chorley, London: HMSO.

Maguire, D. (1990). A research plan for GIS in the 1990's. In Foster, M.J. and Shand, P.J. (eds.), *The Association for Geographic Information Yearbook 1990*, London: Taylor & Francis.

Openshaw, S., Fernie, J. and Carver, S. (1989). *Britain's Nuclear Waste Safety and Siting*, London: Belhaven Press.

Openshaw, S., Goddard, J. and Coombes, M. (1987). "Integrating Geographic Data for Policy Purposes: Some Recent UK Experiences". Northern Regional Research Laboratory Report No.2.

Peuquet D.J. and Marble, D.F. (Eds), (1989). *Introductory Readings In Geographical Information Systems*, London: Taylor & Francis.

Tomlinson, R. (1989). Geographic information systems—A new frontier. In Peuquet, D.J. and Marble, D.F. (Eds), *Introductory Readings In Geographical Information Systems*, London: Taylor & Francis.

19

Monitoring change—Development of a GIS for Dorset heathland

Nigel Brown
Natural Environment Research Council

Introduction

The Institute of Terrestrial Ecology (ITE) is a component of the Natural Environment Research Council (NERC), which also includes organisations such as the Institute of Hydrology (IH), the British Geological Survey (BGS) and the British Antarctic Survey (BAS). Hence NERC's expertise covers the complete range of environmental disciplines, with ITE taking the lead in terrestrial ecology.

In recent years there has been a marked increase in awareness of environmental issues at all levels of society. This increased awareness has brought expertise in terrestrial ecology to the forefront in the quest for sustainable development and environmental protection. In particular, there has been a development in the awareness that man's actions can have ecological consequences on a regional and global scale as well as at the local level.

These developments have occurred at the same time as the potential of geographical information systems (GIS) has gained widespread recognition. As a result, in 1989, the NERC created the Environmental Information Centre (EIC) at ITE's Monks Wood Experimental Station in Cambridgeshire. This Centre now manages some of the most extensive holdings of data on the terrestrial environment in Britain.

Factors affecting environmental change

Ecological systems are complex, dynamic and highly susceptible to environmental changes, such as those brought about by pollution, agricultural intensification, deforestation, urban growth and by global processes such as desertification and climate change. Environmental pollution in various forms has been a general concern throughout the 1970s and 1980s and several recent studies within ITE have addressed aspects of the ecological consequences of pollution (Hornung *et al.*, 1990).

More recently, attention is being focused on the ecological consequences of global warming. ITE is active in the NERC Terrestrial Initiative on Global Environmental Research (TIGER), which will establish a scientific basis for assessing and responding to the ecological effects of global climate change. ITE is also collaborating with the Department of the Environment in a parallel research programme to guide the UK policy response to these issues.

Change detection and the need for information

In order to detect changes in the environment there is a need for "baseline data" to enable questions such as "which are the key species?", "what are current pollution levels?" and "what changes are occurring in species distributions?" to be posed. Long term monitoring programmes such as those carried out by ITE's Biological Records Centre (Harding and Roberts, 1986), and surveys such as ITE's National Land Classification System and Survey (Bunce *et al.*, 1980) are important sources of relevant information.

On a more localised scale, there are individual sites where detailed long-term studies have generated valuable historical datasets. Good examples of these are the Moor House National Nature Reserve (Marrs *et al.*, 1986), and the Dorset heaths, where changes in the ecology of the heathlands have been monitored over many years (Chapman *et al.*, 1989). A number of government departments are participating in the creation of an Environmental Change Network, which will designate a number of such sites, representative of the land surface of Great Britain and NERC has nominated Moor House as a "Lead Site" in this network.

Assuming that data and monitoring programmes are in place it is necessary to construct "interpretative models" of relevant ecological systems in order to understand the mechanisms of change and be able to predict future changes and their magnitude. For example, models can be constructed which describe relationships between environmental conditions—soil type, climate, exposure, grazing pressure, management history—and the botanical composition and successional trends in natural vegetation communities. By building such models into a modern GIS, it will be possible to assess future trends in relation to the rarity, geographical distribution and conservation value of biological species or communities and to make the information and analytical techniques available to a wider audience. These systems can then be used for site management, either by suggesting species suitable for introduction or by indicating how the vegetation is likely to change if the management regime is altered.

Many UK environmental agencies are giving thought to the ways in which our environment, and changes on our environment, are recorded. However, as Professor Last has stated (1981): "without considering how the potentially vast accumulations of data will be handled, this exercise would be worthless".

Current projects using GIS techniques in EIC

ITE possesses a range of GIS facilities and related technology for use in a number of research projects. The GIS include LaserScan (Horizon), ARC/INFO and Tydac (SPANS) whereas related systems include IIS image processing and SysScan digital mapping systems. Current projects employing this technology include:

(1) Global climate change studies
(2) Co-ordinated environmental information in the EEC
(3) European soils mapping (with IH, Soil Survey and the EEC)

(4) UK landcover mapping from space imagery
(5) Biological species distributions and monitoring
(6) UK Critical Loads mapping programme (atmospheric pollution)
(7) Management of map data for the ITE Countryside 1990 Survey
(8) Creation of GIS demonstrators for Dorset Heathland and the Wash.

It is important that these GIS techniques are found meaningful applications to the environmental science being undertaken and are shown to be vital tools in achieving our "environmental objectives". The Dorset Heathland project, mentioned above, is being used as a demonstrator programme for this purpose and this project will now be discussed in more detail to provide an illustration of the active application of GIS in monitoring environmental change.

Dorset heathlands—A pilot GIS development

> "Our lowland heaths are a vital and precious habitat, with some of Britain's most appealing—and endangered—plants and animals, such as Marsh Gentian and the Dartford Warbler. They are the site of many of our most important nature reserves, and at the same time provide eminently accessible countryside, attracting millions of visitors each year".
> Dr Nigel Webb (1986)

The Dorset heaths are of current interest to ITE because they are under increasing threat from urban and agricultural development. Over the past century they have been increasingly fragmented and it has now become important to assess their viability as functional heathland ecosystems. In order to do so it is necessary to characterise the status of the fragments, their shape, size and relationship to adjoining areas.

An existing comprehensive survey/database (Chapman *et al.*, 1989) provides recent detail on species distribution and diversity within an Oracle relational database. This enables detailed evaluation and examination of the ecologically important facet of species diversity as well as recent spatial information on the true extent of the heaths. Historical maps of heathland extent (Good, 1948; Moore, 1961) have been digitised and provide information on the extent of heaths at various dates since 1811. This enables good visual and statistical comparisons to be made and enables quantification of another important ecological facet, ie. the fragmentation of habitats. Fragmentation is critical as many key species such as the Sand Lizard and the Dartford Warbler can only continue to exist within a known minimum parcel size of their required habitat. The simplified maps in Figure 1 give an immediate visual indication of the extent of fragmentation that is occurring. Data of this type can be examined interactively in the GIS along with data within the relational database.

The GIS is also used to overlay these data with other information such as topographic maps and remotely sensed images. Further development of this GIS is leading to an assessment of the present status of each heathland fragment and will help in the exploration of various management alternatives. For instance, one "fragment" of Dorset Heath, Canford Heath near Poole, has been designated a Site of Special Scientific Interest (SSSI) and is one of the larger remaining areas of heathland. Urban development is now a direct threat with a recent proposal to build houses on it. Use of the fully operational GIS will allow for swift recall and analysis of a selected site such as this.

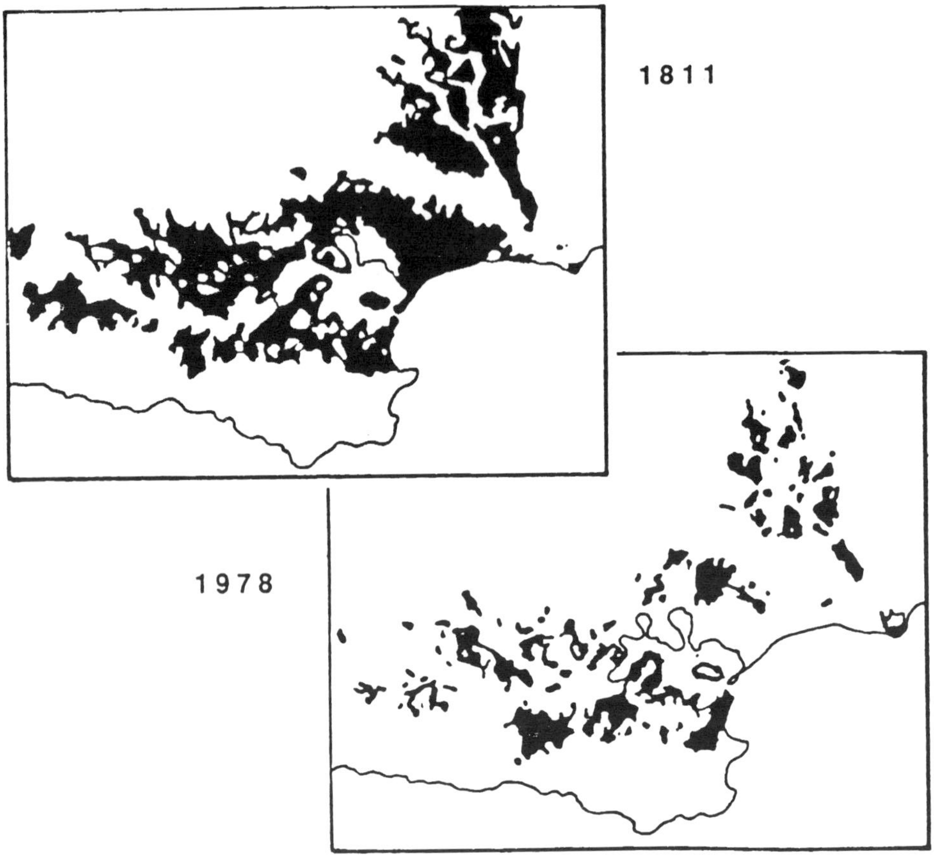

Figure 1. Fragmentation of the Dorset heathland, 1811 to 1978.

Conclusion

We have highlighted here just a selection of issues which now confront the developing use of GIS in monitoring change in the environment. The NERC is confronting these issues from a wide range of interests, from studying individual plant distributions in localised woodlands to more general global environment issues. Projects centred on Dorset to Antarctica, and dealing with subjects from sea currents to ozone depletion, are coming under the scrutiny of environmental scientists using GIS technology in all its forms.

References

Bunce, R.G.H., Barr, C.J. and Whittaker, H.A. (1980). An Integrated System of Land Classification. *Annual Report of The Institute of Terrestrial Ecology, 1980*. 28–33.

Chapman, S.B., Clarke, R.T. and Webb, N.R. (1989). The Survey and Assessment of Heathland in Dorset, England for Conservation. *Biological Conservation*. **47**, 137–152.

Good, R. (1948). *A Geographical Handbook of the Dorset Flora*. Hull.

Harding, P.T. and Roberts, D.A. (1986). Biological recording in a changing landscape. *Proceedings of the Inaugural Conference of the National Federation for Biological Recording*, Cambridge.

Hill, M.O. (1989). Computer programs for identifying vegetation types. *Annual Report of the Institute of Terrestrial Ecology*, 1989. 65.

Hornung, M., Le Grice, S., Brown, N.J. & Norris, D. (1990). The role of geology and soils in controlling surface water acidity in Wales. In R.W. Edwards, A.S. Gee and J.H. Stoner (editors) *Acid Waters in Wales* 55–66.

Marrs, R.H. (1986). The role of catastrophic death of Calluna in heathland dynamics. *Vegetation*, *66*, 109–115.

Marrs, R.H., Bravington, M. and Rawes, M. (1988). Long term vegetation change in the *Juncus squarrosus* grassland at Moor House, northern England, *Vegetation*, **76**, 179–187.

Moore, N.W. (1961). The heaths of Dorset and their conservation. *Journal of Ecology*, *50*, 369–391.

Last, F T. (1981). Introduction. *"Ecological Mapping from Ground, Air and Space,"* Proceedings of a symposium at Monks Wood Experimental Station, edited by R.M. Fuller, November 1981.

Webb, N.R. (1986). *Heathlands, A Natural History of Britain's Lowland Heaths*, London: Collins.

20

Why green?

Mike Foster
A.T. & T. Istel

Instead of using crystal balls to predict the future, strategic marketing people nowadays prefer a PEST analysis. No, not the increase in greenflies due to global warming, but the political, environmental, social and technology influences on a particular market. In the 1980s, technology and social aspects dominated. In the 1990s, environmental issues will begin to come to the fore. Resource depletion, environmental sustainability and social justice will start to become the criteria by which we judge the way we live.

The new decade will bring enormous changes to the way we live and GIS will be an essential tool to help us monitor and implement the radical restructuring of our lives. However, before we go into the role of GIS, let us look at the environmental forces that are driving the imminent huge changes in our lifestyle. The 1980s brought great affluence to 80 per cent of the UK population. Alas though, there is no such thing as a free lunch and this increase in materialistic wealth has blown our ecological balance.

Resource depletion

The invasion of Kuwait by Iraq in 1990 was the first of the "resource wars". It's all too easy as we drive around to forget how quickly the world is running out of oil. There might be a trillion barrels of reserves, but at our current usage of 21 billion barrels per year, that gives us about forty-eight years' supply.

If we were all like the USA where 2 per cent of world population consumes 28 per cent of the oil, the world would run out before the turn of the century. Indeed, despite its huge reserves, the USA will use up its own oil by around 2001. However, an energy conservation policy is currently unacceptable to voters, be they in Michigan, Middlesex or Madras.

Unfortunately it is not just oil that's running low; so too are metals. They are another finite resource and while reserves are more difficult to estimate than for oil and gas, figures of forty-eight years for copper, fifty years for zinc and fifty-five years for aluminium are predicted if current usage growth continues. The speed with which we are using up raw materials is astounding. In the last forty years the world has consumed more resources than in the entire previous history of humanity. Our resources are limited but there is increasing pressure on them through greater affluence and a growing population. However, resource depletion is only one factor.

Environmental sustainability

While the earth has reserves of a trillion barrels of oil, most environmentalists believe the worst thing we could do would be actually to use it, or even half of it. As the United Nations Intergovernmental Panel on Climatic Change (IPCC) reported, global warming is the biggest threat to continuing life on the planet. Carbon dioxide is the prime contributor to global warming and the power station and internal combustion engine are now the leading villains of the late twentieth century.

The 1990 IPCC report called for an immediate reduction in carbon dioxide emissions so we could have a 60 per cent reduction by 2005. Germany has committed to a 25 per cent reduction, but with energy conservation a vote loser, most countries are just endeavouring to stabilise emissions at current levels.

The Meteorological Office has banks of Cray supercomputers predicting gloom and doom resulting from inaction and GIS is starting to prove an excellent tool to enable non-specialist staff to interpret the complex data and predictions. The trends in climatic change will produce a warmer climate globally. Here in the UK we seem set for a warmer climate with more extremes of weather and a much greater rainfall. In 1990 in Worcester, the flooding of the River Severn was the worst for thirty years with people able to windsurf on the racecourse. Yet just three months later, the water company was urging us to minimise consumption as we were in imminent danger of a drought. This see-sawing between extremes is predicted to be more frequent.

Insurance companies might begin using GIS to combine weather data with local geography, soil types and building-type data. This would allow them to offer cheaper prices where appropriate and ensure they do not take another hammering as they did with the 1987 hurricane and the damage caused by the last two dry summers.

An increase in building damage and the threat of flooding for much of London and East Anglia will be important in the long term. However, in the short term, the main impact of a different weather pattern will be on agriculture and food supply. Late frosts, droughts and floods washing away topsoil are already causing problems.

We also have a crazy situation with food distribution. Cheap energy has revolutionised the way we live and we now squander energy at an amazing rate. I live in Worcester adjacent to one of the leading horticultural areas of Britain, the Vale of Evesham. Yet the main food retailers, the local supermarkets, buy their fruit and vegetables nationally so most Worcester residents eat cabbages from Cornwall and carrots from Kent. If you want to buy Worcester Pearmain apples, you have to travel to Birmingham market.

Food for thought

The psychologists who work for the advertising agencies are experts on motivational theory and Maslow's Hierarchy of Needs. We all have a need for recognition, a need to belong to a social group and other psychological needs. The advertisers exploit these with great dexterity whether they are selling pens or Porsches. The 1980s have created a consumer society with a fundamental change in our values and to paraphrase Descartes, "I consume therefore I am".

However, before we reach our psychological needs, we have our basic physical needs such as food. In the UK, the 80 per cent of affluent people take for granted the availability of food and what a fragile resource it is. Many children now believe

Harvest Festival is a sale at Sainsburys, and do not realise that until very recently, ensuring continuing food stocks was the main concern of every society, not just those of the third world. Historically, city size was limited by access to sufficient food for the population.

Plentiful food has been brought about through cheap energy and since we now have to start paying the environmental cost for that energy, it is a situation that must change. One hundred and fifty years ago in Charles Dickens' London, there were cows on Lincoln's Inn Fields to provide the fresh milk. Later, trains brought it in from out of town but the cows are likely to be back sooner than we think.

Towards a New Society

Let us take stock for a moment. Here we are with ever busier lives and life seems so frantic, it's difficult to imagine anything beyond the next holiday. In the background though, global warming and resource depletion are relentlessly winding up the pressure on our current way of living.

Despite the love and peace preached by the New Age gurus, we are in for a bellicose time as countries and corporations seek to protect their ever-decreasing resources. Wars and paranoia over oil and other resources will hit international trading. The breakdown in the GATT talks in 1990 heralded a new era of protectionism.

It is likely then that oil prices will rise initially due to war rather than environmental factors. This will put pressure on transportation costs but we will continue our love affair with the motor car for it is a cornerstone of our consumer society. We will have more crop failures due to the effects of global warming. However, only when these combine with the trade wars to cause famine for us in the West will we start to act seriously about ensuring a sustainable food supply.

Perhaps I am being unduly cynical, but it is a matter of changing our own values. We are a "having" rather than a "being" society and it is difficult to predict the rate of evolution of such *homo* non-*sapiens* behaviour. At whatever speed it happens though, GIS will play an increasing role in this restructuring, in areas such as transportation planning, recycling, agriculture and communications networks.

To combat the oil sanctions of the 1980s South Africa found GIS for transportation planning crucial and its use must increase here. While systems such as Autoroute offer optimisation of long distance routes, GIS will be used for minimising all travel down to an individual street level for refuse collection. In Davis, California, the refuse cart comes daily to collect bottles, garden rubbish and scrap metal in turn—far more energy-efficient than having individual cars going to collection points.

Communications networks will come to the fore as people cut down on their travelling by working from home or a local office. GIS will be essential for planning and managing the highly developed cable networks to every affluent home. It will also be used to manage the network data security and access of employees working from a single neighbourhood office but working for different companies.

The cable system will allow people to order their groceries from home; delivery by a supermarket van to each house is far more energy-efficient than individuals taking their cars to the store. Instead of all the current centralisation, the stores will need to look at what local produce is available and GIS will be essential for this.

With the main emphasis on small local production, there will be a move back from agricultural monoculture to mixed farming. Permaculture or low energy input farming will be the order of the day. GIS will be useful for matching local soil conditions and the emerging new weather patterns to the characteristics of each crop and seed type.

Most of manufacturing industry must move from large centralist production to small, more local units. Planning the best sites for these will again come through the use of GIS. Similarly in recycling, there will be a return to local processing units. Ten years ago there was the Shelton Bar Steel works in Stoke-on-Trent which reprocessed scrap iron from the West Midlands. This did not fit in with the massive plants of the eighties and was closed. Cardiff and Sheffield are committed to recycling 90 per cent of rubbish by the year 2000 and there will be a huge need for local plastic and steel reprocessing units.

In summary, it looks like being a green decade, but how green depends on finding politicians with vision and courage rather than ones who continue to pass the buck.

21

Land use monitoring

Chris Bird
Silsoe College

Introduction

The use of land is monitored, managed and debated by a wide range of interested parties. This article concentrates on land use in the rural context but the debate is no less intense in urban areas. The planning of land use has to take account of economic and social factors as well as environmental and conservation factors. Accurate information on current land use has to be combined with an ability to forecast the impact of land use changes if land is to be managed efficiently.

It is important to understand the different terms that are sometimes employed in discussions of land use monitoring. Land cover mapping can refer to the process of identifying the predominant surface cover for a land parcel. This may fall short of revealing the land use. For example, a football pitch and a cemetery may have the same predominant cover of grass, but their uses are clearly very different. Landscape evaluation refers as much to the visual aspect of the land cover as to the cover itself and so can include such items as building architecture and boundary forms which are not necessarily relevant to the land use. A knowledge of land use implies a knowledge of the human activity on the land and possibly also of the ownership of the land. It can be seen, therefore, that there can be a number of types of geographic data needed in order to gain a complete picture of land use.

In 1990, the RICS responded to the Church of England's report *Faith in the Countryside* by stating that "An integrated policy must be developed to consider all aspects of the countryside, particularly housing, food production, forestry, recreation, nature conservation, rural planning and rural employment". The planning of land use will become increasingly dependent on a range of geographic variables, with the result that GIS technology has an important part to play in the analysis of the data and the provision of Decision Support Systems.

Traditionally, agriculture has dominated the use of land in the UK, but other major rural categories include forestry, water supply, recreation and conservation. Economic developments are leading to a shift in the profitability of different land uses, and changes in the UK landscape can be expected in the future. GIS can play a role in the planning and management of this change, provided that adequate baseline data are available.

Who is concerned with land use?

It is not possible to list every organisation that might employ land use data. Government bodies include the Nature Conservancy Council (NCC), the Natural Environment Research Council (NERC), the Countryside Commission (CC) and the Ministry of Agriculture, Fisheries and Food (MAFF), each of which is currently involved in the development of GIS applications. The Forestry Commission (FC) and the National Rivers Authority (NRA) are concerned, as are the Royal Society for the Protection of Birds (RSPB), the Country Landowners Association (CLA), the Council for the Protection of Rural England (CPRE), and many more. In addition to this, the county council planning departments are concerned with land use, eg. through consideration of planning applications. They are therefore subject to pressure from many of the organisations listed earlier.

The applications of the data can be very different. For example, the NRA is concerned with the nitrate leaching properties of the land cover which are related to possible pollution of the underground water supply. By contrast, the Institute of Hydrology is concerned with the evapotranspiration properties of the land cover, as water released into the atmosphere affects the water balance for a river catchment. Users of land use data can therefore take the entities (ie. the land parcels) defined by a survey and attach very different attributes according to their interpretation of the properties of the cover type. Again, we see a clear role for the analytical capabilities of GIS.

The responsibilities of the government organisations can lead to designations of certain geographic areas which place restrictions on land use changes. The Countryside Commission is responsible for the National Parks and Areas of Outstanding Natural Beauty with regard to landscape conservation. The Nature Conservancy Council is responsible for Sites of Special Scientific Interest. Areas designated as SSSI's carry planning restrictions and the NCC has to monitor these areas and provide comprehensive map data to all other interested parties. MAFF is responsible for Environmentally Sensitive Areas where incentive schemes exist to encourage farmers to enter into voluntary agreements on farming methods. Each of these designated areas carries a requirement for regular monitoring and in each case the responsible body is developing the use of GIS or digital mapping.

How is land use monitored?

The majority of land use monitoring projects today use a combination of ground survey work, aerial photograph interpretation and satellite image analysis. Traditional ground survey techniques are still important and it is impossible to assess the accuracy of other methods without them. However, on its own, ground survey is labour-intensive and costly with the result that sample surveys are all that can be afforded. The use of sampling requires statistical analysis to provide area estimates for the land cover classes which leads to two problems. Firstly, the confidence intervals on the estimates can be unacceptably wide. Secondly, the method does not provide a useable map of the land cover for an area. Where the level of detail required is such that only ground survey work can provide it, then a census survey is employed at a cost. This is the approach taken by the NCC when producing habitat maps based on ecological parameters.

Aerial photography provides a level of detail which is adequate for many land use surveys and the skills of interpretation have been developed over many years. This source of data is particularly necessary when the survey requires details of point and linear features such as trees, ponds, walls or hedges. Many of the county councils carry out regular aerial surveys of their areas. MAFF is currently using aerial photography to provide baseline data on and monitor developments in the ESA's. Silsoe College, on behalf of the Countryside Commission, has used aerial photographs from the 1970s and 1980s to provide census mapping of the landscape of each of the National Parks of England and Wales. Both MAFF and Silsoe carry out ground surveys on a sample basis to assess the accuracy of the interpretation work.

Surveys at a national scale employ satellite imagery if there is to be any attempt to produce a land cover map for a very large area. In this case both ground survey work and aerial photograph interpretation are usually carried out on a sample basis. A combination of statistical analyses and image processing then leads to the area estimates and maps. This is the approach taken by ITE in its *Countryside* 90 project.

An important issue in the monitoring work is that of accuracy. No data is 100 per cent accurate and therefore data should be available about data regarding their reliability. Inaccuracy can be difficult to assess because it is frequently related to a mis-match between the requirements of a classification scheme laid down by field workers and the ability of photographs or images to distinguish features according to their reflectance characteristics. A current area of research is the design of optimum classification schemes for land use monitoring.

Who supplies land use information?

Again, it is impossible to list all the universities, commercial survey companies and government bodies that might supply land use data for various areas. Three UK organisations can be highlighted that have a particular interest in GIS.

The Soil Survey and Land Research Centre (SSLRC), based at Silsoe, has developed a major computer database known as LANDIS—the land information system. Soil data, climatic data and altitude data are held within this system which not only allows access to these data for England and Wales, but enables a user to make enquiries about land use suitability based on models held by the GIS. A consultancy service is provided to land owners on set-aside plans which are a major land use issue at present.

In Scotland, the Macaulay Land Use Research Institute holds datasets in several forms. A National Peatland Database has recently been completed to provide a source of data to inform decisions on land use in the peatlands and to assess peat as a source of fuel. These, and other data, are being compiled in the Macaulay Land Use Information System—a GIS aimed at allowing a wide range of queries on land use issues.

The Environmental Information Centre at ITE, Monkswood, is engaged in collating the many datasets held by ITE, some of which are discussed in Nigel Brown's article. *Countryside 90* is a project which is integrating field, aerial and satellite survey data for the whole country to provide data on current land cover, to identify past changes and to assess the likely future changes.

Future issues in land use monitoring

There is a great deal of activity in the field of land use studies, and GIS is a tool with a high profile in this work. In addition to the organisations and projects described so far, there are projects such as NELUP—the NERC/ESRC Land Use Programme run by the Centre for Land Use and Water Resources Research at Newcastle University, and WALTER—the Wales Terrestrial Database Project, jointly funded by all the Welsh bodies with an interest in land use analysis.

The question of data sharing arises and, on inspection, it becomes clear that there is little co-ordination between the different groups. Each survey works to its own classification scheme which may be similar to, but always slightly different from, another. Worse still, it is possible for schemes to use the same terms but with different meanings. In the near future it is going to be necessary to link together some of this work and sort out the various approaches to data collection.

Finally, the issue of data availability will also be important. Many people have an interest in their environment and wish to take part in the debate on land use and landscape change. Who will be allowed access to the data? GIS provides the technical capability—will the institutions be willing to reveal what they find? The answer is likely to be "Yes, but at a cost".

22

Guarding the water environment with GIS

Keith Annand
National Rivers Authority

The National Rivers Authority (NRA) was set up by the 1989 Water Act in order to separate the regulatory aspects of the water industry from the privatised Water Services Companies. Its prime purpose is to act as an environmental protection agency for England and Wales for the total water environment—this includes surface water, such as rivers and lakes, groundwater estuaries and the coastline. The operational functions are carried out by ten Regions, each of which corresponds in area to the previous Water Authority's area. It is, however, a single national organisation, and is moving towards adopting a common information systems strategy across all the Regions.

On formation in September 1989, the NRA inherited a wide variety of information systems from its predecessors, the ten Water Authorities. Although most of the Water Authorities had made an investment in GIS, this had a largely 'utilities' bias, ie. it was mainly directed to the recording of pipe networks and fixed plant. There were exceptions to this, notably Thames Water Authority's work on river management, and Anglian Water Authority's work on coastal defences.

Although the NRA owns and operates structures associated with the river network, coastline and aquifers, the main emphasis of its GIS use is not associated with these. Whereas a utility is greatly concerned with small scale events, such as street works, the NRA is more concerned with large scale events, such as the extent of an area which has flooded. This emphasis on large scale events leads, perversely, to requirement for small scale maps—typically at 1:10000 or 1:50000, in contrast to the utilities' requirements for 1:1250 or 1:2500 scales.

In common with other environmental agencies, there is a heavy emphasis on impact assessment under a variety of guises, eg. the impact of a flood on a population, the impact of a pollution incident on water abstraction points, or the impact of a flood defence construction scheme on flora and fauna. This in turn gives rise to the requirement for powerful spatial analysis tools, and a need for effective graphical output to present the result of the analysis. It is in this area that GIS has much to offer; very often the NRA's staff are required to argue their case with a developer or planning authority, and pictorial presentation of their investigation's results is much more powerful than presentation in tabular format. For this reason Tydac Technologies' SPANS package is very popular, and is in use in five of the ten regions.

An early initiative, started in the first month of the NRA's existence, was the development of an information systems strategy. This was carried out in the classical way, looking first at business direction and functions, then at the information systems necessary to perform the function, and finally at the technology. GIS was seen as one of the five key systems out of a total of 22 major systems. Significantly, it was the system used in more business functional areas than any other, being the integrating mechanism for otherwise disparate datasets.

The technological solution chosen was a network of minicomputers with each Region linked on a national basis to each other and to Head Office. This poses an interesting problem to the GIS practitioner; GIS solutions have traditionally fallen into two categories—mainframe-based, driving a network of dumb terminals, eg. IBM's GFIS, or workstation/mini-based, serving a local group of users, eg. Intergraph's approach. Fortunately, with the widespread adoption of UNIX-based workstations linked to SQL-compliant databases on server host, the means to implement the chosen technology is now becoming available.

Whilst awaiting the national strategy to be implemented, the Regions have needed to carry out their operational and regulatory functions, and have been using GIS in a number of small systems, some of which are full operational schemes and some pilots. In general, the NRA did not inherit from its predecessors a great quantity of computers. Whilst having the benefit of not tying the NRA to hardware from a wide variety of suppliers, it has meant that most progress has been made with small standalone systems; during the NRA's first year 13 of these were identified in a recent survey, in a wide variety of application areas, which include:

(1) A geographical interface to Water Archive features.
(2) Sea defence management.
(3) Water abstraction management.
(4) Water quality planning.
(5) Rainfall mapping.
(6) Slurry acceptance potential.
(7) Allocation of consents and licences.
(8) Nitrate evaluations.
(9) Fishery and recreational site assessment.
(10) River classification.

On a different scale is the implementation of a river management system in the Thames Region which is implemented using IBM's GFIS product, and runs on Thames Water's mainframe under a bureau agreement. This project has been running for over three years and provides access on a wide area network to a corporate database covering flood defence, conservation and fisheries. In order to provide a map base at an early stage, the 1:10000 scale maps of the Thames and Lea catchments were scanned and vectorised, to give a complete map base from the start of the project. The vectorisation process was necessary only because GFIS did not, until very recently, have the capability of using raster maps. Development on this system will not be taken further until the national strategy is developed in more detail.

A joint initiative has been undertaken by the Institute of Hydrology (IH) and ICL to develop a Water Information System (WIS), and trial of this is being undertaken in the North West Region. This uses a graphics front end on a UNIX workstation, linked to a database which can be implemented in either Ingres or Oracle. The database design profits from previous experience with the Water Archive system, in use in several Regions for the storage of water quality and quantity information.

The NRA is the guardian of the water environment, and the strongest environmental protection agency in Europe. It has enthusiasm and experience in the use of GIS, and has recognised it as being a key strategic information system. The prospects for GIS in the NRA are therefore bright and, conversely, the prospects for the NRA are bright if it grasps the opportunities offered by GIS.

23

Outlook: "Challenging"

Ken Smith
Meteorological Office

Weather affects everybody—it is the business of the Meteorological Office to meet the nation's weather-information requirements. These range from a housewife putting out clothes to dry, to a Tornado pilot planning his sortie, and from a highway engineer concerned with the incidence of fog along a new route, to the Cabinet pondering energy policy in the light of the Greenhouse Effect. Weather is a serious matter; the tragic loss of life on 25 January 1990 shows how dangerous a storm can be, even when accurately forecast.

The Met Office is a classic example of a service-providing organisation within government and so is a prime candidate for Executive Agency status following the announcement of Next Steps. Other options, including privatisation, were considered but the links with military operations argued strongly for remaining within the Ministry of Defence. Thus on 2 April 1990 the Met Office, which employs just under 2,500 staff, became an Executive Agency within the Ministry of Defence. This change of status was not an abrupt, unwelcome imposition but the next logical step in developing our ability to serve our many customers.

Let me emphasise and then illustrate the critical role of IT in meteorology. Our raw materials are observations of the atmosphere which arrive in a constant stream from all over the world; our products are forecasts presented through a wide variety of media, and advice on past weather and climate. Let's follow one piece of information through the system, to see what I mean.

High over the North Sea, a balloon-borne radio-sonde is transmitting values of pressure, temperature and humidity back to a ground-station at Shanwell, Fife. A computer combines the information with the corresponding wind derived from the light-path of the sonde, and automatically generates a message containing the observed data for transmission to the central message switch at Met Office HQ in Bracknell, Berkshire. This switch, based on a fault-tolerant Tandem configuration, operates in store-and-forward mode and within minutes copies of the message will be distributed to offices around the UK. For example, a forecaster at Manchester Weather Centre will be able to retrieve the message both in its coded form and as a graphical display using his MicroVAX-based workstation which holds a rolling 24-hour database of observations.

The Bracknell switch also distributes the message, along with all the others from NW Europe, worldwide via nodes in Washington, Paris and Offenbach. The message formats and telecommunications arrangements are coordinated through the World Meteorological Organisation based in Geneva. OSI protocols are becoming established among the technically advanced Members of WMO, but many links still depend on teleprinter methods. Finally, the message is sent to "COSMOS", the next stage on our data-flow tour.

COSMOS, the central mainframe complex at Bracknell, comprises an IBM 3084 (with a 3081 as backup) which acts as our primary file store, general-purpose machine and front-end to our super-computer, or rather super-computers. A Control Data Cyber 205, installed in 1981, still runs the operational "number-crunching" but is in the process of being replaced by a Cray Y-MP which was installed at the end of 1989. The Cray has 8 CPU's (each approximately the power of the Cyber) with 32 million 64-bit words of main memory and 128 million words of secondary memory. Back to that message from Shanwell: along with thousands of others describing conditions around the world at the land and sea surface and through the upper atmosphere, it has been decoded and stored in our Synoptic Data Bank (SDB) on the IBM. ("Synoptic" means a view of all the data for a given time, as opposed to a view of data for a given place through time.) The observations are available via VDU's and pen-plotted charts to the forecasters in the Central Forecasting Office, and also feed the numerical weather prediction (NWP) models. A technical description of NWP is beyond the scope of this article: suffice it to say that it is based on a mathematical description of how the atmosphere evolves in time at thousands of grid-points distributed over the Earth's surface and at many levels above. Obviously a prediction is only useful if it can be made quickly, and so NWP demands super-computer power. The new Cray will enable us to introduce an improved, higher-resolution version of the prediction model, requiring more computational power but running in about the same elapsed time (a few minutes per simulated day).

When the forecast (out to 6 days ahead) is finished, the results have to be converted into a wide range of product formats for distribution around the world. Because of our proven global NWP capability, Bracknell is designated a World Area Forecast Centre for civil aviation (Washington is the other); we supply data to the flight-planning computers of many of the world's major airlines.

Observations are not only of use in real-time. A long time-series of reliable data for a given location is invaluable to the climatologist advising on the average or extremes of weather expected over the lifetime of a structure such as a bridge. Surface and upper-air observations (like the one from Shanwell) are stored in a permanent archive based on COSMOS known as the Climatological Data Base which supports both scientific work and commercial services.

The Office is also a major centre for meteorological research. Many of the Office's scientists have local computing facilities (typically from the VAX family) which are networked to the major computing facilities at HQ. For example data from COSMOS can be transmitted through an Interlink IBM-DECNET gateway to a micro-VAX or PC on the Ethernet serving the Meteorological Research Flight at RAE Farnborough. These networks need to have considerable throughput; they shift satellite and radar imagery and large files of output from the super-computer.

This year has seen the setting up of the Climate Research Centre at Bracknell, part-funded by the Department of the Environment. The Centre will bring together scientists from the Office, universities and from overseas to work on the problem of understanding the Greenhouse Effect and its consequences for global warming and changing weather patterns. The team of scientists will need extensive computing facilities and so, from 1991, will have dedicated use of a second super-computer, plus workstations and a share of an enhanced mainframe system, to support the running of climate models. These models are essentially similarly to the NWP variety, but include an ocean modelling component and must obviously simulate years rather than days ahead.

Agency status does not change the scientific and technical basis of the work of the Met Office but it does affect the framework within which our activities are carried out. There will be greater autonomy for management, but MoD needs to be satisfied that we are running our business efficiently. Thus new and improved systems for financial management, stock control, cost accounting, invoicing, etc, are required and must be in place during 1990. (These types of systems will be the bread and butter of many readers but are something of a culture shock for scientists!). As a means of collecting and presenting this mass of information to managers, we will be installing the first phase of an office automation system (based on Uniplex) in a few months time. We will continue to use central MoD facilities for payroll, bill-paying and personnel records.

Becoming an Agency also focuses attention on customer satisfaction. There is a self-evident requirement to satisfy commercial customers who have a choice of supplier for meteorological advice, but we need to develop the same discipline with respect to our military and public service customers. Thus one of the consequences of Agency status is a sharper focus on the relationship with MoD to enable us to meet those requirements. MoD also acts as the sponsor for the public services, such as the warnings of severe weather which are issued to the emergency services of support given in the event of air-borne pollution incidents.

Over the last 20 years IT has made a very large contribution to productivity in the Met Office through automation of manual processes such as plotting charts or sending messages. We are continuing to upgrade our computer and communications infrastructure to provide the foundation for further improvements in the effectiveness of our operations. Emphasis is being given to the use of IT to expand our business by delivering the right product at the right price at the right time in the right place. The computer graphics on both BBC and ITN weather presentations show what can be achieved when IT is used imaginatively. We are also using microcomputers such as IBM PCs to support the delivery of products to highway authorities concerned with road gritting, and Apple Macs running desk-top publishing software to prepare the text and graphics for forecasts in the national and provincial press. We are experimenting with the use of a PC network to deliver self-briefing material (as both text and graphics) to RAF aircrew and thinking about how these and other techniques could be used to support commercial customers.

Finally, the success of the Met Office as an Executive Agency involves the satisfaction of the staff. Naturally we feel some apprehension as we enter a period of change and adjust to new procedures for resource management, together with some internal re-organisation. But we also see ourselves as a committed and very professional workforce with many technical strengths, not least in IT, and look forward to meeting the challenge of the 1990s.

24

Potential uses of geographical information systems by environmental pressure groups

Blake Lee-Harwood
Friends of the Earth

One of the working principles of Friends of the Earth (FoE) is that long term environmental improvement is best achieved through the empowerment of local communities on issues of direct local relevance—a sentiment encapsulated in the slogan "Think globally, act locally". Traditionally this has been achieved through methods such as public meetings, leafleting and articles in the local press; methods that are probably little different from those used by the great social reformers of the Victorian age. However, the world changes and pressure groups, if they are to remain effective, must change with it. Geographical information systems (GIS) now offer one of the most powerful campaigning tools in any pressure group's arsenal and future developments will be rapid.

FoE's first foray into the world of GIS was as crude as it was effective. By adapting a system that was used for road vehicle route planning we found that it was possible to enter data on waste disposal sites which could be displayed graphically on maps of varying scales. The product of this early experiment was a series of maps showing the locations of waste tips which the Government had previously identified as posing a threat of water pollution.

The maps were reproduced in *The Observer Magazine* in February 1990 and caused a sensation with nationwide press coverage and questions being asked in the House of Commons. The results, in campaign terms, were a complete success. The issue of polluting waste tips had been placed firmly on the political agenda.

At a local level the effect was even more startling. As part of *The Observer* article we had enclosed a coupon which members of the public could send back to us requesting a map showing information on tips near where they lived. Anticipating a heavy demand we divided England and Wales into 280 areas using postcode sequences. By examining an enquirer's postcode we could identify the appropriate map within seconds and post it out with accompanying information. Such was the efficiency of the system that we answered over 10,000 separate requests for local maps within two weeks.

The system that we developed could also be adapted for displaying other information and could demonstrate in crude terms some of the relationships between the figures displayed, such as the proximity of a potentially leaky waste tip to a public water supply. However, it was an extremely crude tool and in development terms had reached the end of the road for our purposes.

FoE is now considering obtaining new GIS software and appropriate hardware systems with the firm conviction that this technology will play a key role in campaigning in the 1990s. There is an increasing trend for environmental information to be made available in larger quantities and in digital form. Computer databases already exist for surface and groundwater quality in England and Wales, although they are at present rather user-hostile, and the future offers the opportunity for considerable advances. The Environmental Protection Act 1990 requires the setting up of databases for Integrated Pollution Control and local registers of air quality. Local authorities will also have to set up registers of contaminated land. At an international level, FoE is investigating possible access to databases held by the European Commission and the United Nations Environment Programme.

As environmental information becomes increasingly available, FoE could be in a position to enter the relevant portion into a GIS in order to produce campaigning tools of power hitherto undreamed of. GIS analysis could be used routinely in campaigns to produce new information from the analysis of separate datasets. The results of this analysis would be used to generate "products" such as high quality graphics and maps for the media and key decision-makers. Spatial analysis would be used to target key constituencies, industrial sectors, companies and communities. GIS would prove invaluable in identifying breaches of environmental quality standards on a national basis and informing members of the public in the affected locations.

The ability to link geographical information to postcodes will mean that citizens living in an area with a specific environmental problem can be contacted by post and given the facts abut their local environment along with the names and addresses of the regulatory authorities, MPs and councillors. At last, local people can be armed with more of the facts they need to make their voices count in the environmental debates to come. GIS holds out the opportunity for "informed debates" to be genuinely informed and for decisions to be reached in a logical and democratic way.

The opportunities for informing the public are clearly enormous but FoE has another means of entering community debates which GIS could complement. Scattered around the country are over 300 of FoE's "local groups", community groups which set about achieving environmental improvements in their locality. Their activities are diverse, ranging from lobbying their local councils and exposing local polluters to conducting practical projects in recycling. They form the backbone of the FoE and represent an awesome lobbying force if properly coordinated.

The problems raised by the need to service the voracious information requirements of our local groups and to coordinate their activities are enormous. The solutions may well lie with GIS, which offer us the opportunity to send high quality, locally-oriented information to our groups in an attractive and accessible form which can be constantly updated. Linked databases could contain information on the activists involved in the groups and the campaigns they are involved in so that information can be matched to the people who would find it most useful. Ultimately it may be possible to establish on-line access for our groups to some of the system, a task that would be made easier in the first stages by the high proportion of our members who are professionally involved in information technology.

It should be stressed that FoE does not view GIS simply as a means of disseminating information from the centre to the periphery. The passage of information would be a two-way process with information on local environmental problems being fed from our groups and individual citizens back into the system so communities from different parts of the country could exchange information on issues of mutual concern. It should also be added that FoE does not see the utilisation of GIS as being confined to the computer literate. When designing systems great emphasis will be placed on making those systems user-friendly to increase the availability of information to all sections of society.

Information is the raw material that pressure groups consume and information is also the final product. GIS offers us a way of increasing the efficiency of our operations to an extent that we ourselves are only just fully coming to terms with. It is a truism that "information is power", but in a democratic society that power must ultimately reside with the individual citizen. GIS offers us the technological means for achieving that democratisation in a constructive and beneficial way. Perhaps such an advance will actually change the way in which decisions that affect our environment are made and may even ultimately change the way society views this fragile planet that is our home. If GIS can really achieve the benefits that we foresee, it has not come a moment too soon.

For further information on Friends of the Earth call the GIS Information Officer in the Campaign Director's Office on 071–490–1555.

25

Government in action

John Garnsworthy
Department of the Environment

Introduction

This article describes some of the practical applications of GIS and related technologies with which the Department of the Environment (DoE) is involved, mainly through its research programme. A wide range of environmental policy concerns was documented in the Government's White Paper—"This Common Inheritance"—published in September 1990 and GIS have been applied to many of these areas. There are, of course, other central government GIS initiatives on the environment outside DoE but limitation of space prevents discussion of these.

English House Condition Survey (EHCS)

The Department has commissioned the South Eastern Regional Research Laboratory (SERRL) to undertake a research project to link the 1981 Census to the 1986 EHCS with the aims of:

(1) Linking the EHCS postcoded addresses with Census data at enumeration district (ED) level and other relevant small area databases to enhance the EHCS database by the addition of socio-economic data.
(2) Testing the feasibility of identifying areas suitable for renewal or estate action and improving the description of clearance areas using this enhanced database.

This exercise is primarily a pilot for 1991, when the Census and EHCS will take place in the same year. If feasible, the same link between databases will be undertaken to identify areas suitable for area action and compare area characteristics for EHCS addresses between 1986 and 1991.

An enhanced database has been successfully created by linking the 1981 Census and 1986 EHCS and attaching the 'Infolink' database package (containing small area financial data) and the Department's database containing urban/rural classifications. The accuracy of linking EDs to postcodes using various GIS methods has been tested. Results show unacceptable errors when linking individual postcodes to single EDs. Consequently, EDs have been linked to EHCS addresses on the basis of distance radii. The enhanced database now contains two ranges of information with each EHCS address as the centroid and ED data up to radii of 250m and 500m respectively surrounding the address. The feasibility of identifying renewal areas from the enhanced database is being explored.

The use of geographic information systems in the analysis of countryside data

The objective of this project, funded by the Department and undertaken by Bristol University, is to explore the potential use of GIS in the analysis of countryside data. The researchers' brief included:

- to assess user requirements for both data and operational procedures;
- to use case studies to assess the opportunity for setting local impacts in a wider national context;
- to report with advice on the requirements for DoE users of countryside data, recommending what opportunities exist for the use of GIS through demonstrated examples.

Two of the demonstrators, a planning demonstrator and an agricultural demonstrator, are set in Devon and the third will be a national demonstrator.

The planning demonstrator's main objective is to illustrate the potential of GIS in the planning process to provide policy-relevant information. Four positive aspects have emerged. First, is the ability to collate and display data often collected at different scales, for different purposes and by different organisations. Second, as a result of data integration, analytical work is readily undertaken. Third, the speed and flexibility of analysis is increased, and fourth, data on observed change will play an important role in monitoring the effectiveness of policies.

The objective of the agricultural demonstrator is to illustrate the use of GIS to derive new statistics for user-defined areas for input into programmes which monitor the impact of agricultural or conservation policies. The demonstrator will illustrate how GIS may contribute to the analysis of countryside policy matters and show the ability to derive information in tabular or map form.

The national demonstrator will be based on data from the Warren Springs Laboratory. Techniques for incorporating air quality data into environmental assessment procedures for designated areas will be demonstrated, eg. the impact of marginal differences in air pollution on the management of upland vegetation communities. Close association with related research will be established.

Co-ordinated information on the environment in the European Community (CORINE)

CORINE, established in 1985, is an experimental programme of the Commission of the European Communities for gathering, co-ordinating and ensuring the consistency of information on the state of the environment. An essential part of CORINE has involved the investigation of a common computer system for handling data of a geographical nature. ARC/INFO was selected and runs under UNIX on five linked Tektronix workstations.

Three environmental themes are central to the CORINE programme: biotopes (ecological sites) of major importance for nature conservation; acid deposition; and the resources and problems of the Mediterranean region. The UK's main contribution has been from NERC's Institute of Terrestial Ecology which has constructed and maintained a biotopes database holding attributes affecting over 10,000 sites worthy of conservation for their floral and fauna characteristics.

Geological and Minerals Planning Research Programme

DoE is undertaking this programme to provide information needed in developing and implementing policy on minerals and on the significance of land instability and safety issues for planning and development.

Work on minerals includes studies of the extent and quality of onshore mineral resources, which contributes to the information base for Minerals Local Plans, and of offshore sand and gravel resources which might be dredged as an alternative to working on the land.

Studies of land instability and safety have concentrated on a series of national reviews of the nature and extent of potential problems, including: a study of landslides; work on mining instability, natural underground cavities and foundation; and a study of natural sources of contamination (eg. radon, methane, toxic metals). Additionally, demonstration projects on landsliding and mining instability have been undertaken for selected areas. When the series of studies is completed, it will provide an atlas of instability and safety issues covering the whole of Great Britain.

Applied geological mapping studies of selected areas which collate and interpret information on minerals, land instability and other matters have also been commissioned.

All of the recent research has involved the collection and collation of data using computer databases and several have employed digital cartography. Whilst most have not involved GIS, as such, they do provide much data organised in a form which could be introduced into such systems. Some recent studies, however, have been exploring the potential for handling multidisciplinary data sets, including: (a) a study of soils in the Southampton area (described in the following section); and (b) a study of ground conditions in the coastal wetlands of the River Severn which is taking into account both the engineering and environmental implications of development.

Southampton Soils Project

In 1987 the Department commissioned The Soil Survey and Land Research Centre (SSLRC) to examine the potential for computer-derived soils and land use maps in preparing land use planning policies, proposals for development and conservation strategies.

Southampton was selected for the pilot study as a database of relevant geological information and a revised geological base map had been prepared recently by the British Geological Survey. SSLRC has developed a computer-based land information system (LandIS) to organise existing soil and other data. Use was made of these databases within LandIS, supplemented with data from a detailed soil survey. A spatial analysis system (SPANS) was used to manipulate selected groups of data, to provide a selection of interpretative maps. These included themes such as land suitability for individual crops, land capability for forestry and wetland habitat potential. Whilst this approach is a useful tool in developing future policies on options within a given area, the need remains for more detailed surveys at site level for some types of development or uses of land.

The benefit of the technique is that maps can be printed directly from a desktop microcomputer with limited accessories and can be cheaply and easily updated as the database or criteria for selection change. The system is therefore responsive to improved knowledge and to changes in recommended practices and planning policies.

Small Area Health Statistics Unit (SAHSU)

The Government established SAHSU in 1987 following the Black Advisory Group's enquiry into childhood cancers in the vicinity of the Sellafield nuclear processing plant. SAHSU, partly funded by DoE, has the following terms of reference:

- to examine quickly reports of unusual clusters of disease particularly in the neighbourhood of industrial installations and advise authoritatively as soon as possible;
- in collaboration with other scientific groups, to build up reliable background information on the distribution of disease within small areas so that specific clusters can be placed in proper context;
- to study the available statistics in order to detect any unusual incidence of disease as early as possible and, where appropriate, to investigate;
- to develop the methodology for analysing and interpreting statistics relating to small areas.

SAHSU collect data on health "events" (eg. deaths, cancers, live births, stillbirths and congenital malformations), identified by geographical location, and produce event rates and analyses around point sources of pollution. The data are manipulated using a relational database with a geographic structure based on postcodes.

Analysis of land use change in England 1985–1989

The contractors for this research project, SERRL and Halcrow Fox and Associates, will make use of ARC/INFO as an analysis and presentation tool. It will be used for analysing time lapses in recording different land use changes and the implications of these time lapses by mapping various rates of development to allow easy visual assessment of any convergence over time.

ARC/INFO will also be used to undertake a spatial search to identify candidate land use changes records for comparison with known land use changes and to present map results of the analysis where appropriate.

Stock of vacant land in urban areas

In 1990 the Department undertook the first nationwide survey of the stock of vacant land in urban areas of England. Ordnance Survey (OS) surveyors perambulated a selection of 25 hectare squares which corresponded to OS large scale plans (1:1250 maps). A stratified random sample of 25 hectare squares was selected for the surveyors to perambulate and identify vacant land sites recording their size, location and character.

The contractors for selection of the sample, SERRL, used ARC/INFO to identify OS large scale maps (25 hectare squares) with urban land and to identify and select plans according to the sampling criteria. The sample frame consisted of all 1:1250 scale OS large scale plans which intersected with any urban area of 10,000 population or more in 1981. It was then stratified into 63 strata and each stratum was divided into those squares entirely within the urban boundary, those mostly within but on the urban boundary, and those mostly outside but on the urban boundary. ARC/INFO will also be used to present the results of the survey.

ESRC/NERC Collaborative Programme on Handling Geographic Information

DoE is represented on the steering committee for this programme and has particular interest in, for example, the research project: "A Land and Economic Development Information System for the East Thames Corridor". In this research SERRL and SERPLAN are developing a multi-agency, multi-disciplinary GIS/LIS for the East Thames Corridor (ETC). The GIS/LIS will store socio-economic and environmental information on the developable sites within the ETC and monitor land use change on these sites and in surrounding areas. It will be used to carry out research into the determinants of the uptake of development sites in the ETC, the integration of the site-specific remote sensing data with land use change data and the role of management/presentational issues in the acceptance of GIS methods.

The future

A number of potential GIS applications may be identified, not least in areas identified in the Environmental Protection Bill which is a major feature of the Government's programme to tackle the whole range of pollution issues at all levels. One part of the Bill introduces a new Integrated Pollution Control (IPC) system which will bring a cross media approach to controlling emissions, whether to air, land or water, for the most polluting industrial and similar processes. Outcomes of the change to IPC include increased monitoring by industry, in order to prove to the controlling authorities that the requirements of their authorisations are being met, and increased availability of data, from a public register, because of a new provision for full public access to information.

Chapter 26

GIS: A brave, new and greener world?

Roy Haines-Young
Institute of Engineering Surveying and Space Geodesy, University of Nottingham

It is a widely held belief that access to GIS technology can change the way we do things. After all, so the story goes, unless it leads to better working practices, what is the justification for the investment the technology requires? But given the wider use of GIS that we have seen in recent years, have any such pay-offs really been achieved? Nowhere do we need such progress more urgently than in the environmental arena. What evidence is there that the technology has or will lead to more effective environmental management strategies?

This question concerning the effective use of GIS in environmental management is put with particular reference to the UK. It was prompted by a reading of the Government White Paper *This Common Inheritance—Britain's Environmental Strategy* published in September 1990 (HMSO, 1990). The document is significant for the GIS community in the UK because it comes from the same Government department which initiated the Committee of Enquiry into the Handling of Geographic Information (Department of the Environment, 1987). The White Paper provides for the GIS community in the UK a measure of the extent to which the "awareness barrier" highlighted by Chorley has been overcome within a Government department three years on. More generally, it also maps out the wider institutional issues which must be addressed if GIS is to be used as an effective environmental management tool.

GIS: a problem-solving technology

There is little doubt that GIS is an important problem-solving technology in the management of environmental resources. The literature is now rich with accounts of the potential benefits. On the international scene, for example, Davis *et al.* (1990) have recently described how an information system approach can be used in the preservation of biological diversity in California. These authors show how GIS could be used to identify gaps in the network of nature reserves within the state, and to help formulate better management strategies based on a more clearly specified set of priorities. Elsewhere, Jensen *et al.* (1990) have described how the technology can be used to map an environmentally sensitive index for oil spills in an area of south-west Florida. Such map information is essential for contingency planning and for making management decisions after an incident.

In Europe, the CORINE project (Wyatt *et al.*, 1988) has resulted in the basis of an operational system containing a range of information on land, water and atmospheric parameters which can aid the formulation of Community policies for environmental protection, and be used to monitor and assess of the impacts of Community policy. As Rhind (1990) notes, the European Commission is proposing to extend the project as part of a new European Environmental Agency. On a smaller scale within Wales, the WALTER project (Bracken and Higgs, 1990) shows how GIS might be used in the management of an environmentally sensitive area when the issues cut across the remit of several rural agencies.

Despite the progress which has been made, examples such as those described above indicate that many initiatives remain essentially research-orientated. They often address a relatively narrow range of issues, and are often undertaken in isolation from the wider issues of how society approaches the problems of resource use and environmental management. Few projects, it seems, have led to any fundamental organisational change. The lack of significant progress partly reflects the complexity of the environmental agenda itself. Partly, however, it reflects institutional inertia. As the Chorley Committee noted in their report (paragraphs 2.22–2.23), for example, relatively little use of comprehensive GIS has been made to manage rural resources in the UK. This situation was the result, they suggested, not only of practical problems such as the pattern of small land ownership and the cost of systems, but significantly from attitudes within Government departments. Further development they argued, must depend substantially on Government. "A firm lead by the Government", they concluded, "would, more than anything else, provide a powerful impetus towards the more effective use of geographic information." (paragraph 10.4). Since the Environment White Paper aims to describe "how every one can help and what everyone can do" to promote better environmental management it is instructive to consider the extent to which the contribution of GIS is recognised within this document.

On the face of things there is scope for disappointment with the White Paper. The recognition of GIS as an important enabling technology in the area of environmental management does not really emerge in the document. GIS is mentioned only once in the context of research initiatives in the chapter dealing with land use. Although the value of access to information about the environment is stressed throughout the document, the importance of GIS as a means for achieving the integrated perspective necessary for sound environmental management gets little or no mention at all. The worry is, therefore, that despite Chorley, important connections may not have been made.

Information and the framework for debate

The White Paper is, of course, a "strategic" statement by Government. In defence of its apparent lack of awareness about GIS, it could be argued that the specifics of particular technologies are not important. Rather, it is the institutional arrangements which are promoted which are significant, for it is these which determine approaches to problem-solving and ultimately patterns of investment.

Decision-making in the environmental arena, like any other, is dependent on information. Information and the ability to analyse and question the conclusions based upon it are particularly important in the context of environmental management because increasingly decisions about resource use will mean that there will be winners and losers within society. Thus quite apart from judgements about the White Paper in terms of specific policies and technologies, we might also ask whether it deals with the need for good and widespread access to data. Will it, moreover, encourage the integration of data so that the wider consequences of environmental policies can be tested critically?

In the chapter on "Knowledge, Training and Education" (Chapter 17) the White Paper deals with a whole range of data issues. The document accepts, for example, that public debate must be based on sound information and a thorough analysis of the risks, costs and benefits of proposals. It also accepts that Government can help people act in a "balanced way" by promoting good scientific research and monitoring. This requires, the White Paper concludes, the "clear and wide dissemination of the results of research and other environmental data". In a significant passage for the GIS community, the White Paper goes on to state that "the Government will continue to coordinate environmental monitoring with emphasis on the consistent and cost-effective capture, processing, storage and transmission of data". How are such goals to be achieved?

The White Paper argues that the publication of "sound and authoritative" statistical data is essential. It proposes, therefore, that Government should publish statistical reports at regular intervals covering a range of topics relevant to the assessment of the state of our environment. It is proposed that publication of the first report should be within 2 years. In one of the more useful sections of the document, Appendix B, a list of the statistical series proposed for inclusion in the exercise is given. Like a similar appendix in the Chorley Report, at least we can begin to identify data holdings, if not points of contact.

In addition to the publication of statistics by central Government, the White Paper also promotes the right of public access to information held by the pollution control authorities. Clearly, access to raw data, rather than statistical summaries, is essential if policies and decisions are to be tested adequately. but the arrangements for such access are still fraught with difficulties. Although the White Paper notes (section 17.26–17.28), for example, that "HMIP, local authorities, the National Rivers Authority, the Health and Safety Executive and some Government departments keep public registers of environmental and other information", the claim that these "make information available in a form which is easy to use" contains a large element of wishful thinking. Widespread and easy access to public information is still honoured more in the breach than the compliance. Acceptance of the EC Directive which comes into force in 1991 is, therefore, to be welcomed since this will require most environmental information held by public authorities to be made available to the public. In support of the initiative the White Paper says the "Government will propose arrangements, including registers, by which environmental information can be made available in Britain."

The White Paper also asserts that as information becomes available the challenge is to the public and to interest groups to make use of it in a "balanced and constructive way". Access to technologies such as GIS is clearly one way in which we might begin rise to the challenge. In the water industry, for example, one might imagine the situation in which decisions relating to discharge consents might be tested by public access to the data and models on which consent was based. Alternatively, when plans and policies are published by local authorities, the environmental implications of proposals might be scrutinised more effectively if information on the current situation were more widely available. However, we are clearly a very long way from this brave new world in which, even if we have the right to access environmental information, we have access to the technologies necessary to handle it effectively. The challenge for bodies such as the AGI is to explore ways in which the techniques for handling geographic information can be pursued to satisfy not only narrow economic goals in areas such as property and marketing, but also the widest public interest in areas such as the environment.

"Information", it is often said, "is power". In the future, access to technology will be just as significant. The granting of access to information alone is not a sufficient condition for informed discussion. We must also ensure that our public, private and voluntary institutions can join in the debate on at least reasonably equal terms.

A challenge for the future

In the discussions surrounding the Environment White Paper, criticism has largely focused on the specifics of particular policies and on whether intentions can be translated into actions. As Sir Idris Pearce, Chairman of the Royal Institution of Chartered Surveyors, has pointed out, in terms of specific measures the White Paper is only a beginning. "Only when we see a timetable for implementation and a funding plan will it become clear whether the Government's stated intentions will be matched by effective action. For those of us with a more specific interest in GIS it is clear that much the same kind of case can be made."

Despite the lack of specific reference to GIS, the framework which the Environment White Paper sets out for data access is important. The document provides useful ammunition on which future arguments can be constructed. Bodies such as the AGI must ensure that the benefits of GIS for handling and processing these data are understood and receive more tangible recognition. In the short term this can be achieved through inputs into new legislation aimed at environmental protection, and through measures such as those proposed to strengthen the Royal Commission on Environmental Pollution. On the European scene, documents such as the EC Green Paper on the Urban Environment need to be considered, together with arrangements for the development of new institutions such as the Environmental Protection Agency.

More importantly, however, in the long term the example of Government departments looking seriously at the environmental consequences of policy decisions, and of putting in place the institutional arrangements necessary to see that this goal was achieved, would also do much to focus people's minds on what is required. The White Paper leaves in place the coordinating machinery established to produce the document, which includes a committee of Cabinet Ministers to consider environmental policies, and a nominated Minister in each department responsible for considering the environmental implications of departmental policy. Guidelines for policy appraisal are to be drawn up. The White Paper recognises that the right institutional arrangements need to be in place if successful environmental policies are to be delivered (section 18.1). This machinery is now there to be exploited and the opportunities for the GIS community to make its case are, therefore, considerable. Additional change is only likely to occur, however, if the case is argued forcefully, and if tax payers and voters demand it.

Do better data and access to information make for better environmental management? Given the scale of the problems which now confront us we must hope that it is so. It must be recognised, however, that information is but one ingredient in strategic planning. The ability to ask the right questions and the commitment to fund solutions are equally important. Thus involvement of the AGI, and the GIS community in general, in the environmental debate is not likely to be a smooth one. Whatever one feels about the White Paper it is the only one we have—and a start has at least been made. The challenge for the future is to ensure that good intentions in the area of data handling and data sharing, like those dealing with specific areas of environmental policy, are eventually matched by effective government and public actions.

References

Bracken, I. and Higgs, G. (1990) The role of GIS in data integration for rural environments. *Mapping Awareness*, **4**(8), 51–56.

Davis, F.W., Stoms, D.M., Estes, J.E., Scepan, J. and Scott, J.M., (1990). An information systems approach to the preservation of biological diversity. *International Journal of Geographical Information Systems*, **4**(1), 55–78.

Department of the Environment (1987). *Handling Geographic Information*, London: HMSO.

HMSO (1990). *This Common Inheritance—Britain's Environmental Strategy*. London: HMSO, CMD 1200.

Jensen J.R., Ramsey, E.W., Holmes J.M., Michel, J.E., Savitsky, B. and Davis, B.A. (1990). Environmentally sensitive index (ESI) mapping for oil spills using remote sensing and geographic information systems technology. *International Journal of Geographical Information Systems*, **4**(2), 181–202.

Rhind, D. (1990). Global databases and GIS. In Foster, M.J. and Shand, P.J. (Eds), *Association for Geographic Information Yearbook 1990*, London: Taylor & Francis and Miles Arnold, 218–223.

Wyatt, B., Briggs, D. and Mounsey, H. (1988). CORINE: An information system on the state of the environment in the European Community. In Mounsey, H. and Tomlinson, R.F. (Eds), *Building Databases for Global Science*, London: Taylor & Francis, 378–396.

27

GIS development for broad-scale policy applications: The lessons from CORINE

David Briggs
Polytechnic of Huddersfield

Introduction

The CORINE Programme of the European Communities provides one of the most extensive and demanding tests which has yet been undertaken of the development of broad-scale GIS. Established initially as a four year experimental programme in 1985 (Commission of the European Communities, 1985a), it had the remit of "gathering, co-ordinating and ensuring the consistency of information on the state of the environment" in order to guide and underpin Community environmental policy. Subsequently, it was extended for a further two years (Commission of the European Communities, 1990a), and its budget was increased from 4 million to 10.5 million ECUs.

Since then, the longer term future of the CORINE information system has also been assured by the adoption of a Council Regulation setting up a European Environmental Agency (Commission of the European Communities, 1990b; Briggs, 1991). *Inter alia*, this will have the responsibility of maintaining and developing the system as a policy tool. As this indicates, the CORINE Programme may be judged a success. It has resulted in an information system which is already being used for routine policy applications, and which clearly has scope for further development. At the same time, the experience of constructing the system has provided many lessons and shown many of the problems involved. Currently, several other, similar broad-scale information systems are being planned. In Europe, for example, WHO is considering the establishment of a health and environment GIS and in Australia, NRIC and ERIN are embarking on the development of an Australian GIS on rural resources (Johnson, 1989). The lessons to be drawn from CORINE are thus both timely and widely pertinent.

The CORINE GIS

The history, organisation and scope of the CORINE Programme have been discussed elsewhere (e.g. Briggs and Martin, 1988; Briggs and Mounsey, 1989). Details of the GIS design and content have also been provided by Briggs (1990), Wyatt *et al.* (1989) and Whimbrel Consultants Ltd/Huddersfield Polytechnic (1989).

Briefly, the system was initially developed at two separate sites; at Birkbeck College, London, using ARC/INFO (under the supervision of Professor David Rhind), and at CORINE headquarters in the Directorate General for the Environment (DGXI) of the EC in Brussels, using the Siemens SICAD-DIGSY system. Since then, however, the system has been centralised in Brussels, where it runs in ARC/INFO on a series of Tektronics workstations.

The content of the system is outlined in Table 1. As this indicates, a wide range of datasets has been constructed, at a range of scales. Wherever possible, however, datasets are registered to a consistent geographic base, in Lambert Azimuthal projection, based upon the 1:1 million ONC series compiled by the Ministry of Defence, United Kingdom. Large datasets are structured into files based on national areas. The database as a whole amounts to *c* 2000 m/bytes, although clearly data volumes vary depending upon the degree of data compaction, and transient decisions regarding back-up and storage of derivative datasets.

The lessons learned

In setting up any GIS, one of the most fundamental and formidable problems concerns the availability and quality of the input data. This constrains and influences almost every other stage of GIS development and use, including processes of data capture, integration, processing, modelling and display. In the case of the CORINE Programme, this problem was compounded by a number of factors:

(1) the geographic extent of the system was large (*c* 2.3 million km^2) yet the spatial resolution required to meet policy needs was relatively fine (a resolution of <1 km was often assumed);
(2) the range of datasets necessary to meet the demands for information was wide, and often spatially or thematically unrelated;
(3) the time and resources available were limited.

As a consequence, most of the data required for the information system had to be derived from existing, secondary data sources, such as maps, statistical compilations or databases. The completeness and consistency of these, however, were themselves often impaired because—despite attempts by the EC to co-ordinate data collection and exchange in some areas—most member states still act independently in the areas of environmental mapping, monitoring and analysis.

Against this background, it is inevitable that data problems should be paramount. Data availability, for example, was often poor due to the lack of primary survey or measurement in the member states, lack of data processing and dissemination, restrictions of confidentiality, or gaps in available records due to equipment failure and lack of system maintenance. The consequence is that many datasets are either spatially or temporally incomplete.

Inconsistencies also occur in those data which are available. These may arise during almost any stage in the life-cycle of a dataset: during entitation, measurement, processing, modelling, reporting or dissemination of the results. Where secondary sources are being used, data may have been through several such cycles prior to their incorporation into the GIS. Errors and inconsistencies may thus accumulate. In some cases the consequences remain explicit; they may be revealed by differences in nomenclature, classification, measurement units, spatial referencing or sampling

Table 1. Overview of the contents of the CORINE information system

Theme	Nature of the information	Volume of information Description	Mbytes	Resolution/scale
Biotopes	Location and description of biotopes of major importance for nature conservation in the community.	5600 biotopes described, according to approx. 20 characteristics.	20.0	Location of the centre of the site.
		Boundaries of 440 biotopes computerised (Portugal, Belgium).	2.0	1/100 000
Designated Areas	Location and description of areas classified under various types of protection.	13 000 areas described according to approx. 11 characteristics (file being being completed).	6.5	Location of the centre of the site.
		Computerised record of the limits of the areas designated in compliance with article 4 of the EEC/409/79 directive on conservation of wild birds.		1/100 000
Emissions into the air	Tons of pollutants (SO_2, NO_x, VOC) emitted in 1985 per category of source: power stations, industry, transport, nature, oil refineries, combustion.	1 value per pollutant, per category of source and per region, plus data for 1400 point sources i.e. ±200 000 values in total.	2.5	Regional (NUTS III) and location of large emission sources.
Water resources	Location of gauging station, drainage basin area, mean and minimum discharge, period: 1970–1985, for the southern regions of the EC.	Data recorded for 1061 gauging stations, for 12 variables.	3.2	Location of gauging station.
Coastal erosion	Morpho-sedimentological characteristics (4 categories), presence of constructions, characteristics of coastal evolution: erosion, accretion, stability.	17 500 coastal segments described.	25.0	Base file: 1/100 000. Generalisation: 1/1 000 000
Soil erosion risk	Assessment of the potential and actual soil erosion risk by combining 4 sets of factors: soil, climate, slopes, vegetation.	180 000 homogeneous areas (southern regions of the Community).	400.0	1/1 000 000

Table 1 (cont.)

Theme	Nature of the information	Volume of information Description	Mbytes	Resolution/scale
Important land resources	Assessment of land quality by combining 4 sets of factors: soil, climate, slopes, land improvements.	170 000 homogeneous areas (southern regions of the Community).	300.0	1/1 000 000
Natural potential vegetation	Mapping of 140 classes of potential vegetation.	2288 homogeneous areas.	2.0	1/3 000 000
Land cover	Inventory of biophysical land cover, using 44 class nomenclature.	Vectorised database for Portugal, Luxembourg.	51.0	1/100 000
Water pattern	Navigability, categories (river, canals, lake, reservoirs).	49 141 digitised river segments.	13.8 0.3	1/1 000 000 1/3 000 000
Bathing water quality	Annual values for up to 18 parameters, 113 stations, for 1976—1986, supplied in compliance with EEC/76/160 Directive.	2650 values.	0.2	Location of station.
Soil types	320 soil classes mapped.	15 498 homogeneous areas.	9.8	1/1 000 000
Climate	Precipitation and temperature (other climatic variables: data incomplete).	Mean monthly values for 4773 stations.	7.4	Location of station.
Slopes	Mean slope per km^2 (southern regions of the Community).	1 value per km^2, i.e. 800.000 values.	150.0	1/100 000
Administrative units	EC NUTS regions (Nomenclature of Territorial Units for Statistics). 4 hierarchical levels.	470 NUTS digitised.	0.7	1/3 000 000
Coasts and countries	Coastline and national boundaries (Community and adjacent territories).	62 734 km	0.3 3.2	1/3 000 000 1/1 000 000
Coasts and countries	Coastline and boundaries (planet).	196 countries.	1.5	1/25 000 000
ERDF regions	Eligibility for the Structural Funds.	309 regions classified.	0.01	Eligible regions.

Table 1 (cont.)

Theme	Nature of the information	Volume of information Description	Mbytes	Resolution/scale
Settlements	Name, location, population of urban centres > 20.000 inhabitants	1542 urban centres.	0.1	Location of centre.
Socio-economic data	Statistical series extracted from the SOEC-REGIO database.	Population, transport, agriculture, etc.	40.0	Statistical Units, NUTS III.
Air traffic	Name, location of airports, type and volume of traffic (1985–87).	254 airports.	0.1	Location of airport.
Nuclear power stations	Capacity, type of reactor, energy production.	97 stations, update 1985.	0.03	Location of station.

Source: Directorate-General for the Environment (1990)

design. The experience from the CORINE Programme, however, is that they are often accidentally or deliberately disguised; the examples of discrepancies in climate and soils data are described by Briggs and Mounsey (1989) and Briggs *et al.* (1989). Similar disparities occur in the data on air quality, stream water quality and bathing water quality provided by the EC (Commission of the European Communities, 1987). In these cases, erroneous data may be incorporated into the system as though they are pristine and pure (Briggs and Reeve, 1991). The errors are likely to be revealed only in some later application, by which time they may already have been transmitted to other, derived datasets or exported to other databases.

The lessons to be drawn from these experiences are various. In the longer term, there is clearly a need for improvements in environmental mapping and monitoring in many parts of the EC. Similarly, efforts are needed to ensure greater consistency in the data collected. Some attempts have been made in this direction by the EC in the past (e.g. the legislation on the exchange of information on air quality and water quality) but the effects have been disappointing. More rigorous and precise standards are undoubtedly needed. Even so, such developments will take many years to bear fruit. In the meantime it is therefore important to make better use of the data which do exist, or which are becoming available.

In this context, three specific requirements may be noted. The first is to improve and extend the spatial modelling technique, so that gaps in existing data coverage can be filled. The approaches of kriging (Oliver and Webster, 1990) and thin-plate splines (e.g. Hutchinson and Bischof, 1983) clearly offer considerable promise. Secondly, there is a need to improve documentation. At present, the genealogy and history of many datasets is uncertain or unknown. Occasionally—as in the example of the EC Soil Map (Briggs *et al.* 1989)—the available documentation is misleading. Without doubt, more rigorous reporting of data processing would greatly enhance the utility of many datasets. Finally, the scope exists to improve the methods and means of data transfer and exchange. On several occasions in the CORINE Programme the choice had to be made of whether to attempt to obtain and use existing digital datasets, or to recapture the data from their original source. Frequently it proved more convenient to do the latter. In part this reflects the technical barriers of differences in system design; these are gradually being reduced as system architecture becomes more open. More frequently, it was due to the sluggishness of bureaucratic procedures (especially where issues of data ownership were involved). As Burrough (1986) has noted, the adoption of GIS requires fundamental administrative as well as technical development. From the experience of the CORINE Programme this is an area of development in which much still remains to be achieved.

Acknowledgements

The work on the CORINE Programme reported here is the product of efforts by many different researchers over many years. Particular acknowledgement is due to Professor D. Rhind and colleagues at the Department of Geography, Birkbeck College, London, who were closely involved in all the GIS developments referred to here; to Dr J. Maes and Mr V. Schreurs for similar GIS development in DGXI, Brussels; and to Mr M. Cornaert for overall supervision of the Programme. The views expressed here, however, remain those of the author.

References

Briggs, D.J. (1990). Establishing an environmental information system for the European Community: the experience of the CORINE Programme. *Information Services and Use*, **10**, 63–75.

Briggs, D.J. (1991). Towards a European Environment Agency. In Gilg, A. (editor), *Progress in Rural Policy and Planning*, Vol. 1, London: Belhaven (in press).

Briggs, D.J. and Martin, D. (1988) CORINE: an environmental information system for the European Community. *European Environmental Review*, **2**, 29–34.

Briggs, D.J. and Mounsey, H. (1989). Integrating land resource data into a European geographical information system: practicalities and problems. *Applied Geography*, **9**, 5–20.

Briggs, D.J. and Reeve, D.E. (1991). Implications of incorporating environmental data into small area statistical analyses. Paper presented to Technical Workshop on Geographical Database Structures for Small Area Statistical Enquiries, Friday 22 June 1990. Small Area Health Statistics Unit, London School of Hygiene and Tropical Medicine.

Briggs, D.J., Brignall, P. and Wilkes, A. (1989). Assessing soil erosion risk in the Mediterranean region: the CORINE Programme of the European Communities. In van Lanen, H.A.J. and Breght, A.U. (eds), *Application of computerised EC soil map and climate data*, Luxembourg: Office for Official Publications of the European Communities, 195–210.

Burrough, P.A. (1986). *Principles of geographical information systems for land resource assessment*. Oxford: Clarendon Press.

Commission of the European Communities (1985a). Council Decision of 27 June on the adoption of the Commission work programme concerning an experimental project for gathering, co-ordinating and ensuring the consistency of information on the state of the environment and natural resourccs in the Community. (85/338/EEC). *Offical Journal of the European Communities*, **L176**, 14–17.

Commission of the European Communities (1985b). *The 1:1 million soil map of the European Communities*. Brussels: Directorate General for Agriculture.

Commission of the European Communities (1987). *The state of the environment in the European Community, 1986*, Luxembourg: Office for Official Publications of the European Communities.

Commission of the European Communities (1990a). Council Decision of 22 March 1990 amending Decision 85/338/EEC on the adoption of the Commission work programme concerning an experimental project for gathering, co-ordinating and ensuring the consistency of information on the state of the environment and natural resources in the Community. *Official Journal of the European Communities*, **L81**, 38.

Commission of the European Communities (1990b). Council Regulation (EEC) No 1210/90 of 7 May 1990 on the establishment of the Europcan Environment Agency and the European environment information and observation networks. *Official Journal of the European Communities*, **L120**, 1–6.

Directorate General of Environment (1990). *CORINE: Examples for the use of the results of the Programme 1985–1990*, Brussels: Directorate General for the Environment, Commission of the European Communities.

Hutchinson, M.F. and Bischof, R.J. (1983). A new method for estimating the spatial distribution of mean seasonal and annual rainfall applied to the Hunter Valley,

New South Wales. *Australian Meteorological Magazine*, *31*, 179–84.
Johnson, D.B. (1989). The National Resource Information Centre (NRIC)—its role in geoscience mapping and database development. Paper presented to BMR Symposium, Canberra, ACT, 7–9 November 1989.
Oliver, M.A. and Webster, R. (1990). Kriging: a method of interpolation for geographical information systems. *International Journal of Geographical Information Systems*, **4**, 313–32.
Whimbrel Consultants Ltd/Huddersfield Polytechnic (1989). *CORINE database manual*, *Version 2.1*, Huddersfield: Huddersfield Polytechnic.
Wyatt, B.K., Briggs, D.J. and Mounsey, H. (1989). CORINE: an information system on the state of the environment in the European Community. In Mounsey, H. (editor), *Building databases for global science*, London: Taylor & Francis, 378–96.

PART V

GIS and society

Times change and at this particular swing of the pendulum greater accountability is the theme of the moment. This has placed an unprecedented squeeze on resources which in turn has meant that the search has been on to target these better. In the past data that could have been used to target resources has been produced in great streams, but much of it presented in ways that were incomprehensible to all but a few specialists. Enter GIS which has the ability, as a tool, to be used to analyse this mass of data effectively both at a large and a small scale.

The ability to integrate different datasets to test an hypothesis on deprived areas or to provide command and control for the police or ambulance services is proving invaluable. This ability to integrate diverse datasets, coupled with the ability of a GIS to produce a readily understandable graphic, should ensure an expanding market in the service of society.

28

Working for a safer society

Anna Cross
University of Newcastle-Upon-Tyne

Working for a safer society

During the last decade, public awareness has been heightened considerably by media coverage of very sensitive, yet potentially devastating, issues concerning the environment that we live in. Global warming, the destruction of the ozone layer and incidences/causes of all forms of cancers are just a sample of some of the topics prominent in today's society. The result has been an increase in demand for possible solutions and the adoption of any technology or methodology which may serve to improve our knowledge.

Since "most human activity depends on geographic information: on knowing where things are and understanding how they relate to each other" (Chorley, 1987), it would seem that such issues lend themselves ideally to a geographical information systems (GIS) approach. The location of certain factors has always been considered a very powerful and reasonably acceptable surrogate for other unmeasurable and/or missing variables of interest. GIS allows the storage, manipulation and combination of this information, with the additional advantage (over traditional methods of analysis) of a very effective communication medium, that of the map.

The potential is reflected in the ever increasing number of GIS-orientated meetings, and in the proliferating projects which combine environmentally-related subjects and GIS technology. Common examples of ongoing research are; energy 'planning', flood 'modelling', metal pollution 'monitoring' and nature 'management'.

Case study

GIS technology has been implemented, and is under evaluation, in a study to identify excesses of childhood cancer in Northern England, and in turn establish whether they may be related to the local environment.

This was brought about by both mounting public interest and medical uncertainty as to the possible causes. GIS was perceived therefore as a new and potentially useful tool to tackle this problem and hopefully provide new insights into the aetiology of these cancers. It has led to the development of several environmental databases, including those most favoured, such as the siting of nuclear installations, background radiation levels and so on. More importantly though GIS also allows one the freedom to go beyond prior theory, because it can work within a data-rich, theory-poor framework, such that any element of interest that is mappable can be incorporated into the research, so one can also look to vaguer possibilities such as the effect of smoke levels, heavy metal pollution, etc.

This is by no means a one-off attempt to employ GIS in this type of approach. With environmental health a major policy issue in the late 1980s, members of the European Economic Community formed a consultation of relevant bodies, including the World Health Organisation, to discuss the implementation of GIS to tackle a very similar problem, be it on a much larger scale.

> Target 19: "By 1990, all Member States should have adequate machinery for the monitoring, assessment and control of environmental hazards which pose a threat to human health, including potentially toxic chemicals, radiation, harmful consumer goods and biological agents".

It can be seen therefore that many academic and public bodies are joining the GIS bandwagon, with the sole aim of being able to develop systems which will both integrate their vast amounts of geographical data and ultimately use it to improve upon their existing knowledge and working practices. However, GIS lacks sophisticated spatial analytical capabilities. Unfortunately, unless this is solved the users' expectations of GIS are never really going to be met by the present state-of-the-art GIS reality.

The future

Ideally, end-users are seeking analytical functions which will allow them to answer fundamental questions such as:

- Why are there excesses of childhood cancers here as opposed to there?
- How can I predict criminal behaviour, in order to enforce some precautions, or heighten public awareness?
- Where should one locate radioactive waste dumps, in order to minimise risks to society and our environment?

However, whilst promotional literature for GIS software claims 'advanced analysis functions', their interpretation is vastly different from that of the user, ie. 'abilities to generate buffer zones around features, remove common boundaries between area features and pattern area features' (definition from a well-known vendor brochure). In fact, this amounts to no more than data manipulation and map transformation techniques, which will not serve to answer the questions posed above.

Thus, while GIS can accommodate the data explosion, effectively manage databases and map any results to such queries, it does not hold the key to get from the data storage stage to the required results. Hence, many users, irrespective of their research bias, are now asking 'Well we have all this data, now what are we going to do with it?'

Preliminary steps towards answering this plea have been developed as a result of the research on childhood cancer in Northern England. The first was that of a Geographical Analysis Machine (Openshaw *et al.*, 1987), which was employed on the existing point cancer data, before the actual implementation of GIS technology (although it is still directly applicable to GIS analysis). Basically, its function was to look at the population at risk and the cancer incidences, and highlight those areas where there appeared to be an excess of cases.

The second method, Geographical Correlates Exploration Machine (Openshaw *et al.*, 1989) constituted a progression from the latter, in that it took the environmental coverages as developed and stored in a GIS and searched for possible relationships between the point cancer data and that of one or more human/physical aspects. Both methods are chronicled in further detail in other articles.

It is accepted, though, that these methods are in their infancy and are still largely descriptive in nature. However, they do go some way to relieving the onus of simply 'eye-balling' patterns in the data, as offered by GIS map-based analysis. They serve, therefore, to highlight and subsequently restrict research to more specific areas of interest, providing an essential step up the GIS analytical ladder on which further developments can be made.

Conclusion

It can be seen that the implementation of GIS technology is becoming increasingly widespread over many organisations and disciplines. Its applications are numerous, because it lends itself to anything and everything which is potentially mappable. However, whilst this data explosion stage is occurring, there is little thought (or, for those further along the process, the seeds of thought are only just being sown) as to how we can actually use these datasets for further benefit.

The foundations are already there, with the number of adopters illustrating how useful the basic GIS toolbox can be in society. The future, however, requires to build upon this base to enable expectations to become reality. In the case of the European-scale health GIS, emphasis needs to be upon data sharing and compatibility. A greater awareness also needs to be given to error within, and as a product of, GIS, establishing some form of quantification to reduce the possibility of uncertainty when the results are to be used in important decision-making situations. Finally, the integration of sophisticated, generic analytical functions would enable the data within a GIS framework to be used to its full capacity, such that its role becomes the formulator of action within society, as opposed to simply mapping it as it presently stands.

References

Department of the Environment (1987), *Handling Geographic Information*, Report of the Committee of Enquiry, chaired by Lord Chorley, London: HMSO.

Openshaw, S., Charlton, M. and Wymer, C. (1987), A Mark I Geographical Analysis Machine for the automated analysis of point pattern data. *International Journal of Geographical Information Systems*, **1**, 335–358.

Openshaw, S., Cross, A. and Charlton, M. (1990). Building a prototype Geographical Correlates Exploration Machine. *International Journal of Geographical Information Systems*, **4**, 297–312.

World Health Organisation (1988). Summary Report: Consultation of Environmental Health Information System in the European region, Berlin (West) 21–25 November 1988.

29

Towards more healthy living

Jerry Fishenden
Health Education Authority

The Health Education Authority (HEA) was established in 1987 by the Government as a special Health Authority within the NHS. Its objective is to encourage the population of the UK to adopt a healthier lifestyle. Such a complex and important task requires a demanding combination of statistics and information on the demographic, morbidity/mortality and general lifestyle characteristics of the whole country. Such information is required in the correct form before it can be of much use in strategic and operational planning and implementation.

The HEA (and its former incarnation as the Health Education Council) has always had access to standard research information such as the General Household Survey and OPCS data. In addition, it commissions its own research on health behaviour and awareness. Such data, however, is clearly often of use within each of the districts, enabling them to prioritise and focus their work on health issues relevant to their own needs. The HEA is working at a national level, so it needs to be aware of national trends and needs, as well as local requirements. It thus has a broad and exacting requirement for information across both levels.

Working through partnerships, such as with Queen Mary College, the HEA has already begun to map relevant information on a national basis built upon such locally-oriented information. Its recent publication of 'Mapping the Epidemic' was a graphically presented map of the nation, by health authority district, of coronary heart disease (CHD). This was the first such publication of its kind and although it added nothing directly to what was available locally (districts know their own locality's statistics), it was the first clear attempt by the HEA's research section to map the pattern of morbidity and mortality across the whole geography of the UK.

While such publications have their uses (the document was generally well-received), clearly they can only present a snap-shot, static picture of one moment in time. Future trends will require further reprints, showing changes over time (and, one trusts, an eventual reduction in the incidence of CHD). A more effective way of capturing, storing, analysing and accessing such data is clearly available through the use of computer-based systems.

In 1988 the Authority selected a parallel-processing computer system to fulfil its central computing requirements, permitting high-speed manipulation and interpretation of large amounts of data. With such power available, it now has a unique opportunity to collate a comprehensive national database of relevant, geographically-oriented statistics on all areas of the nation's health in which it is active. Furthermore, a pilot scheme to link in all district health authorities is underway, a plan which will eventually provide on-line access to health promotion and education services within the NHS. Cross-analysis and interpretation of regional and district-based demographic, geographic and health data such as age, sex, social class, ethnic origin, income group, occupation, place of residence, etc, will enable the HEA to fine tune its campaigns and distribution, with greater ability to target appropriate materials.

The NHS information base is rapidly expanding as the Government pushes information technology (IT) within all of its functions. By the end of March 1991, all GP surgeries should be computerised. Similarly, work is in progress on Family Practitioner Committees (FPC's) and within hospitals to integrate all information systems. The HEA's IT strategy embraces these developments and it is intended to achieve, and provide, access, where appropriate, to other NHS.

One of the most labour-intensive tasks to date has been the transformation of data from standard statistical systems, such as SPSS and SAS, to graphically presented formats. While there are tools available, they require considerable time in manipulating data into the final desired format. It is in this area that there needs to be most change. Spreadsheet packages such as Wingz have shown the potential for graphically stored information to become dynamic, manipulated by background hyperscripts (essentially sophisticated macros). In such packages, it is possible to define user 'buttons' that when 'pushed' on-screen, can pull in relevant information in a graphically-orientated format while at the same time providing access, if required, to the raw data underlying such representations.

Another possibility is the increasing availability of interactive audio/visual compact disk (CD) systems. With the appearance of the Commodore CD-TV at home consumer prices and later the Philips CRDI system, the potential for using CD-based information systems is finally becoming a reality. In fact for some time the hardware available has been outstanding, of a high performance and unrealised potential. And it is this latter that is the stumbling block. The quality of software is several years behind the hardware platforms now available. Professional awareness of the potential of the new technologies lags even further behind that.

The real challenge for the IT professional in the coming decade is to push the software to a far higher level of functionality and to educate specialists in such fields as research in the new methods available for accessing, manipulating and presenting data. There still remain too many systems generating page after page of columns and rows of data in endless tables of information. Users require, and should demand, better. The new graphical methods of presenting data are an opportunity we should seize. The Authority's ability to affect real change in the health of the nation will in part rest upon its ability to maximise both the accessibility and the content of its information. The health of the nation is too pressing an issue for the Authority not to take advantage of these new technologies and it will not be left behind in this challenge.

30

Geographic information systems for ambulance services

Ian Van Creveld
West Midlands Ambulance Service

The objective of the West Midlands Ambulance Service is to respond with the nearest, suitably equipped ambulance which is available, quickly and safely to an incident, with the appropriate trained crew. Current resources provide up to 55 fully equipped front line ambulances at peak periods, supported by fast response cars and motor-cycles manned by paramedics. These are based at 25 locations. Contact is maintained with staff by use of VHF radio, both mobile and hand held, also mobile telephones and alphanumeric pagers. There are BT private wires to all stations as well as hospitals, Police, Fire and Airport controls.

The West Midlands Ambulance Service deals with some 700 emergency and urgent incidents every day. The service covers an area of 345 square miles serving a population of 3 million. Particular problems which arise within its area are the Midlands Link Motorways M5, M6, M42 and M54, including the part of the M6 motorway from Spaghetti Junction through to Rayhall—the busiest section of motorway in Europe (170 000 vehicles a day, but 20 000 per hour at peak times, according to West Midlands Traffic Police data).

In addition to three airports (including one of international rating) there is also a high pressure (1600 psi) pipeline complex carrying diesel, petroleum and aviation fuel only three feet underground throughout the Midlands. With two nuclear waste trains passing through the area each week our major disaster plans are always at the ready.

The old solution

Previously all requests for emergency ambulance transport were dealt with on preprinted forms. These were handed to an officer who decided on deployment of vehicles. With the aid of a glass-covered vehicle schedule and chinagraph pencils he marked the location of vehicles. A coloured push-button light system gave him only three states for each vehicle, Available On Station, Busy and Available On Radio. The system was time-consuming and also generated a lot of paper. An information retrieval system gave the nearest three station location to any incident.

Present method

Calls from the public and GP's are received via an ACD (Automatic Call Distribution) telephone system and are input direct into the computer by staff using colour terminals. The advanced computerised Command and Control programme will cross-reference the telephone number and address then display the incident with up to twenty four of the nearest available vehicles and their current status.

Next development

Within the next twelve months a vehicle tracking system will be installed in the Emergency Control. With the aid of radio beacons any vehicle can be located. This will enable the Command and Control computer to give the optimum response to any incident based on the actual locations of the vehicle and the incident. Each address will be cross-referenced by its location to give a grid reference number. The Command and Control computer will then give the computerised Mapping System the location information and instruct it to display an icon at the relevant position. The progress of any vehicle can be monitored throughout an individual case. An overhead video projector will display on an eight foot by six foot wall-mounted screen the whole of the West Midlands area complete with all calls that are in progress. As the Command and Control computer knows the location of the incident and the Mapping System will display the vehicles' movements, as ambulances arrive at their incidents the systems can automatically provide status updates for statistical purposes.

Conclusions

Once all three systems are up and running and interfaced with each other, there will be a greater utilisation of resources, better monitoring of vehicle movements, and ability to identify vehicle locations when assistance is required urgently. In addition there is a facility to archive data and re-present it in various forms, either as plotting one vehicle over a twenty four hour or seven day period, or displaying the same type of incident within an area over a predetermined time period. When combined the system will be capable of providing sufficient information to enable the West Midlands Ambulance Service to produce in-house their own Operational Research programmes.

Appendix

West Midlands Ambulance Service Emergency Command and Control

Equipment

CPU	2	Bleasdale 740mb drives with the UNIX operating system. 1st machine copying to 2nd machine via Ethernet connection
Terminals	20	Microcolour Graphics type M2320

GIS	4	DEC Colour VDU type VR290-D3
	2	VAX 2000 drives running Laser Scan software
	2	VAX 3200 drives running Laser Scan software
	1	Barco 400 overhead colour video projector
Ancillary	2	Philips 512k PCs type CM1480
	2	Mannesmann Tally 85 printers
	1	Hitachi Laser disc drive type MSCDEX

Contacts

BM Johns
Regional Chief Ambulance Officer
West Midlands Ambulance Service
Fourth Floor, Falcon House
6 The Minories, Dudley
West Midlands DY2 8PN Telephone: 0384 455644

W Whitehouse
Information Technology
Address as above Telephone: as above

A Sheehan
Emergency Command and Control
Address as above Telephone: as above

Profile of the West Midlands Ambulance Service

The West Midlands Ambulance Service was formed in 1974 following the National Health Service Reorganisation Act. Previously the area was covered by a number of smaller ambulance services, which were formulated following the 1946/7 National Health Act when the Minister of Health placed the onus on the local Authority to provide ambulance services. The WMAS covers some 325 square miles incorporating the former West Midlands Metropolitan County Council area and parts of Staffordshire and South Staffordshire. It is divided into two divisions and is served by 25 ambulance stations, some 1064 staff and 381 vehicles for a population of three million.

It takes almost fifteen months to train an ambulance person fully to accident and emergency standards, which is undertaken at our own purpose-built training school.

There are four controls, three of which deal with routine out-patients and one for urgent and emergency calls. During the period 1988/9 the following statistics are available for numbers of patients conveyed: emergency, 115 081; urgént, 88 020; out-patient, 912 351; giving in total 1 114 452. With the population of Birmingham being just over one million, it is the same as every man, woman and child being conveyed by ambulance once in the year.

Service vehicles completed 5 730 971 miles. This is equivalent to driving the family car around the world almost 235 times. The vehicles consume in the region of 470 000 gallons of petrol costing almost 900 000 pounds. The Central Emergency Control attends some 700 calls a day and with passing calls out, contacting doctors or the police, a call is answered, on average every 30 seconds of the day. With a budget of almost 18 million pounds a year the WMAS is the second biggest and busiest ambulance service in Great Britain.

31

GIS and society: A local government viewpoint

Eric Cripps
Cumbria County Council

Introduction

Local authorities exist in a financial climate which expects increases in the provision or enabling of services with steadily decreasing resources and with ever greater restrictions imposed by central government on how they might spend revenue or invest capital.

An example might be the situation for most Local Education Authorities (LEAs) in regard to the Local Management of Schools (LMS). In the LMS case, in one way or another the greater part of the education budget administered by LEAs is devolved on a formula basis to schools whilst the remainder is retained by the LEA to spend on its own initiative, though it may prioritise services, such as educational psychology and welfare which are administered by the authority rather than by the schools. Even if any argument that the overall funding of the education service is inadequate is discounted, there are clearly major statutory limitations on the extent to which local authorities may allocate their resources.

Another example is in the provision and co-ordination of community care in Social Services. Here, there are ever increasing demands for services to the growing numbers of elderly in the community, the increasing numbers of handicapped people discharged from residential care to live and be cared for in the community and so on, and yet the financing of "Griffiths" has been postponed until the next Parliament. There is increasing material deprivation and social disadvantage especially in our cities and urban areas, growing numbers of single-parent families and associated social problems, all of which impose demands on shrinking levels of public expenditure. Road building and maintenance and property maintenance is inadequate—many school buildings throughout the country are in a state of disrepair.

This sample from the list of demands on local authorities has to be set against relative shortfalls in most revenue support grants settled by central government on local authorities. There are only two ways for a local authority to bridge the gap between income from revenue support grants (plus the national business rate) and its budget; increase the Community Charge or reduce expenditure. For political reasons, increasing the Community Charge unnecessarily is usually not on. Service cuts are also highly unpopular among both politicians and Community Charge payers. The demands on local authorities seem impossible to meet and mean "more for less".

Such demands require both severe prioritisation of resources and shedding the provision of services from local authorities to private companies and voluntary organisations.

Increasing demands and shrinking resources also mean a healthy re-orientation to "value for money" in the delivery of community services. Relating the value of services output at quite small geographical locations to the cost of their inputs in revenue expenditure becomes very important. The particular measure of accountability of local authorities brought about by the current gearing of the Community Charge to local government expenditure also brings about a less healthy preoccupation with the value of localised outputs of services compared with the revenue costs of such services. On one side of the political divide there is a concern with "are my constituents getting their money's worth of the share of the local authority's resources for which they are paying?"; on the other side there is a concern with "my constituents cannot afford to pay the Community Charge anyway, so I must keep the cost to them of local authority services down".

Modern day requirements for prioritisation, cuts, and value for money in service delivery bring into sharp focus the need for the strategic targeting of local or community resources.

Communities, especially in a county like Cumbria, tend still to be defined spatially and in most areas social phenomena such as deprivation and many social categories which demand services manifest spatial variations. It becomes essential to understand the local geography of resource allocation—whereabouts the local authority spends its money—and of local need and demand on which services need to be targeted.

The role of GIS and the business need

Strategic information for the targeting of community resources thus has a clear spatial dimension. It is as important to know where services are needed and where resources are spent as how much and on what. In local authorities, two forms of GIS meet the needs generated for strategic management information in this context.

Small scale GIS for strategic management

The first provides for the integration of social and economic data with small scale geography and for quite powerful spatial modelling and analysis. The geography is mainly comprised of unstructured polygons in a hierarchy of administrative areas, such as the County, Districts, County Divisions, Wards, Parishes, Enumeration Districts, and School catchment areas. There is also point data describing attributes of different sets of point locations. The geography also usefully comprises notional catchment areas based on Thiessen Polygons or their quadtree version or voronois.

The data files of this system contain attributes for each of these locations or areas which describe the demographic categories of their populations, births and deaths. These attributes also record appropriate variables from the small area data in censuses. In the files, there are also data derived from these census variables such as standard scores for "material deprivation". There are also strategic data from an authority's administrative files, such as the proportion of children receiving free school meals, and by location or for small area, the numbers of elderly in care, the numbers of children in care and other social service clients. Other data attached to these locations are children recorded in Area Health Authorities' Child Health Registers.

For many of these locations and areas there are also measures of employment and unemployment. There will also be the forecasts of the values of some of these variables as well.

Other important attributes of locations will be all of the individual properties, schools, colleges, residential homes, day care centres and so on which the local authority owns and which are the physical locations of most of the resource inputs to the various parts of the geographical area administered by the authority. Very important ones are the amounts of capital and revenue expenditure which the authority makes at each location throughout its area.

Indispensable tools in this system are the central postcode directory and the postcode address file, which enable most of the entities at the focus of most strategic analysis to be located, provided that they are postcoded. Point locations can be defined to about 100 metres and the contents of ward and other areas can be generated either via gazetteer type tables or point-in-polygon routines. The fact that, in this type of system, tolerances in the location of data of about 100 metres are acceptable, says a great deal about the broad brush nature of the strategic targeting of community resources.

The range of "geographies" essential to this kind of system was described a little earlier. Optional geographical features, again in the form of unstructured vectors, of great use for illustration and for orientation purposes in mapping, are entities such as roads, lakes, urban areas, and in the case of Cumbria land above 300 metres, all of which are available from small scale digital maps such as the OS 1:625000 map.

The main forms of analysis undertaken in this system involve the thematic display of spatial distributions of single variables, multiple overlay techniques, single and multi-map spatial modelling of variables, and two map area analysis, mainly mapping functions of attributes of one map into another. Spatial query is another very important function of this type of system, but a crucial feature of the GIS in this strategic area is its ability to generate polygons, such as Thiessen polygons or voronois around point entities. Other extremely useful functions are those of creating spatial buffers around points (circles) as well as along arcs (roads in particular).

The range of data outlined, and relating to different geographical entities, can be integrated together to perform a variety of market area analyses in the public sector. These analyses provide essential strategic management information highlighting where resources ought to be targeted, or for answering spatial "what ifs" in response to targeting options. In a later section, some Cumbrian examples of applications of this type are given.

This first kind of GIS is typically housed in a research and intelligence unit of a local authority, charged with providing timely management information for strategic decision support. Although there is a need to standardise data recording and particularly the locational referencing of key strategic administrative data, a central research and intelligence unit can by and large be asked to get on with the application of a GIS of this kind together with the consequential data acquisition, the provision of skills of analysis and running the system.

Analyses from the system might be required by a County Council as a corporate entity or by service departments—Educational, Social Services, Planning, Highways—but in this instance the departments themselves do not need to run or know too much about the GIS. The outcomes of applications of a system of this kind are corporate, not the system itself.

A large scale corporate system for day-to-day business

This is not true of the second kind of system, which is focused more on day-to-day business and administration and is essentially a corporate system.

At the heart of the system, is the large scale corporate digital map. For most local authorities, this will comprise the OS 1:1250 maps for urban areas, the 1:2500 maps for developed rural areas and the 1:10000 maps for the remoter rural areas. Together these form the continuous, seamless, scaleless map which provides the up-to-date corporately managed map base for GIS applications in departments, probably within a peer-to-peer network of systems. The corporate map may be required to provide for straightforward automatic mapping applications in some locations in the network.

Some GIS applications will require fairly simple map data. A number, however, will require fully structured or topological data. A role of the keeper of the corporate map is therefore to structure unstructured map data for those applications which need to be able to do polygon processing or network analysis.

These latter applications are clearly in highway maintenance, school transport, school catchment area analysis, property terrier and particularly property management. In this system, attribute files are maintained and probably held within user departments.

There is a greater requirement for accuracy in the location of attributes. In such systems, there will be a much more extended set of descriptors for each entity. More stringent standards with regard to locational referencing, and textual and graphical data transfer, need to be applied throughout the system.

The principal applications of these large scale systems are at the departmental level. The targeting of resources, in many cases, becomes a day-to-day management decision. In highway maintenance it may be concerned with identifying lengths of the road network in a part of the county, which qualify for the short term expenditure of available money, and scheduling them with lengths of highway in other parts of the county map. In the Education department, it may be that there is a requirement to integrate data on children at particular locations, with highway network data, with bus schedule data and to route the children to school. This is day-to-day resource planning at a large geographic scale.

Applications in Cumbria

Cumbria's applications have all been with a system of the former type, with small scale unstructured map bases and with issues of a highly strategic nature in resource allocation. The software is Tydac's SPANS and is run by Cumbria County Council's Corporate Information Unit (CIU). The County Council is currently developing a strategy for the larger scale system, which it sees as only possible in Cumbria by the middle 1990's, although a start on it is highly desirable in 1991.

SPANS has been in operation in the CIU for two years and has been used in a large number of analyses of geographic distributions of social and economic data and of service data during that time. Only a few examples can be discussed here, with a particular focus given to its use in targeting the County Council's resources.

Recently, the CIU collaborated with the Education Department in a study of ''Social and Educational Disadvantage''. The main purpose of the study was to prioritise the allocation of the Cumbria LEA's Initiative Fund and other central spending by the LEA on school psychology services, educational welfare and nursery education, in favour of the socially disadvantaged areas of the county.

To be able to do this, the study had to measure the statistical relationships between distributions of "material deprivation", the consumption of free school meals (a reasonable proxy for the distribution of income support) and educational under-attainment measured by tests in primary and junior schools and GCSE results in secondary schools.

The correlations and regressions which established these relationships were of course computed with conventional statistical routines. The manipulation and integration of variables into compatible geographies were crucial GIS applications and essential prerequisites to the statistical analysis. The spatial mapping of the analysis, though significant in presenting the results, was less fundamental than the ability to aggregate and allocate census enumeration district data (in attribute records tagged to ED voronois) to polygonised primary school catchment areas. With these new spatial distributions of census variables, appropriate standardised scores of deprivation for school catchment areas could be computed and related to the in-school data described earlier and defined by point distributions.

From this analysis, about 320 areas in the county could be ranked on a scale from poorest to most well-off small areas within the county. Similar analysis was done for Social Services and related to the distributions of elderly population and other potential client groups in Cumbria's population. These distributions of clients came from Social Services' actual caseload and the provision of various services. The spatial imbalances of the probable need and supply of various social services could then be studied. They were also mapped.

These kinds of study, currently an important aspect of the work of the CIU, look at the spatial distribution—whereabouts in the county (measured as the needy groups in small areas) the needs are for major services.

The supply side is looking at spatial distributions measured by annual expenditure on employees' wages and salaries and other revenue expenditure made by all committees throughout Cumbria. Each physical cost centre within the County's accounting system is being located to within 100 metres via postcodes and the CPD. Attribute files recording categories of revenue expenditure for voronois defined around the point locations for these cost centres are created and include expenditure which may be recorded as being spent at the headquarters of the County but has to be spatially distributed to these notional sub-areas.

These spatially defined inputs can then be compared with the spatial measures of need described above with any spatial variables describing consumer preferences at various locations throughout the County and with spatial distributions of service outputs. These can all be attributes of point locations, or polygon sub-areas of the County.

Conclusion

In very recent years, there has been a growing application of GIS and spatial analytical techniques to market research problems. These have added to a much longer established catalogue of applications in different aspects of environment analysis. Market research approaches using socio-economic variables are highly appropriate to targeting resources in the public sector.

32

Real-time vehicle tracking

Alexander Shaw
RAC Motoring Services

Technology in RAC patrol vehicles

The C.A.R.S. system (Computer Aided Rescue Services), the RAC's command and control system, is the medium by which breakdown calls are received, and appropriate resources are allocated, for breakdowns suffered by members anywhere in the UK and N. Ireland. This system went live at its first location, Stockport for NW England, during 1987 and after a period of progressive phased implementation covered the UK by 1989. The system allows the RAC to give a much improved service to members and to respond quickly to all calls for assistance.

Underpinning this improved service are two main features:-

(1) Improved information about the location of a breakdown. This is still the most troublesome feature of calls for roadside assistance in that the member may have only a hazy idea of the precise location, or may have difficulty in its description. (Consider for a moment some of the answers that are received when you ask a stranger for directions!). This problem is addressed by the use of a structured data input and cross-referencing to the Post Office PAF (postal address file). Provided the member knows on which road or street the vehicle is located and the name of the nearest adjoining street then a precise Ordnance Survey reference can be determined. A large gazetteer has been compiled in which use can be made of road names (e.g. A64), local landmarks (e.g. outside the Pig & Whistle), RAC and AA call boxes, car parks, etc., and the telephone number being used by the caller may also be used to give an indication of location.

(2) Improved management of resources. Basically having patrols at the right place at the right time makes a considerable difference to the quality of the service, so managing the location of the patrols is very important. Vehicle tracking technology provides a means by which this location data can be much improved.

A market leader in the field of vehicle tracking technology is the Datatrak system. This utilises a network of low frequency navigation transmitters and a locator unit carried by the vehicle. From the signals received from each transmitter within range, the unit can compute the Ordnance Survey grid coordinates of the vehicle and this data can be passed to the radio system for re-transmission. The data can be used by the receiving control system to pin-point to within 50 metres the location of each vehicle being tracked. The system was developed originally for security purposes so that the progress of vehicles transporting high value loads could be continually monitored, and indeed has had a number of successes with hi-jacked loads being recovered and criminals apprehended. Datatrak Ltd. is a joint venture of George Wimpey and Securicor.

The tracking technology is shown to best advantage when used in conjunction with an interactive mapping display. This allows the vehicle location to be superimposed on the map and this display can be readily understood by the operators. This will show not only the precise location of the vehicle, but the speed and direction of travel will also be evident.

A number of RAC patrol vehicles in the West Midlands area have been fitted with the Datatrak locator units and antennae. Tests were carried out to verify the accuracy of the location information and the equipment has been found to be very reliable. The RAC already has a sophisticated radio system with an MDT (mobile data terminal) in the cab of each of its rescue vehicles, via which breakdown details are passed to each patrol. Particular emphasis is placed on the use of 'data' for radio communications, rather than the less cost-effective 'voice'. The Datatrak unit can interface with the MDT and patrol location information can be included in all 'status' communications. Therefore the despatchers will know the precise location and activity of each patrol. This is particularly useful when taken in conjunction with the OS reference worked out for each breakdown location as it will allow the despatchers to display a map with the locality of the breakdown, plus any nearby patrols. This will assist in determining which is the most suitable resource to allocate to the job. This will not be used with every job as very often the decision is quite straightforward. However, there are many cases when the decision is complicated by various geographic features—rivers, motorways, etc., where such a display would prove to be very effective and reduce the incidence of time-consuming mistakes. The routes available to the patrols and the likely travel times can readily be estimated, for example. Also the display will indicate to the despatcher the geographic distribution of patrols over the area for which he is responsible. This will allow better coverage to be maintained and reduce the tendency to congregate the patrols in a small geographical area.

The patrol is the front line in the rescue services provided by the RAC and as such constitutes a very important corporate asset. Whilst part of this asset is the equipment and expertise embodied in each patrol, its location and management are also very important. To date the CARS system assumes that the patrol is at the location of the last reported breakdown or at a location as advised by the patrol (it is not always possible to remain at breakdown locations, motorway breakdowns being an obvious example). Vehicle tracking allows the locations of each patrol to be validated and reduces the tendency to wander 'off station'.

Where the volume of work exceeds the capacity of the available patrol resource, then external agents may be used by the RAC. The agents are usually small independent businesses who specialise in vehicle repair and recovery and are trained and equipped to the required standard. They are often used by all the breakdown clubs/organisations and are particularly effective at times of heavy demand, or in sparsely populated areas, where patrols would be under-utilised for most of their time. Fitting a tracking unit within agents' vehicles would make the despatchers aware of an agent's breakdown vehicle in the vicinity of a breakdown. This will allow more effective utilisation of this resource. Secondly the despatcher will be able to monitor progress on a breakdown to which an agent's vehicle has been sent and will know in particular the 'at scene' times.

The explosion in the use of mobile telephones has greatly increased the level of communications with motor vehicles. Whilst this may be a boon to the driver when a breakdown occurs it is still essential that a precise location be given and therefore the driver must know where he/she is (not always the case particularly on a dark/wet night), possibly even having to leave the vehicle to ascertain their location (a singularly annoying situation in inclement weather). This is a particularly difficult situation with lone female drivers and whilst the motoring organisations are prepared to search for motorists in distress, the vaguer the location then the wider the search area and the longer the potential waiting time. Although not yet commercially available as part of the mobile telephone service, the technology to address the above problem is available. The cellular radio network could use the radio telephone transmission as a beacon and calculate a location from the reception characteristics at surrounding receivers, or at least determine from which 'cell' the transaction is emanating. This has been done with stolen vehicles with mobile telephones, but a "where am I?" service is not yet generally available; no doubt suppliers will respond to demand.

For optional efficiency of the rescue service, precise knowledge of the breakdown location must be known, together with the location and status of resources needed to tackle the breakdown. The location and management of resources can be improved by the use of vehicle tracking technology, interfaced to an interactive mapping system. Mobile telephones may be valuable in determining the location of breakdowns should there be sufficient demand for this service.

PART VI

GIS and the infrastructure

Those businesses whose major assets are lines, rails or pipes have been involved in a major move from government-owned to private business. One of the biggest tasks has been quantifying these assets, many of which are buried. On top of these business needs are external pressures in the form of the customers and environmental pressures, not perhaps from our own government but often from the EC.

All these forces dictate the use of GIS at two levels, on a macro and micro level. These approaches are being explored cautiously as the industry learns the capabilities of the new systems and the price of making wrong decisions. Datasets have also become important with boundaries and gazetteers featuring in all plans. Disappointingly, this will lead to much duplication unless some form of central brokerage can be set up for what is after all bureau information. The government has a role here as a priming force to start the flow of data.

The Public Utilities Street Works Act (PUSWA), due to get Royal Assent this year, will be a potent force in that it will allow the use of electronic messaging to exchange notices (the notification of an intent to dig a hole in the road) and create a Computerised Street Works Register (CSWR). The CWSR will be a log of all intended work in the highways based on an Ordnance Survey Grid Coordinate (OSGR). Access to this database will be perhaps by road name or OSGR. Importantly it will act as an impetus for the statutory undertakers to get their plant records up to date.

33

The road more travelled

Ralph Robbins
The Automobile Association

1990 was the year when the diverse parts of the GIS industry began to appreciate the other parts that make up the whole. From my perspective as a data supplier I have been disappointed for many years at the lack of understanding by users and software suppliers of the costs of sourcing, editing, compiling, structuring and presenting data. This failure to recognise fact has resulted in the cost of maintenance not even being considered. Perhaps it is a necessity of evolution that these things take time. Perhaps the proliferation of seminars, conferences, meetings and visits is gradually having an effect.

It is not only users who are recognising facts of life. Data suppliers are having to appreciate that if they want to sell their product, market forces apply and a realistic pricing policy is essential. The advent of PC-based systems presents an opportunity for much more widespread use and appreciation of the power of GIS. There is also the ever-present fear among data suppliers that this popularising effect will make their data more prone to piracy. Examples are available of a cavalier attitude to data ownership by some software suppliers and rampant piracy by users if the opportunity presents itself.

One event worthy of note was the highly successful PANDORA project. The British partners in this EC DRIVE programme project, the Automobile Association and Ordnance Survey, created separate datasets of road networks and road, traffic and service attribute data that were effectively for the in-car navigation systems suppliers. Field trials were accomplished although not witnessed by the data suppliers on the grounds of commercial confidentiality. Favourable reports were issued. The application to continue the research into maintenance procedures by the triumphant teams was disappointingly rejected by the DRIVE office. Surely an opportunity missed! The mainly happy result by the PANDORA participants in 18 months would appear to indicate that we are ahead of our European colleagues in this field who have 3 years to complete the same exercise. Their projects are still running with, apparently, mixed progress.

I think it has to be stressed that data suppliers are as diverse as the subjects they cover, therefore one can only speak from one's necessarily narrow viewpoint. However, from where I sit it is clear that vendors who formerly purveyed their wares in paper form to a different market are learning fast. They are learning to respond to the new markets' requirements of subject and detail, to present data in the required format, at an appropriate price. They are also learning that lead times acceptable in paper publishing are not necessarily acceptable for electronic publishing. They are learning that professional representation is a requirement, consultation is vital and response needs to be fast. Above all they are learning that their market in electronic publishing is not the fag-end of paper publishing or a nice little earner on the side, but is a powerful and growing market in its own right which if cultivated and serviced in an efficient and professional manner will in a few short years provide rewards hitherto unsuspected. The product range is growing fast and the quality is excellent. Try us and see.

34

Tapping into GIS in Yorkshire Water

Jim Osborne
Yorkshire Water

Introduction

Yorkshire Water's business

Yorkshire Water is one of ten water services companies set up by the Water Act 1989. The Company provides approximately 4.4 million customers with water supply and sewerage services over an area of approximately 14 000 square km. Following privatisation, it now operates within a strict regime of regulation and price control and is monitored closely by the Director General of Water Services. The industry is having to respond rapidly to the introduction of higher standards of water quality and increased awareness and sensitivity to environmental issues, whilst trying to remedy the dereliction of its assets following years of under-investment.

In the lead-up to privatisation, the industry produced asset management plans to establish the level of investment required over the next 20 years to achieve declared standards of service. Routine investment planning is now based directly upon the use of these asset management plans.

Business objectives and requirements

Within the scope of its business, the Company is seeking to maximise the use of its existing assets and achieve the most cost-effective investment in new assets necessary to achieve its declared standards of service, and identify new business opportunities to generate an acceptable rate of return for its shareholders. Both requirements result in an increased demand for more effective information about the location, condition and performance of its asset infrastructure, the nature and extent of service problems, and the wider needs of its customers. With the bulk of its assets and customers geographically dispersed, there is a clear potential for the use of GIS.

Scope of this article

This article reviews the status of GIS within Yorkshire Water and presents a subjective view of the issues involved in progressing with its wider-scale development and implementation.

Information and GIS strategy

The Company is a major user of technology within many aspects of its business. The scope of its computer systems incorporates the traditional billing and financial applications but the Company is increasingly adopting computer and communications technology within its operational and engineering activities, e.g. telemetry, digital mapping, and CADD. Historically, computer systems have been commissioned, developed and used as discrete applications. During recent years, however, a policy of integration has been pursued in response to several internal and external factors: the user community is demanding "one-shot" data entry, i.e. data is entered into the Company's network of computer systems only once; within the private sector, there is now a specific need for consistent and auditable reports to external monitoring organisations; and the opportunities presented by the technology itself. So, whilst there is a clear potential for GIS, the specific need is for "integrated GIS".

Practical adoption of GIS to date

DM/FM strategy

With the bulk of the Company's assets being underground, in the form of pipe networks, it is essential to have accurate map-based asset records as a basis for organising routine operational work and for assessing the need for new investment. The Company also has a statutory duty to maintain accurate records of its underground plant.

In 1987 a 5-phase strategy was adopted to introduce digital mapping (DM) for water main and sewer asset records. It was based upon the following principles: during the initial phase, "invest small, think big" and grow from a basis of knowledge and experience; use DM to provide the necessary foundation on which to build an integrating environment (DM/FM) to extract and analyse data from a range of application systems and present it more effectively through a "mapping window"; and carry out major reviews after each phase. It is also characterised as evolutionary rather than revolutionary.

Phase 1

The initial phase was structured to gain experience and improve organisational awareness of the potential of the technology, anticipating that it would be difficult to present a traditional cost/benefit analysis in support of the following phases. The project had many facets, of which the technological elements were only part, e.g. system specification, system procurement, development and implementation, personnel selection and training, organisation, data conversion, acquisition of OS Maps.

In 1989 the first major review was completed. Whilst confirming that the objectives, structure and elements of the original strategy were sound, it did identify a number of constraints upon the rate at which to progress:

(1) GIS products are not yet sufficiently mature and reliable to support the development of wide-scale operational facilities within an integrated GIS environment of the type required by Yorkshire Water;

(2) the Company has a limited capability to collect geographically referenced data of a quality consistent with wide-scale investment in GIS,
(3) equipment is relatively expensive but reducing in price in real terms,
(4) OS digital map data is not readily available throughout the Company's area,
(5) the necessary expertise is at a premium,
(6) it is doubtful if any GIS companies and existing software products at present can support a fully operational service within a large corporate enterprise.

In particular, it was noted that there is a world of difference between the capabilities of the technology, as witnessed at a slick demonstration, and the ability of a GIS company to harness that capability within a major implementation and thereafter to provide a reliable operational support service.

Clearly, many of these constraints are characteristic of any new and evolving technology, but they are all manageable. Fundamentally, the issue becomes one of how quickly to progress rather than one of whether or not to do so at all.

The situation in 1990

In January 1990, an updated DM/FM Strategy was approved and sanction given to proceed with the next phase. The principal elements of the next phase are to establish the necessary basis and then demonstrate a capability to convert approximately 14 000 water main records and 13 000 sewer records by 1996, and to migrate to a second generation DM system to take advantage of more capable and robust software products that have become available since the initial system was commissioned in 1987.

Towards the end of 1990, steady progress is being made to increase conversion capacity whilst specification of the second generation system is underway. Unlike the initial system, the second generation DM system is being structured more around a number of "building blocks" or principal layers within a hierarchy of map-based records systems:

- map management system
- water asset records system
- sewerage asset records system
- boundary management system.

These in turn will be supported by a "common services module", which will incorporate plotting and gazetteer facilities that will be shared by all GIS applications. In respect of the gazetteer, the strategy provides for a corporate gazetteer to be established and maintained through the DM system and used by several existing non-graphical applications, and all future GIS applications, whether or not they are map-based.

The way ahead

DM/FM strategy

The Company is continuing with its evolutionary approach to the introduction of GIS, at a rate which is consistent with the capabilities of the technology and the organisation's capability to adapt to its introduction. During the period 1990–92 the bulk of available effort will be given to laying essential foundations, primarily through the introduction and extension of digital mapping records systems and a corporate gazetteer, as the basis on which to build subsequent GIS applications.

Organisation and data quality

1990–91 will see the introduction of a new organisational structure that includes sections and posts directly and wholly responsible for the management of data quality. As recommended in the review of the initial phase of DM/FM strategy, a regional team will be created to maintain map-based records. They will be required to provide a service to agreed standards with access to their records being made available through a network of distributed workstations.

Research and Development

Proposals are being considered for a number of R&D initiatives which include an examination of the feasibility of using Global Positioning Systems for data capture of point locational data, an assessment of the improved effectiveness of a GIS user interface to an existing application, and the potential uses of scanned data.

Summary

It is now accepted that digital mapping and GIS technology offer significant opportunities in the longer term for improving overall effectiveness in managing the Company's asset infrastructure. Experience indicates that there are many short and medium term issues, often non-technical, that need to be recognised and addressed in progressing with the development and implementation of digital mapping/GIS. The major Key Success Factors are cited as follows:

- a recognition and acceptance that the reliability and quality of data is as important as the technology, and, therefore, a culture change may have to be engineered throughout the business before major investments are committed,
- organisational change needs to keep pace with the introduction of the technology,
- there needs to be a detailed knowledge and understanding of the actual capabilities of the technology within the potential areas of application.

Clearly, there are many aspects about developing and managing a GIS environment that can only be learnt effectively through direct experience. At some point, therefore, it becomes critical for a business to make a start and then to progress at a rate consistent with the overall capabilities of the products and market place, matched to the needs of the business.

35

British Telecom—Capturing the lines

John Falconer
British Telecom

British Telecom's Access Network valued at over 3.5 billion includes well over 30 million cable pairs radiating from telephone exchanges. The cables containing these pairs are carried by nearly a half million kilometres of underground duct and more than four million telephone poles. Much of the detail of the Access Network is recorded as plant on map and schematic drawings which are stored and maintained by traditional drawing office techniques. In recent years British Telecom has improved the range, quality and speed of its customer service through the modernisation of its network and widespread introduction of computer-based planning and operational systems. Increasingly quality of service and efficiency benefits are being achieved through integration of our many computer systems. Existing Access Network records do not generally include new work until it is completed or are frequently not accessible on the required timescales. The limitations of existing records hinder the effective management of an annual investment of hundreds of millions of pounds in the growth and modernisation of our Access Network.

The existing record comprises over 650,000 map-based and pseudo-geographic schematic sheets plus an estimated 12 million duct space records (schematic plus alpha-numeric record of cable fill in duct bores). These records are used by our own staff to plan and carry out work on the network and to advise contractors, local authorities and other utilities of the position of our plant. Within British Telecom we have well over 1,000 drawing office staff supporting more than 2,000 planning staff working on enhancements to the Access Network. Based on a recent business survey and considerable existing knowledge of GIS technology, new plans for computerising the Access Network records are being considered. A clear message from our initial work is that a full vector data capture, particularly of schematics, is too expensive and cannot provide a clear payback on the investment. To optimise the data capture process, the initial capture is expected to be a mixture of raster scanning and pure alpha-numeric data. This alone will improve the accessibility of the information and protect the existing document base against further physical deterioration. As required, items of plan on the raster image will be linked through vectorised nodes to the alpha-numeric data. Similarly all new work and updates arising from physical surveys will also be recorded in linked vector form. In time the creation of vector and linked alpha-numeric information will make duplicate information on the raster image redundant so that it can be deleted or archived. A key feature of the data capture will be the addition of confidence attributes which will allow the quality of data to be considered before it is used. The new vector/alpha-numeric linked data will not take the same form as the existing record, removing in particular much of the existing duplication between, for example, plant on map and schematics. In particular we anticipate that in the full vector environment schematics will be machine-generated.

Later phases of the work are expected to include interfaces with our Customer Service System (CSS), workforce and stores management systems. 1991 is expected to see piloting of data capture and initial applications software.

36

Management of the implementation of a digital records system

Mike Ives
British Gas plc

Introduction

The British Gas Digital Records System (DRS) has been evolving over a number of years and has already passed through the study, trials, evaluation and supplier selection phases. The project is now in its implementation phase where the management issues are complex and are likely to apply to most utilities from small local operations to large national organisations such as British Gas and British Telecom. The implementation phase covers the data take-on and the operation of the system with the aim of obtaining the benefits as early as possible.

Digital Records System

The initial phase of implementation of DRS covers distribution operations for the low, medium and intermediate pressure gas supply systems. British Gas is organised into twelve regions with some 100 district offices. For digital records, each region will have its own system comprising a central processor with the master regional database and user workstations in district offices. A key aspect of the system is that it incorporates a Distributed Data Management System whereby operational copies of the data are held at each district to provide users with fast, secure local access to information. The central regional processors are also linked to the regional mainframe computers to provide access to the engineering and corporate databases. DRS is a national system whereby the overall management is centrally controlled in the areas of configuration management, software customisation, training development, support and map supply. Regions are responsible for the efficient operation of the system.

At present, initial equipment has been delivered to eleven regions and the remaining system will be installed shortly. The data take-on task is substantial with some 225 000 km of mains to be digitised over 90 000 map sheets. A phased implementation is being undertaken with each region undertaking its take-on a district basis. The whole conversion is planned to be completed by 1995.

Configuration management

Digital records utilise equipment which is in a rapidly changing area of technology. A key issue is the management of product enhancements and substitution. Comment has been made about making progress with a lengthy data conversion exercise when the base technology is changing. Operational phases of an implementation demand stability yet there is the need to take advantage of the additional benefits of new equipment. Equipment supplied to the initial district will be different from that supplied to the last one; a difference which will manifest itself in issues of training users and supporting installations as well as in the human considerations of one having a better facility than another.

Changes and evolution are not just considerations for hardware but equally apply to software. In DRS there are three software layers; the base operating software, the mapping application software and the customised software, and upgrades. All software needs to be carefully managed and tested before implementation in a large distributed configuration. The key consideration is the provision of continuous operational facilities for users. System linking provides a further area of consideration as evolving communications software needs to be compatible across processors from different manufacturers.

Software customisation

Software customisation takes the basic system components and creates an efficient, secure operational system. For a large organisation, centrally developed software has the advantage of providing uniform systems and hence uniform procedures across the organisation. It also has to take account of the differing viewpoints across the regions to gain user acceptability, but it must avoid the nightmare of trying to provide all things to all men.

A key aspect for software development is the nature and location of the users and the tasks they will perform. DRS is an operational system whereby the predominant use will be records management to support engineering operations. Currently these tasks are undertaken by district clerical support, tracer/records clerks and draughting staff; these will be the majority of users of the system. In addition the system will also provide important facilities and services for engineers in respect of gas system management and analysis. DRS will support a large user base of widely differing needs and capabilities. The majority of users will be in district offices which are typically between 20 and 100 miles from computer services expertise which is located in regional offices. The net effect of these considerations is to impose strict requirements for the user-friendly nature of the system and also on the data and system integrity.

The customised software must protect the user from the technical complexity of the system. This degree of customisation increases the size and complexity of the software which imposes further management considerations. A phased development has been undertaken to provide a manageable, early implementation. Typically the phases address map take-on, new geography, mains and plant take-on, data updates and user systems. Each software release must be managed from an installation point-of-view as well as the development and user viewpoints. For digital records, software is installed at over 20 sites and this will grow to over 100 sites in the next few years.

Distributed operations at district level have major advantages but also impose their own additional considerations. Distributed data management for DRS is not just a question of sending the maps and mains records to each district system. It must also consider the linked (alphanumeric) databases for mainframe systems. The high degree of customisation of the system leads to DRS containing extensive management control facilities which again are required at both regional and district offices. Consequently, distributed data management is not straightforward and is further complicated by the fact that the system must cater for changes in the organisation. Typically these could alter the location at which tasks are undertaken, a regional task could be distributed to districts and vice versa, and also the areas covered by district change. The net effect is that the software and distributed data system must provide for any task being undertaken at any site and that the map coverage of an area could change; the system must automatically undertake the necessary adjustments. Finally distributed data management must control the write access to the system so that users in different sites sharing the same map can both undertake data updates in a safe, controlled manner.

A major issue for any system is data integrity. DRS provides extensive checking for data input and for mains digitising; this includes a cross-check with data from the engineering database on the mainframe computers. It is also important to provide validation controls for system supervisors and managers.

Training

Training has to address all aspects of the system and its operation taking account of the potential users' wide range of skills and expertise in the operation and management of computers. These users can be categorised into general users, records maintenance, records supervisors and managers, computer and communications specialists, and regional project management. Each category has its own activities associated with the operation and support of the system. It is important to recognise the ongoing needs for training as the tasks of users alter, and also to accommodate the increasing numbers of users and user turnover.

For a large scale implementation the numbers involved in the different stages of the project increase dramatically. During the trial and selection phases of the project the numbers actively involved will be of the order of 20. This increases to above 100 during data take-on and will exceed 1000 during the operational phase. Whilst the numbers increase rapidly, the level of computer and system expertise decreases, which imposes requirements of the user friendliness of the software (and also the training).

DRS training has to cater for the differing needs of the users and it has to be effective so that the system can be operated in an optimum and efficient manner. The basic philosophy adopted is that training courses and materials are produced nationally. These are then used to train regional trainers who will in turn train the users in the regions. Training for the majority of users must provide comprehensive background material and instruction into the use of the system. A further vital category is that of efficiency factors where certain skill practices provide productivity gains. These hints and tips are built up through experience and an effective method must be found to disseminate them throughout the user population.

In the UK mapping users are fortunate in being able to utilise the Ordnance Survey national grid. Even starting from this advantageous position, map users need considerable training in the map-related issues of the system. Most users are familiar with the Ordnance Survey large scale maps but the concept of the 37 Ordnance Survey 88 feature codes is alien to them. Having mastered this concept of data coding and manipulation it is necessary to consider the time domains incorporated in the system:

- speculative, for new developments;
- approved, for projects;
- as built, both Ordnance survey and British Gas own-drawn geography, i.e. temporary data from projects before update; and
- history.

Each of these areas replicate the OS88 feature coding. The introduction of digital map updates leads to a whole range of issues relating to the replacement of old Ordnance Survey and project geography with officially updated data. This may require adjustment to areas of data which were manipulated during the project stage. The proposed introduction of early update of kerb data, Rapid Response Survey, will increase the complexity of the user manipulations even though it will improve the operational use of the system. Finally the user needs to be able to work within the security considerations of distributed data management tools and reports provided by the system.

The training implications of map and geography manipulation cover a wide range of activities and new areas of concept. The issues raised for map and geography areas are mirrored for gas mains and plant, and also the operational use of the system. In addition the supervision equipment needs to be covered by training.

The wide range of training requirements needs to be satisfied by an equally wide range of training techniques, facilities and materials. Background courses are required to introduce new concepts, for example basic map issues. The use of the system is covered by courses which combine presentations, demonstrations and hands-on experience. Mains and plant training also needs to consider records preparation; for British Gas this includes the alignment of records with data in the mainframe engineering database.

Finally it must be recognised that the introduction of this area of new technology requires more than just training in the operation of the system. It also requires a whole range of new skills and techniques which will need to be built up over a period of time, needing active management steps to ensure that the skill level is retained and not diluted through absences or complacency.

Support

Support covers the critical areas of fault reporting, diagnosis, rectification and associated management tasks. The support facilities must be aligned to the skill levels of the users in a similar manner to that of the customised software. In particular, support needs to accommodate both system faults as well as user operational issues. It is important that the user gets a fast service from the help desks but also is kept informed of progress on the particular matter.

In the particular case of British Gas, the support organisation also deals with the delivery of new software, upgrades and fixes, and the supporting materials. This area rapidly becomes complicated with the increasing number of sites in operation across the regions.

Summary

The digital records system is large and complex and it requires a commensurate level of management. The system and its management facilities are geared to support the users in their productive operational use of DRS. The software facilities are designed to remove the technical complexity from the user tasks; in a similar manner the management, control and validation facilities are designed to assist managers. It is vital in such a system that management is provided with feedback on operational usage and productivity.

37

Keeping on the rails

Simon Lofthouse
Penn Communications

London Underground is one of the world's largest urban rapid transit systems. It serves an area in excess of 400 square miles, which includes 273 stations, and it carries approximately 815 million passengers per year. A key requirement for London Underground's successful operation is the provision of accurate and timely information to enable the correct management decisions to be made at the appropriate time. The Company's IT strategy is based upon creating a standardised mechanism to increase accessibility to this information. Due to the diversity of the Company's data and the large number of people that require access to it, finding a single system that can store, analyse and present this information in an intuitive, easily accessible fashion is an extremely complex task. An area of technology that is presently being investigated by London Underground is GIS (geographical information systems).

London Underground spends approximately £500 million per year on major engineering projects and general maintenance. In order to manage and coordinate this vast amount of work the Company had previously relied upon traditional paper Ordnance Survey maps. For many years, maps were used, extensively, to plan and review work both internally and externally with the public utilities. However, the inherent drawbacks of these maps, such as scale conversion, overlap and general wear and tear, often made them difficult, inaccurate and time-consuming to use.

In order to solve these problems London Underground decided to digitise its maps. Utilising state-of-the-art mapping technology, the Company converted all of its 1300 traditional maps into a digital format that could be viewed on a computer. Not only did this improve accuracy and efficiency, but it also substantially reduced administration time and cost.

Having produced this geographical information system for its Civil Engineering and Permanent Way* Departments, London Underground began to look at the system, and the presentation of information in a geographical format, in more detail.

It was soon realised that by extending the geographical information system to incorporate an additional range of digitised information, including Company assets—stations, signals, platforms, escalators, etc.—and public utilities—gas mains, electric cables, sewers, etc.—a powerful system could be produced capable of managing a range of operational aspects.

The concept of overlaying information over a digitised geographical map was, finally, further developed to include real-time information such as closed circuit

*Permanent Way is in charge of all matters that relate to the track.

television (CCTV), train movements and fire alarms. The resulting system became known as SMS (Station Management System). Based around the latest workstation technology, geographical, graphics and database software, SMS enables station staff, line staff, management and the London fire brigade to call up on screen all relevant drawings, maps and information needed to operate a station on a day-to-day basis during a crisis situation.

A UNIX dual screen, colour graphic workstation provides the constant information. Ordnance Survey maps of the surrounding area and utility services, station layouts and textual descriptions are shown on one screen, whilst the other displays the real time information through dedicated windows. This will include the status of fire alarms and escalators, train arrivals and closed circuit video pictures.

By using a 'drill down' method, SMS can be used to scan down from a geographical view of London, focus in on a station's specific location and move down from street level through all the public utilities, layer by layer, to form a complete picture of the whole station in relation to its surrounding area.

At its most extended level, SMS provides London Underground with the means to identify and locate all of its assets quickly. It can assist with the collection and assimilation of operational data and provides a common interface with the emergency services and public authorities. Its incident simulation facilities also provide a valuable training tool for operational staff.

Russell Fawcett, head of IT, London Underground, comments:

> "SMS is a fine example of our innovative use of computerised geographical information. However, in our view, geographical information is only one part of the jigsaw. If it is to be used to its greatest effect it must be used in conjunction with the far wider facilities of graphical information in general.
>
> We have used SMS to prove the technology and we now lead the world in computerised geographical transport management systems. There is, however, still a long way to go, and many new applications are still to be investigated."

SMS was piloted at Green Park Tube Station and the results are still being evaluated. However, computerised geographical information is used on a day-to-day basis by London Underground and the Company is committed to extending its use in the future.

The author acts as P.R. consultant to London Underground.

PART VII

GIS and the financial and commercial world

Over the last decade GIS has primarily remained in the domain of large public sector organisations, utilities and research institutions. However, with the rapid development in PC and workstation products—particularly those which come value-added in the form of up-to-date data—business professionals involved with finance and commerce are finding GIS to be an attractive decision-support tool. In this section we have grouped together those papers which address the application of the technology in the financial and commercial world. A range of interesting topics are discussed from housing development to store management.

38

GIS with reference to housing development

Steve Gill
Midsummer Computing Exchange

This review of the use of GIS with respect to public sector housing separately considers the control and management of housing and housing development. Firstly, the management and maintenance of existing housing stock is examined, then consideration is given to land registration and the development process. The use of GIS in these three separate stages of housing development is, these days, a feasible proposition and has been demonstrated in both pilot schemes and live projects.

In 1988, the last year for which figures are available, the estimated public sector housing stock in Great Britain, managed by local authorities, new towns and housing agencies, was approximately 6,172,000. Each of these properties has associated information to which the landlord needs ready access. This information, being related to a postal address or precise geographic location, is spatially related and can thus be incorporated into a GIS and subsequently referenced via a unique key.

There are a number of different methods used to create the key or unique property reference number (UPRN) for each property. Typically they are based upon postcodes and national grid coordinates. These schemes create a reference number for the property which is unique throughout all of Britain. However, other schemes can be used to generate a reference number for property within a small locality. For example, the Milton Keynes Development Corporation GIS project uses a numeric district code which, when combined with a numeric building site code and a plot number, yields a unique property reference number.

Each of these properties needs to be related to a base map; usually an Ordnance Survey map. The maps themselves are held in the system in either raster or vector form. Vector map data gives greater versatility than raster data and is therefore perhaps preferable as the map base. However, vector data is more expensive and it is still not yet available for all urban areas of the country. Raster mapping may thus be the only option for certain authorities and more especially those with large rural components.

Once a base map is in the system and an overlay has been prepared which contains details of the property owned by the landlord, then he is in a position to carry out queries via the graphics. If further themes or overlays are entered into the system, it then becomes possible to check relationships between different datasets. For example it would perhaps be possible to establish relationships between maintenance costs and the incidence of vandalism within different localities. Before the emergence of GIS the preparation of plans displaying such disparate datasets was both time-consuming and expensive.

In 1987 approximately twenty per cent of the housing stock in Great Britain required repair or improvement. This figure gives some indication of the amount of maintenance work carried out on behalf of public bodies. Much of this work is let to tender with the consequent need for plans and schedules of repair. If these are prepared using GIS technology then they can be prepared more accurately and at less cost than in the past. They can also be rapidly reproduced if necessary. Housing-related landscape contracts can similarly incorporate map sets prepared from a GIS. Organisations using this technology not only benefit from cost savings but also have access to more accurate data—for example mowing areas can be calculated rapidly and accurately using such a system.

Most of these contract plans would be prepared using the standard OS 1:1250 mapping as a base. However in 1989 the OS produced a subset of their 1:1250 digital map called OSLAND. This dataset was prepared for only a small area of the country and contained only property boundaries and the boundaries of other large features such as road carriageways and lakes. The boundaries within OSLAND have all been generated or inferred from associated OS map sheets. The boundaries therefore do not necessarily represent the legal definition of a property or feature. Thus OSLAND is perhaps best suited for work related to property transactions where precise legal boundaries are not required to be shown.

Perhaps the most onerous use of digital maps within the housing field is the preparation of a digital property register. HM Land Registry have just conducted an initial pilot scheme whereby they have entered the register for Peterborough into a GIS. An allied use with a differing emphasis is the preparation of plans for conveyance purposes. It is very easy to assume that one can take the existing plan for the conveyance and digitise it onto the digital map base using grid co-ordinates taken from the plan. Unfortunately very few conveyance plans have been prepared on an up-to-date map base, many of them having been hand-drawn in the last century. In such cases it is not unusual to find spatial errors of five metres when a boundary on a conveyance is superimposed onto the current map base.

Each boundary has to be entered into the system such that it is in its correct spatial relationship to all the surrounding features, whilst simultaneously maintaining the accuracy of the plan. Having entered the property boundary into the system, one has to be aware that if the OS carries out a re-survey then a feature, previously used to establish a spatial relationship, may be moved in position. There are very few major errors in recent OS surveys but positional changes must be expected where the prior survey is relatively old. A strategy is thus required to identify and perhaps modify boundaries which are based on this earlier data.

Turning now to the development process, many people involved in, or on the periphery of, GIS will have seen vendors demonstrating the capabilities of their systems to cope with zoning queries and to identify land ripe for development. GIS are ideally suited to these manipulations. Within the public sector, an authority's land terrier monitors land for development. A public body which installs its land terrier on a GIS will thus receive a much improved turn around in the response to land ownership queries. Using a manual system most queries take a full day to answer, whilst those concerning estates may take in excess of a week. A GIS however will answer all these queries within an hour—including the preparation of the necessary drawings. Once land ownership information is available on the GIS then the local authority can also record and monitor the flow of PUSWA information and monitor the opening of roads.

Recently GIS has started to become evident in the sale of housing. If details such as house type, price and floor plans are entered into a small GIS then the queries of potential purchasers can be answered via the system. Now if this data is combined with a 3D model of a development then the vendor can demonstrate a "walk-through" of the estate right into the particular property the potential purchaser is considering. Although this technology is presently expensive, there is currently such a pilot scheme being prepared for an estate with high land values.

There are definite advantages to be gained from the use of GIS technology in the housing field. However authorities should be aware of the differing criteria to be used for data capture, depending upon whether the correct legal boundary is required or whether its approximate position, as defined by features present on the OS base maps, will suffice. In the former case a higher level of interpretative skills and more time to enter the data will be required. For the perceived benefits to be realised there must be a real commitment by the organisation as a whole to the project. If the skills required to implement a GIS are not available within an organisation then, for the system to be a success, they must be bought in from outside.

The 1990's will see more local authorities moving into the GIS marketplace and the emphasis will shift away from public utility applications. The GIS marketplace will become more dependent on spatial analyses of area features. This increased awareness of spatial analysis requirements will result in greater computing power being available for the same real cost.

39

Computers in commercial property research: A case study of Prudential Portfolio Managers

Paul McNamara
Prudential Portfolio Managers Ltd

Introduction

This article outlines the various uses to which computers are put in the commercial property business. Following a brief review of the range of functions for computers across the property industry, a more detailed examination of one major financial institution is carried out. Particular attention is paid to two research-related initiatives recently undertaken by one of the UK's leading property investment organisations.

Main uses of computers in commercial property

It is clear from the recent 'Property Computer Show', held in October 1990, that computers are now being used in almost every aspect of the commercial property business from helping in the mundane management of individual properties to the strategic planning of investment portfolios. Taking the landlord function first, systems have been developed for property management. This requires combining details of ownership, occupation and leasehold structure with information on the timing of rent reviews and lease renewals. These link directly with systems of property accounting which issue rent demands, calculate fees and log expenditures. In some local authorities, databases of information of individual properties have been linked to Ordnance Survey mapping systems. There are now systems available to link building plans to these land terriers.

Within a property agency, the main commercial developments are systems which store information on the characteristics of properties to which a client's requirements can be matched and a shortlist of relevant properties for inspection produced.

Turning to more analytical applications, there is a wide range of packages designed to evaluate the financial worth of property assets. These calculations, though simple in conceptual terms, can be extremely complicated in practice. Commercial properties often include a multiplicity of leases of different types and lengths, with different conditions and valuation practices, all of which need to be brought together through valuation packages. Performance analysis systems, developed to examine the performance of individual properties and sets of properties, are more complex, particularly those that attempt to attribute the performance to given features of the properties under examination. However, a range of such systems is now in existence.

All of the above systems are employed within individual organisations. However, there are systems in operation which serve the property industry as a whole. For example, the Investment Property Databank (IPD) is a major database storing information on individual properties (£40 billion value in total). This database is used to monitor performance in the property market generally. (A second database with similar objectives is about to be launched, namely The Property Research Database.) There is an on-line property press cuttings and information service (FOCUS) which stores information on property items from across the UK. Contextual socio-economic data can also be extracted from this system as can estimates of the level of enquiries for different types of premises in centres across the UK. Specialist databases can be bought on retail users across the UK (SAMI, URPI), offices in central London (APR) and planning applications (Glenigan Leads).

Prudential Portfolio Managers Ltd

Prudential Portfolio Managers Ltd. (PPM) has a specific range of interests which it uses computers extensively to satisfy. PPM is a fund management organisation. Its Property Division (PPMP) is therefore interested in obtaining good investment performance from investment in commercial property within predetermined bounds of risk. This is done primarily through the purchase and sale of existing buildings. However, the Property Division also has a significant commitment to property development. It husbands many hundreds of properties in six different funds which together have a value of nearly 5 billion. These funds are of very different sizes and have very different objectives.

The company is therefore both an investment organisation and a major corporate landlord. Investment strategy is determined through a regular review of changing market conditions in all the various investment media such as gilts, UK equities, overseas equities and commercial property. From this analysis, monies are injected into or taken away from the property 'exposure'. Investment policy is implemented through the sale of existing assets and the purchase of new assets.

As will be appreciated, PPMP has both landlord and investment management roles to fulfil. Not surprisingly, therefore, many of the systems outlined above are in place within PPMP. The Property Division has invested significantly in the development of systems for the management of its own stock and the evaluation of development and investment opportunities. Given the in-house commercial agency function (not to be confused with Prudential Property Services, the residential estate agency chain), there are also systems for storing information on comparable market evidence.

Over and above this, PPM has established a research team to analyse property market conditions and make recommendations on fund strategy. This team has built a wide variety of systems to attribute asset performance to a property's locational and physical characteristics, to test the efficacy of different fund strategies, and forecast future rental growth and total performance in the UK property market and its component local markets. It is with reference to two initiatives undertaken within the property research team at PPM that the rest of this article relates.

Property Research Information System (PRISM)

There are two main outputs from the property research team at PPM. First, there are forecasts of the likely performance of the commercial property market. Second, there is advice on the actions or strategy the funds should adopt to take advantage of these forecast conditions given a known level of cash flow and a current exposure to different property markets. The forecasts by the research team cover the national market, regional markets, and sub-regional or 'town' markets, and are produced quarterly for the three main sectors of the market, namely shops, offices and industrials.

There are intentions within the team to produce forecasts for retail warehouses and the emerging 'B1' mixed office and production units. Leisure and residential investment sectors are not forecast explicitly, although a watching brief is kept.

Problems of data assembly

Clearly, the production of forecasts on the likely level of rental growth and total performance of these different types of commercial property asset requires a substantial process of data assembly. Before describing the system that has been developed to handle this, it is worth dwelling briefly on the practical and conceptual problems associated with assembling such data. Given the house philosophy that it is essential to see property in context with real-world social and economic processes, forecasting requires extensive information on the economic, demographic and social processes operating in the areas being considered. Information is also required on the property markets in these areas, especially rental levels, rates of growth in rental levels and the yields which are being used to capitalise rental flows to obtain market prices.

Given the very different socio-economic processes affecting the different sectors of the property market, a wide range of data has to be assembled. For example, employment estimates and forecasts at higher level SIC codes and other basic economic variables have to be gathered to estimate demand in the industrial and office sectors. Population estimates and forecasts of consumer expenditure are crucial to understanding demand changes in the retail sector.

Information on the future supply of commercial space is easier to obtain because of the need for the vast majority of new space additions to pass through the land use planning process. Historic and current supply is, ironically, more difficult to estimate. The property research team at PPMP makes use of aggregated government statistics which were published in a variety of forms until (disastrously) they were cancelled through cost cuts in 1986. More recent information has been assembled from a private sector collection agency which has gathered information on planning applications across the UK since 1986.

Information on property rents and yields comes from a variety of private and public sector sources and has to be treated with great care. Sources include published indices and series (usually from chartered surveying firms), databases of actual properties and, given the size of PPMP's property funds, the considerable information held in-house. Care is required when assembling information because the different sources often use different conventions for data collection and presentation. Some rental series relate to actual properties (with all their peculiarities), others relate to standardised, hypothetical properties. The rental level quoted may relate to provable rents (i.e. that which has been known to have been achieved in the marketplace) or open market rents (relating to that which would be achieved if a deal was to be done at the time of data collection). Some rental series use best rents for a local market which means that the location to which the rent point relates will change from entry to entry. Other series use a fixed rent point. All of these nuances have to be recognised by the analyst.

Problems of data analysis

Clearly, the problem for the commercial property researcher is to assemble data in a meaningful way so that data from different sources relate to the same geographical entity and land use type and can be used for forecasting rental growth and total performance in the future. This presents both practical and conceptual problems.

Conceptual problems emerge when one considers the nature of property markets. When forecasting, say, the 'Newcastle industrial market' or the 'Manchester office market', what do these property market definitions actually mean? Similarly there are classic geographic conundrums about what catchment area it is relevant to base retail analyses upon.

A wide variety of practical problems stem from the uncoordinated way in which social, economic and property data is collected in the UK. Employment data relates to Amalgamated Office Areas, recent planning application data relates to town definitions determined by the company concerned, population forecasts relate to counties and old metropolitan borough areas, rental information relates to 'towns' or 'centres', however defined.

More varied still is the way in which data directly related to property classifies land uses. Each separate data source seems to create a different land use classification all of which need to be brought together in the most consistent way possible for appropriate forecasts to be produced.

The PRISM system

This lack of data compatibility has been a major stimulus for the property research team at PPM to establish an integrated database to bring these disparate sources of data together to permit meaningful analysis. The PRoperty Information SysteM (PRISM) comprises a number of data modules and aims to achieve a number of objectives. Data modules include population data and forecasts, employment data and forecasts, planning applications, data on PPM managed properties and property market data. The aims of the system include provision of consistent and comparable data for modelling locations across the UK. There is also a set of operating objectives to allow flexible reporting, capacity for system evolution, speedy usage and multiple access by different researchers.

The main function of the system is to link the various strands of data together so that complex data selections can be made. The data can then be brought down into either an econometrics package, a performance analysis package or a more flexible spreadsheet facility. The two prespecified packages allow regular research requirements to be satisfied. The spreadsheet package allows substantial flexibility for subsequent data manipulation and examination after selection. A typical enquiry would be to identify the historic data on the demand, supply and rental performance of office property in a given location and load them into an econometrics package for the production of forecasting equations. Another would be identifying those areas in the UK that have experienced, say, 5 per cent population growth per annum over the 1980s, and examining the performance of retail properties in those areas.

PRISM works by ensuring that all the data modules placed on the system have a clear spatial identifier, property sector identifier and temporal identifier. Underpinning this, a thesaurus has been established to link the various spatial and sectoral definitions so that the locational or land use definitions used by different data sources and suppliers can be translated into an in-house set of areal and land use classifications embedded within the system. Postcode based areas, 'catchment areas' and administrative areas have been linked. Data is stored at its lowest spatial level and aggregated up.

PRISM allows bespoke 'area profiles' to be defined combining any set of locations. These can be used to structure subsequent data enquiries and bring together sets of data for newly defined areas such as the M40 corridor. This use of profiles liberates the research team from the straightjacket of the varied area classifications from other sources and facilitates more meaningful analysis. Once established, these profiles can be used with the same speed as any more conventional area definitions like district councils or postcode areas.

PRISM is not solely for the use of the research team. The particular importance of PRISM to other parts of PPMP is the on-line facility through which relevant planning applications from anywhere in the UK can be reviewed. Investment surveyors can, with great ease, identify competing developments in areas they are considering or identify, nationwide, the development activities of rival companies.

As yet there is no mapping facility attached to PRISM although this is under careful consideration. At the moment, most of the analysis revolves around gathering information to investigate a given location. Mapping facilities could become an important research stimulus for looking at the point data held on areas and, more importantly, individual property assets and discerning patterns and possible causal factors in their performance. However, there has been one major research project carried out at PPMP which has made very great use of mapping systems and it is to this that I now turn.

Examining the future pattern of demand for office space in central London

As suggested earlier, forecasting likely future supplies of property space is not too difficult to do. The greatest difficulty comes in forecasting the demand for space. Fortunately, in an office market like central London, forecasting demand for space is helped by relatively good information on employment growth supplied from a variety of sources. However, the major problem for the analyst is where, within the complex and interrelated functional networks operating and evolving in central London, that aggregate demand for space will be expressed. In recent times there always seem to have been a number of conflicting hypotheses in currency on the future geography of office demand in central London. These range from a mass exodus of firms fleeing exorbitant accommodation costs to an assurance that traditional foci for specialist employment would be maintained irrespective of changing cost and demographic environments.

As a result of these uncertainties PPM, as major holders of commercial property in central London, and Stanhope Properties plc, as major developers of commercial property in and around the London area, combined to design and commission a survey of occupiers' views about their preferred locations in central London. A market research company with experience of interviewing senior decision-makers in central London (SRU Ltd.) was recruited to conduct a questionnaire survey of a structured sample of office occupiers across central London. The sample entailed over 200 detailed interviews. The results of the exercise remain confidential. However, of interest to this paper is the use of a geographic information system (GIS) to interpret the data received.

Each interviewee was asked a range of factual and attitudinal questions about the parts of London they were located in and other areas they might consider occupying. This type of enquiry is not unusual to surveys of occupiers carried out by chartered surveying practices with a specialism in central London. Where this particular survey differed was in the request made to interviewees to describe, on a series of maps covering central London, inner London, the south-east region of the UK and the UK as a whole, the 'ideal', 'acceptable' and 'wholly unacceptable' locations for the conduct of their businesses. By describing three areas explicitly, a fourth area, namely 'generally unacceptable' locations was discerned. Therefore, the exercise resulted in a series of over 200, four-tier, contour maps describing the future locational preferences of a wide variety of office users based in central London.

The aim was to aggregate the mapped data to give a picture of overall patterns of locational preference. Because different types of user occupy very different types of offices as well as locations, aggregations for individual occupational sectors gave indications of the types of office that would let well in different areas within central London.

The mapping work was undertaken by Dr. Ian Heywood and colleagues at Salford University Business Services Ltd. using the SPANS Geographic Information Systems Software (Tydac Technologies Inc.) on IBM hardware. Each individual map was digitised as was a rudimentary street pattern of central London. Once done, the GIS was able to produce aggregate maps as required for any subset of the population of interviewees.

The resultant maps gave some strong indications about the extent to which traditional areas would retain their appeal to land users. When coupled with detailed information about the future pattern of the supply of office space the prospects for rental growth and the total performance of different parts of central London office market became clearer. This and other information is of direct relevance to the future development and portfolio rationalisation programmes of the companies concerned.

Conclusion

There is a wide range of applications for computer technology in the field of commercial property. PPMP, as a fund management organisation, has embraced computer technology in both the management of its business and in the research field. In particular, an integrated database capable of bringing together data from a wide variety of published and in-house sources, in a compatible form suitable for innovative research work, have been pioneered.

To date PPMP has not viewed the attachment of mapping facilities to their PRISM database as cost-effective. However, with the assistance of Salford University, PPMP has engaged in research on the central London office market which has made extensive use of a GIS as a primary research aid.

40

Store planning using GIS

Carmen Elliott
Research Planning Manager, Debenhams plc

Introduction

Over the past year, Debenhams has been using a geographical information system for assessing potential store locations, monitoring the performance of existing stores and assisting in store development. This article describes the applications and benefits of the GIS in retail planning.

Store location

Debenhams currently operates some 80 stores throughout the UK in major cities or county towns. Since 1985, we have opened 12 new stores each of which was subject to extensive research to determine its viability. Opening new stores and extending existing stores are such crucial decisions that it is vital to evaluate a store's market potential and competitive position. The increasing use of computerised techniques over the past few years has provided retail management with quick and accurate information to assist in making decisions.

With the growing amount of information and need for ever more detailed research, our old methods of largely manual data manipulation and analysis were becoming unduly time-consuming and cumbersome. We felt what was needed was a system that could provide flexible handling and storage of large datasets. The particular benefits of a geographical information system, apart from the obvious ease of access to information, are in the assessment of potential new sites—what is the population within the various drive-time bands and catchment areas; what is the demographic profile; what is the retail potential in the area; how will the catchment area overlap with other existing store catchment boundaries? These questions can be asked of several sites at the same time to identify the one with the best potential.

The GIS

The geographical information system we use is GEOPiN, a version of ARC/INFO software which was customised to allow for a system of menu-based routines. We identified a number of tasks that had to be performed by the database system to meet the distinct stages in our store location analysis. These included:

(1) - Catchment area definition
- credit card customers
- drive times
- other methods;

(2) - Area profiles
- census
- PIN;

(3) - Retail potential
- store mix.

Defining catchment areas

In assessing a store's potential and the associated market we have first to define the catchment area. For existing stores, we use our store credit card penetration levels to define "primary" and "secondary" catchment areas. We assume that the whole population in the primary area are potential shoppers at the store. For the secondary area we make the (not unreasonable) assumption that the influence of the store will diminish with distance and hence use a percentage weighting of the population in the secondary area. A simple program allows for the penetration of credit card by postal sector to be determined, thereby allowing for primary and secondary catchment areas to be flagged. Figure 1 illustrates the primary and secondary catchment areas around our store in Kings Lynn.

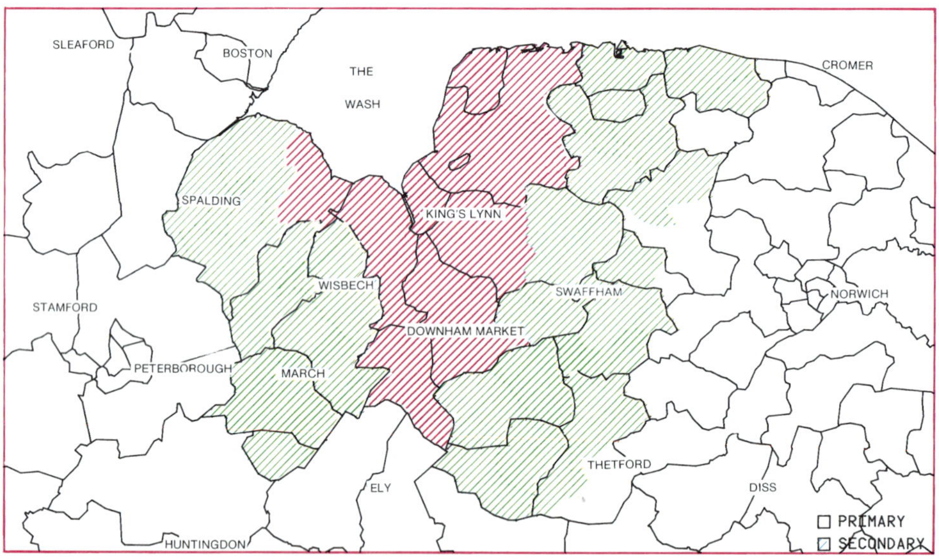

Figure 1. Primary and secondary catchment areas around Kings Lynn.

New stores present a different problem. We have no established customer database and the GIS present the best alternative with the use of drive times. The drive time is basically a boundary determined by the distance that can be travelled out to a shopping centre in, say, 20 minutes by car along roads radiating out from the centre. Assumptions are made in this program about the surrounding area (rural/urban) and implicit in this, the speed characteristics for the different types of roads, vehicle flow conditions and junction delays. In this case, the GEOPiN system presents us with the flexibility of defining catchment areas where lack of data

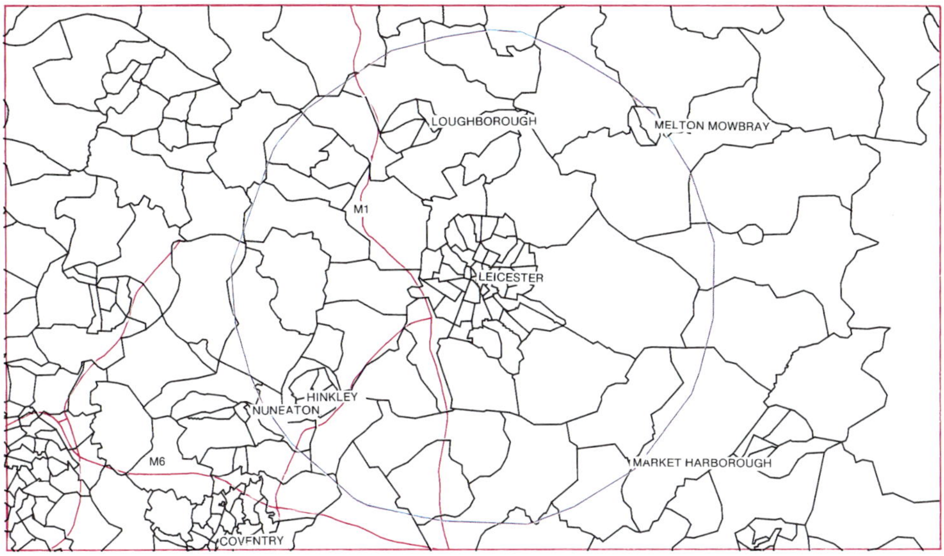

Figure 2. Drive time band generated for 30 minute drive from Leicester.

provides little in the way of alternatives. Figure 2 provides an example of a 30 minute drive time band around Leicester. The system is also flexible when we need to input some subjective element to the catchment definition. For example, the presence of competing centres would restrict the catchment boundary and in these cases we override the drive time data where it seems appropriate. After defining the catchment areas we look at the population figures to assess the demographics and the potential spend on the merchandise sold at the stores.

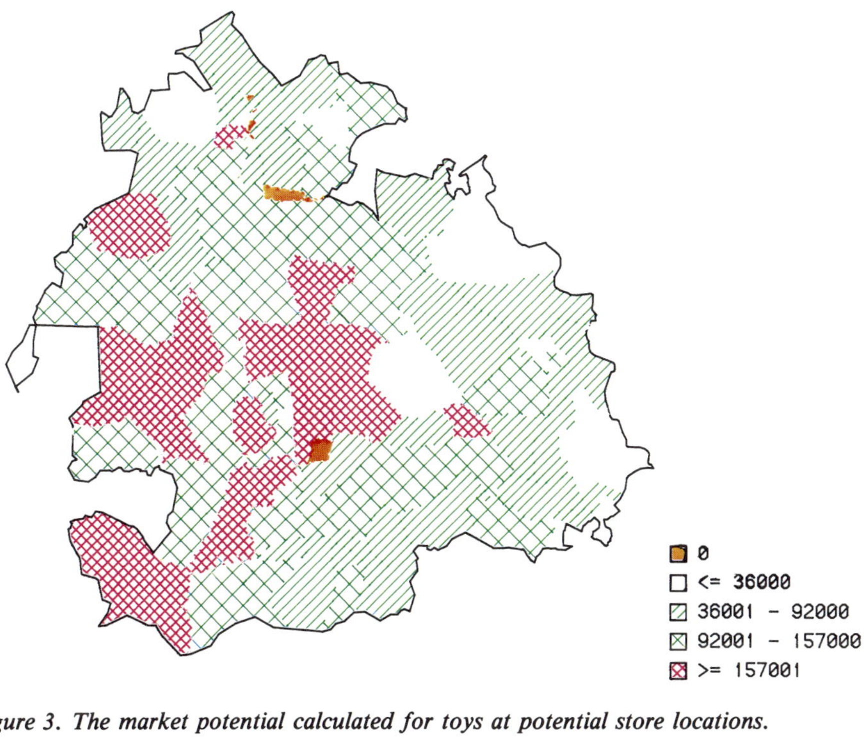

Figure 3. The market potential calculated for toys at potential store locations.

Area profiling

As the socio-economic and demographic profiles of customers within different catchment areas vary, it is important for the product mix to be considered separately for each store to ensure an appropriate allocation of space. To be able to profile the demographic makeup of our catchment areas, the demographic variables are held at postcode sector level for the UK within GEOPiN. This is a definite advantage over the old system where we could only get census variables for larger areas, e.g. district or county level. Together with PIN (Pinpoint Identified Neighbourhoods) profiles, the GIS offers a sophisticated system of allocating the store product mix to match the customer profile within the catchment area around the store. We also use the Census and PIN profiles to cross reference any other *ad hoc* market research surveys that we do. For example, if we needed to conduct research among young couples with children, we can tap into the database to narrow the research sample size.

Retail potential

A program within GEOPiN, the retail potential report, provides estimates of the annual spending potential within a catchment area for the merchandise groups that are sold by the store. The retail potential reports have two main areas of application. They help to provide us with a preliminary evaluation of any new location as well as a means to consider and monitor the performance of existing stores. The market size figures from the report give us an indication of the potential sales available in the catchment area and this enables us to determine the store merchandise mix and the space that should be allocated in the new store. We can do this for a number of reasonable catchment scenarios to identify the one with the best potential. In the case of existing stores, analyses would indicate market shifts over time and can help guide the merchandising mix on offer. Figure 3 shows the market potential for toys in a catchment area.

Current issues

A fundamental input to the analyses described above has been the 1981 Census and with time the data becomes outdated. Although there are recognised methods of projecting future population levels, not everyone agrees on the best method. There needs to be more work done in this area if we are to make the most of GIS.

Conclusion

The next decade should see continued growth in the application of GIS in the retail scene. The development of precise targeting, down to postcode level, offers retailers more detailed information than can be gleaned from simply using the ABC socio-economic divisions. The increased competition over the next decade, for a diminishing number of ideal sites, is going to require much more skill and effort and the GIS has a relevant role to play in this area.

41

Assessing the potential of branch outlets using GIS

Alan Howe
National & Provincial Building Society

A year ago National & Provincial (N & P) came to the conclusion that it needed to know far more about its branch network so that decisions on their operation could be taken with a degree of certainty. GIS was used to produce scientific analysis on all N & P's 318 branches. The GIS project comprised data on branch catchment areas, including sales potential, population and customer profiles. Categorisation according to needs, such as refurbishment, relocation, changes in staffing levels and appropriate mix for product sales targets, were all seen as key objectives.

N & P has a customer database file held on a Unisys A17 Mainframe. The file has been postcoded and deduplicated which means that a customer only appears once regardless of the number of accounts held. Because many building society accounts are held jointly, there is a differentiation between the first and second named, with the first named being deemed the primary holder and, in the case of share accounts, holding voting rights.

To supplement the management information which the customer database can produce, a PC version of Pinpoint's GEOPiN system was purchased to run on an Olivetti XP9. This is a 386 machine with a maths co-processor and 320 megabytes of disk for storage. This amount of disk was necessary as large chunks of the main customer database are selected and downloaded daily to the PC to be plotted and analysed.

Because all our customer records are postcoded, it is a relatively easy task to manipulate this data and to produce maps displaying the present coverage of branches and product penetration. The catchment areas are further refined using a drive time module to give an effective range to the branch that a customer could cover within a given time. The time is dependent on the size of the town/city and working patterns. Once the final catchment area for each branch is defined, GEOPiN reports are produced calculating the population using census data supplied within the system. These are categorised into socio-economic groupings A, B, C1, C2, D, and relevant age groups in line with N & P's target audiences.

Data supplied by Financial Research Surveys (FRS) is also held on the PC for 40 key financial products, including mortgages, insurances, card holdings, etc., and accessed by GEOPiN to provide potential penetration of specified products within each area. Data on competitors is entered, for instance, to show the locations of their branches, thus completing the missing pieces of the jigsaw.

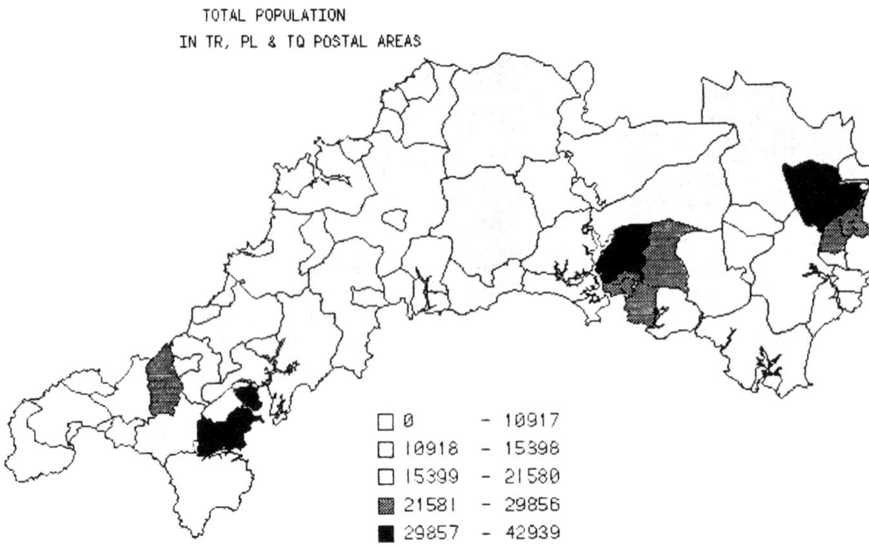

Figure 1. Total population distribution of the south-west.

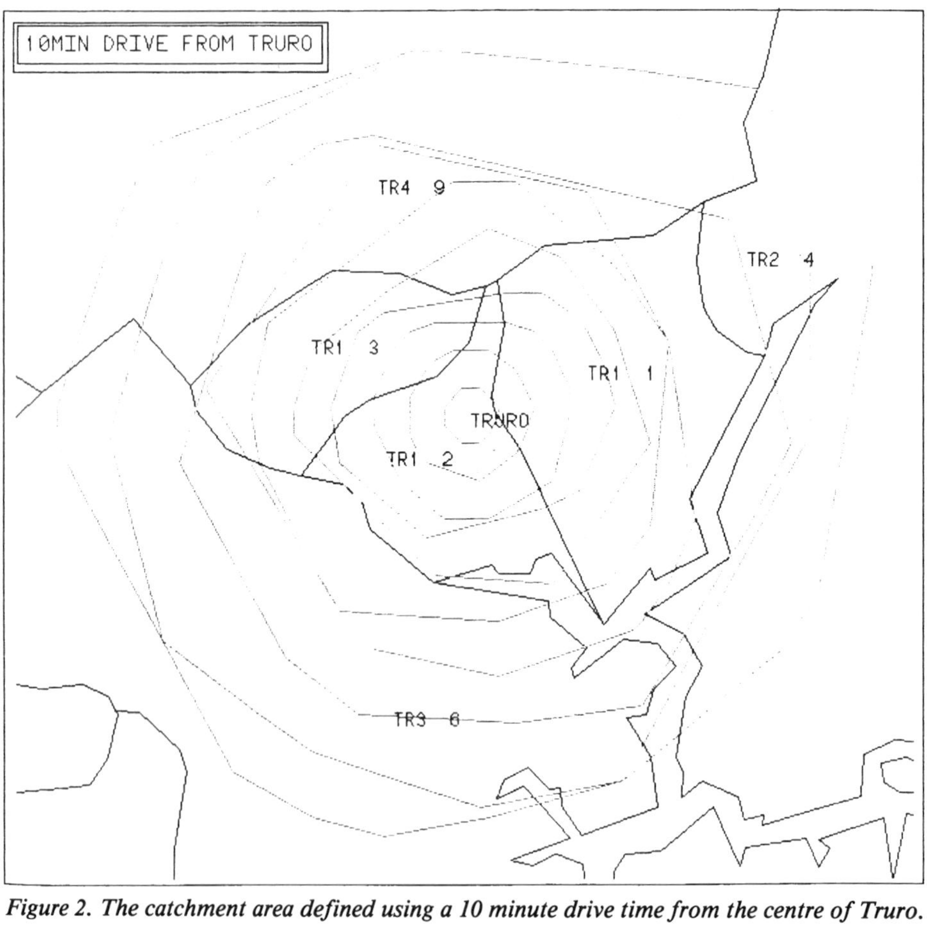

Figure 2. The catchment area defined using a 10 minute drive time from the centre of Truro.

All these results are combined onto a database with actual branch performance details. Each item is given a weighting according to importance to give a "final score" for each branch. Once this figure has been calculated, the data are sorted into order using the "final score" as the prime key. Plotting actual performance against branches' trading areas potential has resulted in the classic dogs and stars of the Boston Matrix.

GIS has helped management focus on the needs of individual branches. Among other things, this will help increase performance, identify necessary staffing levels, provide input for cost/benefit analysis on refurbishment and reduce the risks often associated with relocation. As customers are geographically based, information should also be geographically based. Computer assistance is essential as dealing with a large customer database and mountains of data by manual methods is not feasible. Knowing more about customers' working patterns, shopping preferences, together with detailed analysis of the places where they live, enables N & P to create facilities to service their needs better. This applies not only to existing customers but to potential customers as well. The need to understand those people who do not buy is just as important as those that do.

GIS data, therefore, influences the siting of cash machines, the target audience for telemarketing, requirements for a direct sales force and their individual sales areas, customised products—the list is almost endless. The three examples of graphical output, from the South-west, are shown below and on the previous page.

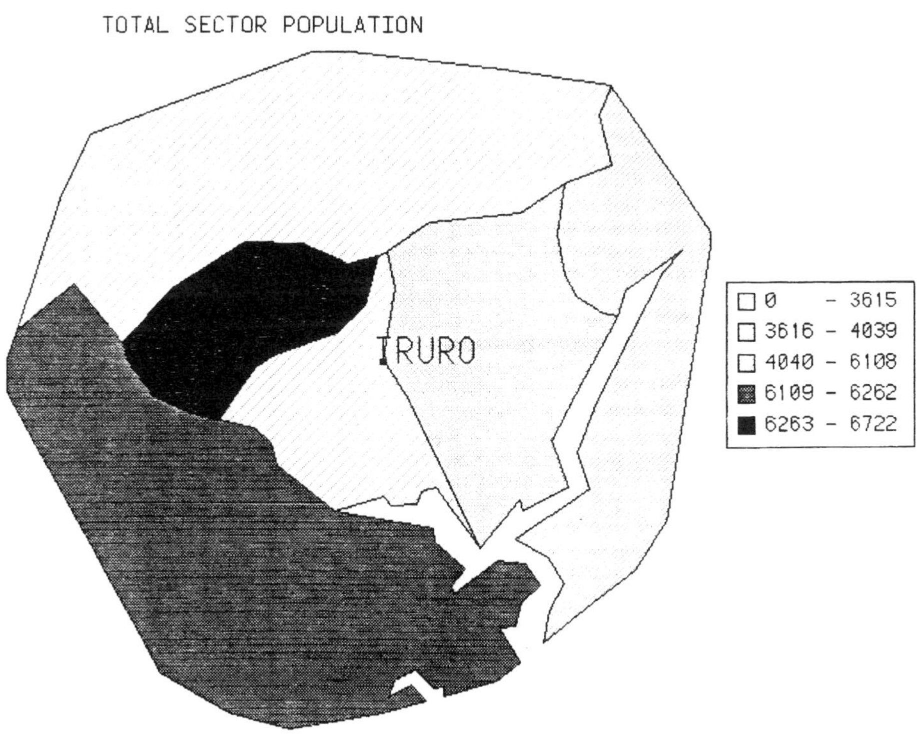

Figure 3. Population within the 10 minute drive time boundary for Truro.

42

GIS in the commercial and financial world

Jonathan Reynolds
Oxford Institute of Retail Management

Retailing

The last eighteen months have not been easy for the UK retail sector. The explicit target of government economic policy and no longer a fashionable stock, the sector has experienced a sea change in terms of growth rates, by comparison with the consumer boom period of 1984–88. No longer can multiple retail businesses rely upon consistent performance from a national roll-out of a standard fascia with a standard product range. Store location and market planning departments within retail companies were becoming very adept at using a straightforward set of methods and techniques to identify new sites. Such techniques included:

- site screening (the identification of an initial set of priority locations for a store development programme);
- catchment area definition (the designation of trading catchment areas around a store, with the help of customer data or expenditure estimates and local demographics);
- store portfolio evaluation (the comparison of trading catchment areas for existing outlets of a retail business).

All of these activities deal with the collection, analysis and dissemination of spatially referenced information; all are therefore susceptible to the integrating capabilities of GIS applications. Yet whilst three in four UK retailers claim to have a store location or planning unit, few have investigated the possibilities of GIS, and a mere handful are actively using GIS. Why is this?

For many, the in-house procedures in use during the 1980s were developed incrementally and tailored to the particular needs of an individual business or roll-out of a single retail format. The purchase of an integrating system to replace these well-honed procedures, and the upheaval in established practices, data processing and validation, was considered neither practicable nor cost-effective. For those companies who chose to rely largely upon external sources of information and analysis from the growing number of geodeomographic data companies, the purchase of a bespoke, full-blown GIS represented a significant increase in the scale of investment in normally underfunded departments. Moreover, additional professional staff would need to be recruited to manage the system.

Further, the emphasis within many retail businesses has now changed from bullish expansion to cutting the cloth to fit; managing within conditions of considerable uncertainty. Is it therefore the wrong time to be advocating the kinds of new systems investment which GIS represents? On the contrary, it is the retail company which innovates and invests during times of slower growth or decline in sales, which will emerge the stronger when economic conditions improve.

However, retail companies' needs are changing in two distinct ways: in the way they address the external environment and the way the internal environment of the firm is managed.

The external environment

In less certain times, retailers must move to monitor and forecast trends in the external environment with greater rigour. The out-of-centre retailing we see around us today, for example, was largely a consequence of broad urban population movements in the 1970s, combined with significant increases in personal mobility. In the next decade, however, competitive advantage is more likely to accrue to those retailers who move from deriving development and marketing strategies from purely demographic trends towards those who are able to manipulate spatially-referenced lifestyle and other behaviourial information in increasingly sophisticated ways. As if to reinforce this shift, the Henley Centre for Forecasting has identified the local dimension to economic and social change in the UK as an emerging theme for the 1990s. There are likely to be many new challenges of this kind.

The internal environment

Presently centralised retail organisations will find it difficult to "think locally" until they become better aware of their own businesses and harness the currently discrete datasets which exist within the business. Using information technology will be necessary to develop new linkages in the retail value chain, between data building blocks in merchandising, store planning and market research, for example. Skills and experience hitherto locked within one division of the company become more marketable, and a better case can be made for the purchase of management systems, including GIS. Within conditions of uncertainty, IT may also be required to rationalise and utilise increasingly complex information flows within the retail operation—developing an understanding of product and product-range as well as store catchments, for example, driven by point-of-sale and -purchase data.

Subjecting real sets of local market data to analysis is one way of exploring the scope and potential of GIS to respond to these new needs. Regrettably, much of the raw data needed for such analysis is necessarily confidential to the companies concerned. However, the Oxford Institute of Retail Management, working with a number of retailers, developers and geodemographic data companies, has built up some extensive sets of information relating to local UK markets which illustrate both some of the new challenges which retail companies face, together with the assistance GIS can offer to other parts of the retail business. Two of these cases are shown by way of conclusion, very briefly, below.

Case A: Modelling new shopper behaviour

Figures 1 and 2 show the early results of the modelling of new shopper behaviour in response to the opening of one of Europe's largest out-of-town shopping centres, at Meadowhall in Sheffield. The project uses ARC/INFO to capture and analyse all spatially-referenced demand and supply data connected with the new centre.

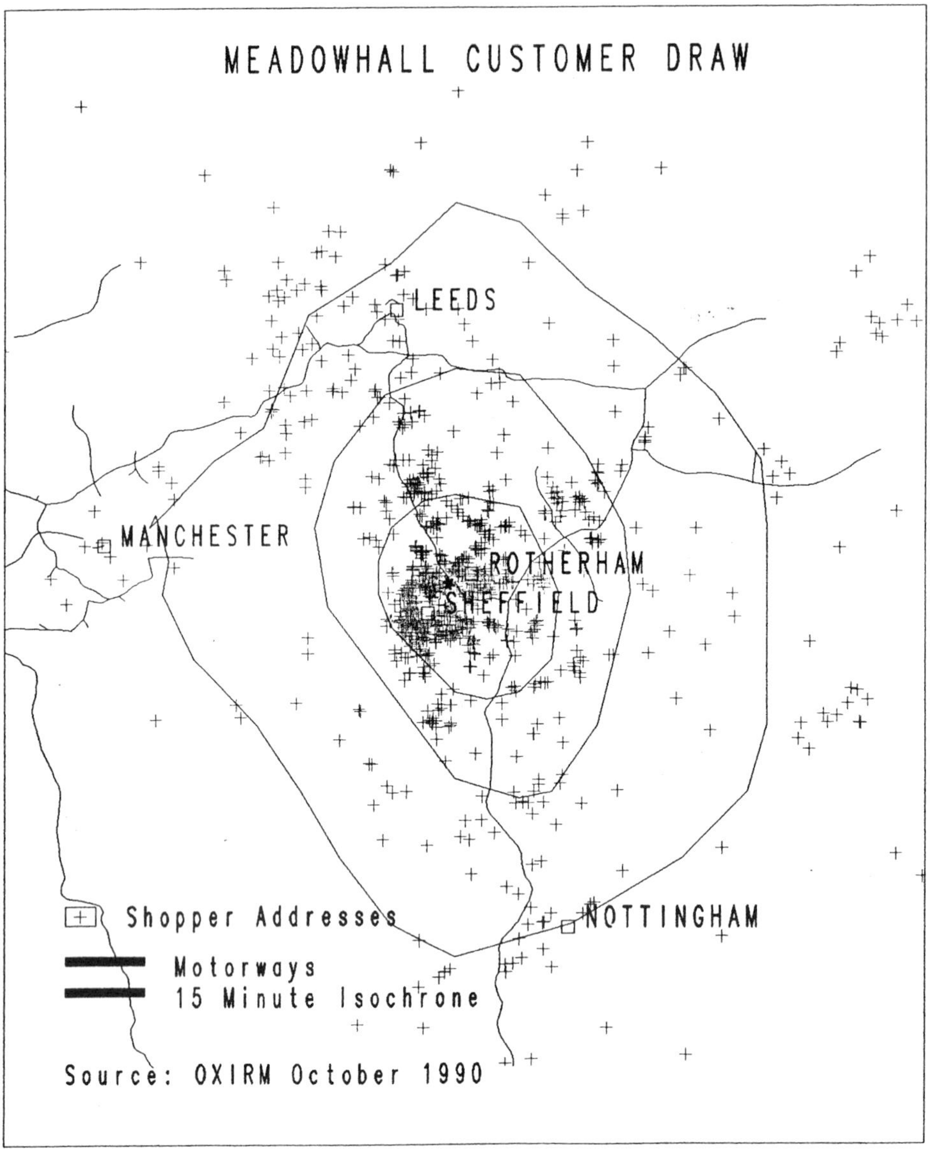

Figure 1. Draw of shoppers to the Meadowhall Centre, Sheffield (central star).

Figure 1 shows the clear early and extensive attraction of the centre, based upon in-centre interviews. Fully one-third of all spending and a quarter of visitors are coming from beyond 45 minutes from the centre. The draw along motorways M1, M18 and M62 is equally pronounced. Figure 2 applies a standard geodemographic discriminator (supplied by Pinpoint Analysis) to provide a rapid 'snap-shot' of customer characteristics compared with the background population. The centre is clearly attracting households from neighbourhoods characterised by high proportions of young, upwardly mobile families and affluent families. Such analysis provides a significant guide to the likely level of future retail investment in the centre.

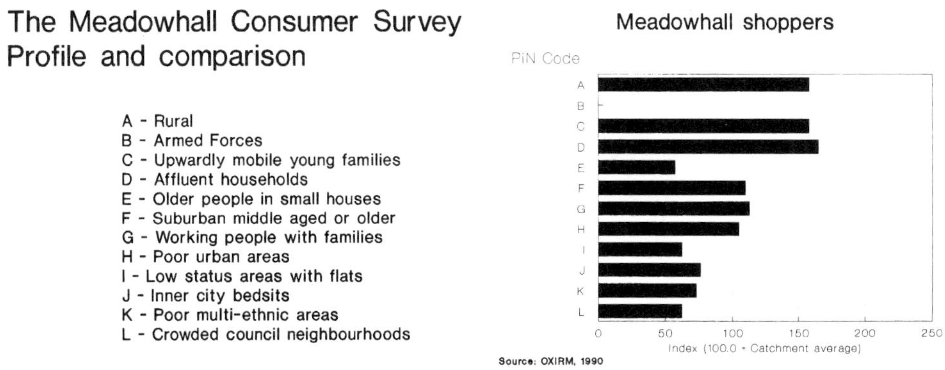

Figure 2. Survey of Meadowhall Centre shoppers, based on standard geodemographics.

Case B: Ranking local labour market vulnerability

By its definition, the retail sector is tied much more closely to local markets than many others, and is consequently much more vulnerable to small area trends in labour availability. Given the shifting environment for the recruitment and retention of labour, if a company was better able to identify those stores within its portfolio where the labour market was tighter, recruitment policies and resources could be appropriately targeted.

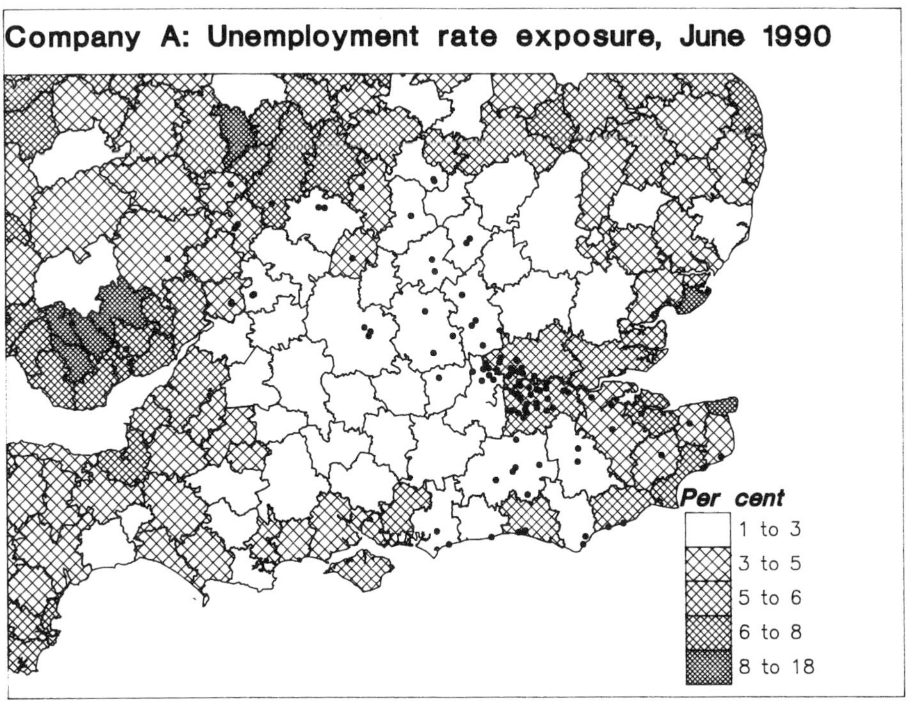

Figure 3. Rate of unemployment (per cent) and store locations, southern England.

Figure 3 shows that some of this potential is already achievable using existing manpower information systems (specifically NOMIS, the National On-line Manpower Information System) in concert with a commercial GIS, ARC/INFO. The figure shown is part of a larger project assessing the exposure of a grocery retail business to changing local labour market conditions. In the figure, the rate of registered unemployed in June 1990 for travel-to-work areas in the UK is displayed alongside store locations. In combination with a series of related variables, the system reports a quantified expression of labour market vulnerability in Company A's stores.

Acknowledgements

OXIRM gratefully acknowledges the support of Meadowhall Centre Ltd and Pinpoint Analysis Ltd in this research.

PART VIII

People and GIS

In this section we have brought together papers on selected human issues that confront a manager bringing in a GIS. The bigger the implementation the more important become the issues. The main issues discussed here are project management (S. Astbury), training (J. Petch and R. Haines-Young), the selection of the right sort of operatives (A. Tait) and the implementation of GIS in various bodies. It is getting these issues right through proper preparation and planning that a GIS implementation stands a reasonable chance of success.

Perhaps more importantly, for the longer term at least, is the need to train the right sort of people. Only people with a good base of numeric and geographic skills will be able to realise the powerful tool that GIS can be (it is easy to see today the lack of understanding that people have in the information contained in a map and in their ability to use one).

43

The importance of people in implementing a successful GIS

Peter Elkins
IT Southern Ltd

The importance of people—oh what an opportunity to be trite, and what temptation to enlist those endless platitudes. Perish the thought! Let's turn a penetrating stare towards the importance of people. If we really do want geographic information systems (GIS) implemented and established as a bed-rock in Information Technology, at the end of the day, "people" must be the most important focus of attention.

I have approached the topic not in the generic sense, but from the stand-point of many years spent implementing and building GIS at Southern Water, an organisation which has gained immense experience over the last eight years and at the time of writing has some 50 networked SUN workstations providing GIS and its support based on Alper Systems Records software.

It is noted that people can and certainly do behave in qualitative and emotional ways. In this respect, success or failure in the implementation of GIS will be by virtue of the way the needs of people are handled. In GIS, whether as user group corporately, or in an august and enthusiastic body like AGI, people are of key importance in maintaining geographically based information as a driving force.

Let us reflect upon organisations and the decision process, confidence and perception, the human resource and cultural change, and probably most importantly the fact that GIS does support real world tasks.

Decisions, decisions, decisions

Automated decision-making is not yet a reality. As people, we are still all decision-makers whether Chairman or Excavations Supervisor, and we all need information support in taking those decisions. Although we live in the age of the "flat" organisation, with business organisations, utilities, and some Local Authorities losing the multi-level pyramid look, the interaction between people in the decision-making process has not lessened. However, information by space and time, the stuff of GIS, is becoming a powerful unifying principle; a common facility available to all organisations and one which can streamline the decision processes providing we understand the human activities of the business in the first place.

There should be an emphasis on the role of GIS in top-level management decision-making. This in turn will establish a better understanding of overall decision support opportunities available to an organisation.

Hype

Confidence is a valuable commodity when gaining the support of people. This can never be more true than with implementing GIS.

At times, "hype" in GIS has not helped; it has excited people to expectations in excess of reality, and to believe in rates of expansion greater than those which could be matched by software. This phenomenon, which we have seen generated from both within the organisation and by vendor communities, has disturbed the confidence of senior management and end-users of GIS alike. Recognition of human predispositions to doubt and suspicion, particularly when the understanding of GIS is underdeveloped, is critical to the final success of the project. Involve people with every step taken; help them understand what the real offerings are at any stage in the development, listen to them and try to accommodate their views. Memories can be short, so keep reminding them.

Perception and respect

People perceive; they do not always rationalise the objective merit of a GIS facility presented to them.

At Southern we have found that to misunderstand a current perception can lead to ill-timed and ill-designed developments. When reviewing user satisfaction, there is a need to listen carefully; there is need to see developments through their eyes, to listen to their requirements, but not necessarily to adopt the solution they propose.

Respect for users' opinions and perceptions is critical in planning, designing and implementing systems. The only GIS facilities surviving, in the long-term, will be those which have resulted from respecting the users' views. This will be so, even though changes in those views can be expected. People are important; their views, even if they appear unnecessarily conservative, must be taken into account, documented and actioned where appropriate.

Human resources

The introduction of a GIS can streamline the human activities in many areas of business. But what of the prime mover, the human resource? Take the data conversion people; in-house or by contract, their dedication to quality and efficiency will continue to be depended upon.

At Southern, the 30-strong data conversion team are now into the fourth year of travail, having devoured some 12000 km of water mains, 600000 water features, thousands of scanned maps, and about half of the 24800 km length of sewers. With this achievement comes the attendant responsibility for maintaining the integrity and currency of deployed data. These people have onerous and repetitive tasks; tasks with which they may become easily bored. Crucial to success are the selection procedures for these key individuals; we must place people to the task with precision. Once recruited, they must be trained to the required standards and their interest retained by continually creating challenges and rewarding their successes.

Cultural change

Culture, in our context, is about people and how they create working arrangements and processes to deal with ever-changing business objectives. The vision generated by the application of geographic information systems, is now a potent force for those who will be instrumental in actioning cultural changes. The "culture steerers" will be GIS decision-makers; they are important to the long-term success of implementations. They will ask, quite rightly, "Why are we spending the money?" Their full understanding of the long-term possibilities must be maintained if they are to be a vital driving force in our organisations. People, however, have short-term memories and a continued awareness programme is important if this is to be countered. AGI has, I believe, a central role to play here. Continued support from those with authority for cultural change is most critical to success. Keep the "champions" briefed, keep them nurtured.

The end-users

End-users of GIS, wherever they sit in an organisation, must be regarded as the reason why it has all happened. GIS provide both infrastructure and end-user facilities and it is this complex balance of functionalities which can make selling to the user difficult. We must remember that systems without end-users will be of academic interest only.

Key users in the front line of an organisation can be unforgiving, if functionality offered does not meet their expectations. Their acceptance of facilities is critical to the further stages of implementation which will yield the major benefits. The operational user can make or break further developments and we should all respect this powerful position. It is salutary to reflect on the question as to who we, the system providers, are providing for! GIS deployment is not merely an interesting intellectual exercise and such a view is likely to spell disaster. A key message is to listen, provide feedback to designers, and improve. A full explanation of the cost implications must be available, naturally.

Feedback

Be proactive, review now! People are now committing their futures to GIS especially in the utilities. Can we make sure that the managers of those trail-blazing, first GIS users, interpret correctly the signals of both enthusiasm and disquiet from their staff? In the process, let's ensure we always give that objective view, which will allow an accurate evaluation of the GIS's use.

Demographics

The importance of people is never more noticed than when they are unavailable. We are probably all familiar with the sense of shock when it is realised that there is no one to perform some critical task. Demographic changes are now stalking us all insidiously. Predictions persuade us that the early 1990s represent a significant "low" for "people supply". How can we operate an organisation, a business, with fewer people? Clearly the importance of people will be raised to an even higher level and the issues discussed earlier become even more focused. Lean organisations could make heavy demands on geographic information systems; are we ready for the challenge?

Conclusion

GIS are based on sophisticated technology and will be capable of delivering increasingly powerful solutions to business problems. Success can be guaranteed only if the human resource, with its frailty and qualitative behaviour, can be accommodated. The importance of people remains paramount.

44

How to prevent project management

Stewart Astbury
Byers Engineering Company

Introduction

AM/FM/GIS projects are complex. There are many things that can go wrong in the course of implementation. Some problems occur because of inadequate planning, others because of poor execution, and yet others because of fallacious thinking. Typical problems in each of these areas will be reviewed and the likely consequences identified, along with recommendations.

Fallacious thinking

Fallacies are erroneous beliefs that are held in the form of principles or purposes and are used to guide reasoning and decision-making. They are particularly dangerous because often they are not explicitly stated. It is well worth the time to address key concepts in the project's scope and the feasibility documentation. Following are examples of some of the most common fallacies.

(1)	**Fallacy:**	We should implement all functionality simultaneously.
	Consequence:	Overcomplication of development and cutover to new system; blown budgets and schedules.
	Recommendation:	Design in detail, implement in phases. Get some benefits soon.
(2)	**Fallacy:**	We must choose the system that meets the standard architecture requirements of our information systems group.
	Consequence:	Selection of a system that may require significant development time and cost.
	Recommendation:	Balance IS needs with a system that meets user needs.
(3)	**Fallacy:**	We can get our contractor to do things for free—so they can get the big prize down the road.
	Consequence:	Products hastily prepared, more in ''demo'' mode than functional.
	Recommendation:	Obtain the developmental resources needed to do the job right the first time.

(4) **Fallacy:** We must use strategic thinking to position this technology best.
Consequence: High level lip service with a lack of content; no direction for project.
Recommendation: Match the technology to the needs of the end users.

(5) **Fallacy:** Data capture is a copy job/our records are in good shape.
Consequence: Not designing a realistic and effective conversion plan; data that do not support key applications.
Recommendation: Data capture is a process of building an intelligent database from numerous records; a detailed design is needed to ensure a quality database.

(6) **Fallacy:** We can let the pilot project produce the design.
Consequence: A waste of time and money; no usable product; disgruntled users.
Recommendation: Develop the requirements and design; code and test the software before starting the pilot.

(7) **Fallacy:** This technology is user-friendly; it will be a big success.
Consequence: Unhappy users clinging to the old system; inability to "cutover".
Recommendation: Perform organisational impact study; determine effect on users; provide training and preparation for new systems.

Planning

Detailed planning requires hard work. It is tempting to gloss over key aspects and hope for the best. Following are several areas where this approach can lead to severe problems:

(1) Failure to perform a realistic cost/benefit analysis. At the end of the day, the true costs will be accounted for. Additional budget requests are painful. Enough said.
(2) Failure to project adequate staffing. This may be particularly tricky in the areas of vendor support and user operations/maintenance during the learning curve.
(3) Failure to perform a complete design. New ideas will surface which may cause changes to the data structure. Programs must then be modified, databases recompiled, conversion reworked—all at considerable expense.
(4) Failure to consider the impact on the organisation. Users will not be able to use new technology; border skirmishes will break out between departments; old systems will remain in place.
(5) Failure to obtain upper management commitment. The money won't be there when needed. The prudence of the entire undertaking may be questioned.
(6) Failure to anticipate costs to customise the package. This arises from a failure to get a realistic understanding of user needs and wants. This costs money, takes time and causes hard feelings between customer and supplier.
(7) Failure to determine computer resource requirements. Slow response time,

inability to transfer files, lack of storage space...these excuses will get old quickly from a user's viewpoint.

Implementation

The problems which crop up during implementation are the worst because they are the most urgent and visible. The sins of the past require retribution at this time. This is the project phase to make certain everyone knows what they need to know.

(1) Training—many jobs will be changed. Procedure manuals, training, and on-the-spot support are essential.
(2) Backlog—because of the engineering and construction completed during the conversion period and the initial learning curve, significant backlog posting will develop. Better to plan on using contract support than never to get the database current.
(3) Politics—the project manager must be sensitive to the needs of all users, technical support and management. It is worth much to listen and fulfil their needs. Here one must be careful to walk a tightrope over project scope and user satisfaction.
(4) Project tracking—it is essential always to know the status of costs and schedules. A ready (and honest) answer goes a long way to maintain project credibility.
(5) Project milestones—success comes a step at a time. There should be a number of planned "little victories" to keep morale up.
(6) Test plan—software systems frequently have errors—better for the project team to discover them than the users.
(7) Communication—realistic, weekly status reports to all concerned will build better tolerance for the eventual problems that occur. Most people are reasonable if given time to adjust to a change.

A final word

A clear understanding of the overall problem addressed and how this project will solve it is absolutely essential. An ill-conceived, unnecessary undertaking cannot be managed, no matter what the skills of the project manager. The project's basic purposes should be reviewed constantly and all change requests evaluated in their light. Lastly, always ask the question, "Does this idea make our work more productive and save money?" If the answer isn't a clear yes, the idea should be re-thought or abandoned. Managing an AM/FM/GIS project is challenging but achievable. It simply requires a clear focus, detailed planning and vigilant monitoring on the part of the project manager.

45

Training for GIS

James R. Petch
University of Salford

and

Roy Haines-Young
University of Nottingham

The dominant theme of articles on training in previous *AGI Yearbooks* (Gittings, 1989; Unwin and Dale, 1990; Willis and Nutter, 1990) and of a host of other articles on education and GIS (see for example Westcott, 1989; Heywood and Petch, 1990) has been the so-called skills shortage. This article deals with the same issue but comes to a somewhat different conclusion about what to do about it. Its objective is not so much to suggest that more should be done to reduce the shortage but rather to open up further the debate on what GIS education should consist of.

Unwin and Dale (1990) initiated the debate in the 1990 *AGI Yearbook*. They outlined the two principal curricula in GIS, from NCGIA and AutoCarto, but raised two critical concerns. Those were, first, that the material of the syllabi were content-driven, being teacher-centred rather than student-centred and, secondly, that there was a danger that GIS would become a subject to be taught rather than a means of doing things (and thereby become sterile and marginal). These points need amplifying and reinforcing and to be responded to with some concrete suggestions.

Amongst the educational difficulties cited by Unwin and Dale (1990) are:

- differing conceptions of the field, which makes definition of any formal educational aim difficult;
- different student needs in education and training, which leads to a variety of possible modes of delivery.

What strikes us is that these difficulties seem to be a consequence of believing that GIS should be taught in one way. The futility of this assumption is apparent simply by asking who is going to use GIS education and for what? The answer begins with A for Agronomist and ends with Z for Zoologists. Each group will have its own problems and own methods. These can be translated into issues of data structures, system architecture, data functions, display media and so on. It is clear therefore that no GIS curriculum can suit everyone. In addition applications people do not necessarily need to know about systems and vice versa.

This situation prompts a second question: is there a core of knowledge or a set of skills which do need to be learnt by any GIS student? It is clear that not all parts of the NCGIA and AutoCarto curricula are necessary but it is not at all clear which parts are. We suggest that only one section of each curriculum is essential but that there are two areas not covered by either which are also essential. The first is the area on cartographic and spatial analytical concepts which deal with data properties, spatial presentation of data and spatial operators. The second is the architecture and spatial analytical functions of different systems.

This may seem a rather thin list but it needs to be judged in the context of the answer to a third question: how can these GIS skills and knowledge be used? The answer to this question can only be 'ask the expert in a subject where GIS is to be applied'. The points and issues of many of the subject areas in the GIS curricula mentioned here, and in GIS texts, require explanation or responses specific to applications areas. Areas such as 'institutional aspects of GIS' and 'applications of GIS' from the AutoCarto curriculum clearly fall into this category. But there are other areas such as those about spatial concepts, or about spatial operations on points, lines, and surfaces, which only make sense to non-GIS specialists when dealt with in terms of their own systems of interest. What must be remembered in curriculum design is that abstract GIS ideas and methods are of little use or interest to people who want to use GIS to do a job. Additionally, they are difficult and tedious to learn or master when there is no context or direct use. In addition so much of GIS methods and operations is software-specific. Take digitising for example; digitising into ARC/INFO, SPANS or SYSTEM 9 is quite different conceptually and procedurally. The fact that some topological properties are preserved in one system but not in another makes some spatial operators impossible in one and trivially easy in another. Questions of error are use-specific. Input data and output data issues are generally use-specific. In other words, teacher-centred material which does not take account of problems is unlikely to provide the basis for a successful course, especially in non-academic environments. The answers to these problems of what to teach must be given by a person who understands an application area and specific GIS software.

The parallels drawn by Unwin and Dale (1990) between quantitative methods in geography in the 1960s and GIS in the 1990s are striking. It is easy to see now that quantification *per se* is meaningless. So it is with GIS. We all need to gain good skills to do better what we are already doing and the acquisition of skills allows us to do new things. But just as a pure statistician alone cannot answer questions about research problems so a pure GIS expert cannot do spatial analysis. GIS is part of the scientist's, engineer's or manager's toolbox.

Currently there is the idea that GIS training should to a large extent involve the coverage of abstract ideas about data, computing systems, spatial analysis and a host of general topics such as organisational implications. Perhaps this is not so dangerous when there are so few skilled people and when demand for these skills is so high. However, such a situation has to change as GIS become widely used and applied. The general purpose curricula will become less effective and less relevant as people who need education become the designers and operators of systems. These people will function within a single organisation and use perhaps one GIS. Therefore, they will have limited but specific educational needs.

The types of GIS educational programme which should take the place of general curricula are perhaps nearer to courses such as the diploma at Silsoe, where GIS is taught as part of a Master's degree on Land and Water Management. Such integration is vital if the real position of GIS in a discipline or practice is to be appreciated.

The key problem of GIS is not a technical problem of data and computer systems but the problem of translating a job into a spatial data system.

There is a fourth question about GIS curricula on a particular issue which nobody seems to address: is a GIS geographic or just georeferenced? Of course georeferencing is the essential component of GIS. It is important because of the power that mapped data gives in terms of inspection, investigation and interpretation. But is there not more to it than that? Any old time geographer will give the answer—map skills. Maps are not simple graphic displays of information. An experienced worker can show great skills in using maps. These skills of design, analysis and interpretation are a real craft not accessible simply by looking at maps or producing graphic displays. The power of the map is in the richness of information it contains and displays. Rudimentary map reading and map-making skills are taught less and less in schools and universities but these are fundamental skills for GIS. What GIS has to offer is diluted without the ability to "read a map." Which GIS curriculum contains this topic?

References

Gittings, B., (1989). Education and Training—The Missing Link? In Shand, P. and Moore, R. (Eds) *AGI Yearbook 1989*, London: Taylor & Francis, 323–324.

Heywood, D.I. and Petch, J.R., (1990). GIS Education: a business perspective. Paper presented at Conference on GIS in Education, Leicester, 1990.

Unwin D, and Dale, P., (1990). An educationalist's view of GIS. In Foster, M. and Shand, P. (Eds) *AGI Yearbook 1990*, London: Taylor & Francis, 304–312

Westcott, B., (1989). IT—The Skills Shortage, In *Proceedings of Conference on Information Technology for Local Government 1989, The Challenge of Change*, Blackpool, England, 32–33.

Willis J. and Nutter, R.D., (1990). A survey of skills needed for GIS. In Foster, M. and Shand, P. (Eds) *AGI Yearbook 1990*, London: Taylor & Francis, 295–303.

46

Recruitment and selection for GIS data capture

Anne Tait
Manpower UK Ltd.

Introduction

This article reviews some basic issues in the field of recruitment and staff selection with regard to GIS. (The term GIS is used commonly through the article to describe spatial information systems whilst recognising that there are alternative terms which could be appropriate.) Information Technology should be designed for the benefit of its users. This article suggests that GIS personnel should also be selected for their ability to add value to expensive data capture activities. There is evidence that specialised recruitment and selection techniques from the temporary staffing market can enhance this process.

It is clear to anyone who encounters GIS-related technology that it is an exciting concept that will launch business management into the next decade and beyond. GIS has arrived over a decade after the initial impact of Information Technology on the business world. It is yet another technological leap which many organisations will take to gain a sophisticated commercial edge. The introduction of new technology has an impact on the culture within an organisation. GIS technology is no exception bringing with it additional considerations because it demands "new" skills.

The complexity of GIS technology and the type of information it deals with require not only high initial expenditure on hardware and software but also large-scale data capture/conversion tasks. Equipment is of no value to the organisation until the data for its operation has been processed. The quality of this data will form the foundations of the corporate database; it must be accurate, and is dependent on the quality of the personnel employed. Most organisations will be looking for a method of data conversion that provides them with quantifiable value-added results and tangible returns on the high initial investment for equipment. Unfortunately thoughts about human resourcing issues tend to be reactive rather than proactive, stimulated at the moment when beneficial returns become the focus within an organisation.

Data preparation/capture/cleanse and maintenance

Most large organisations have moved towards investment in new technology over the last two decades, from electronic typewriters to international communication links. Modern office equipment has alphanumeric capacity and progressively, graphics. GIS technology adds the dimension of a movable, geographically-referenced background which creates the need for additional operator skills.

Entering information into the database requires the conversion of geographically-referenced material and alphanumeric data from a huge variety of sources. This work involves not just electronic conversion, it demands interpretation, sorting and interrogation before it can become a formative part of the hard copy source. Often quantities of this work will be done manually not electronically.

Volumes of corporate data collation and conversion are considerable by anyone's standards and consist of layers of information relating to one asset. Once plotted, there may be additional data to be entered, sewer records, flow variations, customer information or service/maintenance updates. The latter require constant revision through the whole data capture exercise.

Data preparation and conversion projects will ideally only be performed once if the essential accuracy needs are addressed. This will require a number of people to complete the tedious task of collating and sorting information from reference manuals, index cards, maps, etc., the final output from which will become an integral part of the corporate decision-making mechanism. If these people are not personally committed the quality and accuracy of the results will suffer. Only when a bad decision is made on the basis of inaccurate and possibly misleading information will the full impact of the situation become apparent. This could be days or months into the future and may give rise to critical repercussions.

Recruitment

A data capture exercise in alphanumeric information is, by comparison to a GIS data conversion exercise, straightforward for staff to perform. There are data conversion and VDU skills readily available for purchase in the marketplace. The concepts of alphanumeric electronic data transfer have become well-established in the modern business world. GIS utilises more complex applications than normal office automation procedures. Its corporate users are rarely those from the standard office environment where telephones, fax and PC's are items of common equipment.

GIS kit seems initially to require "experts" who have always worked in the planning, engineering or other specialised technical worlds. However, asked how he or she feels about digitising all day to transfer information from hard copy maps, a cartographer will inevitably say that the work is monotonous and repetitive but simple to execute. Technical specialists will identify that the interpretation of the information being transferred is the skilled aspect of the work.

GIS data conversion exercises can be divided into two tiers. Firstly, preparation of the data and interpretation of the information. Secondly, accurate conversion of the data through a systemised approach onto the electronic database, paying attention to detail and adhering to predetermined corporate objectives. Once the parameters of a data capture exercise have been defined by the organisation, and a brief given to individuals performing the data conversion/capture activities, induction and training will be required.

Increasingly organisations are meeting the staffing demands of large single projects by the use of cost-effective contract staff teams. With GIS data capture/conversion more organisations are beginning to consider this option but finding the right contractor can be difficult. There are two main problems. Firstly,

GIS operator skills are not readily available in the marketplace. The skills for digitising tend to be linked currently with other specialised skills. Secondly, designing a job specification for GIS data capture is a complex and time-consuming task made more difficult if non-technical staff are to be appointed.

Where do GIS data capture operators, who can input data onto GIS equipment to specified standards, come from? Data capture certainly describes the activity in general terms but how can digitising be identified at this stage to the uninitiated? These issues are no less pertinent when directed at organisations wishing to utilise their own personnel.

Inevitably, when existing staff within an organisation are utilised for GIS work, a reduction occurs in the overall productivity of the organisation. Often this will affect the quality of the data capture exercise because people are being de-skilled and de-motivated. This creates a poor scenario for quality work, exacerbated as project size increases.

The only method, at present, for GIS data capture teams to gain experience with spatially-referenced information and GIS hardware is to "learn as they go". This applies to both in-house and contract personnel. Performing basic survey work quantifying street furniture and digitising directly from validated hard copy sources does not require a specific recognised qualification. Training is bound to be required and should incorporate aspects of team-building and a clear idea of the project objectives. Basic data capture work is boring and repetitive and a good team spirit will help, as will a sense of direction.

Long-term strategic human resourcing should acknowledge that GIS data capture projects will eventually become a thing of the past. Strategic planning could then transfer contract staff with idiosyncratic knowledge of the database to roles where they can maintain and manipulate the final database as required for the corporate good. Personnel recruited should be selected for their potential to form part of the corporate staffing on a full-time basis after this hectic first phase.

Staff selection

Teams of people with no prior experience in GIS can quickly develop skills in digitising, survey work and the interpretation of geographically-related data. Their personal interest, confidence and enthusiasm grow as they make a significant contribution despite their unqualified and varied backgrounds. Careful selection ensures that staff are committed and they can contribute a huge amount to the success of the project simply by being interested at the appropriate level. This is clearly a good basis for quality work.

Successful teams have been assessed for their natural abilities using a battery of tests (developed by Manpower UK Ltd.) to meet task definitions rather than job descriptions. The results of these tests provide a profile of the individual's natural abilities and, combined with a specialised interview selection programme, give a performance prediction. They have also all been assessed for their keystroke speed,word-processing skills (where relevant), administrative capabilities and supervision needs. Careful selection of staff for these projects can be a major factor in guaranteeing quality information in a database.

GIS equipment tends to be much more complicated than a single PC and keyboard. Anyone moving from standard office technology will have transferable skills but interpreting map-based data is not something a word-processor or VDU operative would normally be required to do. Basic alphanumeric work is more straightforward than digitising, in most GIS operators' opinion. Therefore, this work often represents a welcome learning curve for them and can be perceived as a positive opportunity in their careers.

Despite the use of sophisticated selection and testing techniques for recruitment to these projects, personnel remain cost-effective compared to the more expensive technical staff with whom they work, as are contract staff who have a clear idea of the work, the project objectives, the deadlines and the time commitment required of them. Such a customised staff group will usually be more motivated than in-house staff, co-opted for the duration, utilising only a small percentage of their skills. Using in-house staff from an organisation can mean that whilst they are doing basic data capture their own work is not being done. It can also mean that highly paid individuals are being asked to do low-skilled work when their contribution could perhaps be used more effectively in the supervision of others.

Project planning

Whether using contract or in-house staff, avoiding poor quality work should start early in the planning process. Ownership of the project and all its individual facets and phases must be clearly defined to get the best out of the people involved. This requires an approach that starts at the end-user requirement and works backwards to the Board. During the course of such an activity, people, systems and corporate objectives receive a healthy review. If managed correctly, personnel in the organisation can benefit immensely from this costly and labour-intensive activity (which normally involves a full cross-section of the organisation's management hierarchy and technicians, and relies on hands-on capability and theory being constantly linked).

Unfortunately the decision-makers are rarely operators themselves. Their technical understanding of the machinery is high but they have not experienced digitising for any length of time. This creates a "grey" area which can have adverse effects on the planning of a data capture exercise. Communication can further be hindered by differing knowledge levels concerning the practicalities of basic GIS data conversion.

There are some organisations who have planned their approach to data capture to meet carefully-planned criteria. They have logically identified their corporate needs and are able to look at the market and demand a specific service with regard to contract data conversion staff. They have also addressed the motivational needs of the teams and made their training investment in personnel with predicted performance levels.

The use of contract teams facilitates on-site data conversion activity on an organisation's own equipment. Task control remains with the organisation, as does quality control.

The fundamental principle of this approach is that an organisation acknowledges responsibility for the validation and interpretation of its asset information. The information, therefore, remains corporate. (That is, the people who decide on the characteristics of the database are the people who will continue to use it after data capture is complete.)

An organisation must understand that, during the conversion of the data, unique and complex issues are going to arise. If the staff involved are accurately selected and have ownership of their work, their motivation will add unquestionable value to a project.

The important components of all these processes are the people and the consistent quality of their work. Behind every corporate GIS are the hands and minds of the people who feed it information. A corporate GIS will only be as good as the personnel selected, at every level, to use it.

47

Integrated GIS

Roger Morgan
Thames Water Utilities

The definition of GIS produced by the Standards Committee of the AGI includes the words "The primary function of a GIS is its ability to integrate data from a variety of sources."

There is little doubt that GIS have come a long way over the past few years, but progress to date may not have put us in the best position for the future—put another way, it might have been better if we had not started from here. So where have we gone wrong? GIS have largely developed from mapping systems and until recently little attempt had been made to integrate them with other corporate systems. The result is that they are still largely independent—tacked on as an afterthought when they should be absolutely central to the corporate information of the business. There may be cases where standalone systems are justified, but not within most utilities. Nobody would take very seriously the idea of fitting the QE2 with an outboard motor.

Why are we where we are today? Utilities have always had a need to keep a variety of information, both geographic and non-geographic. Traditionally the former were drawn maps, sketches, etc., showing the geographic location of assets and little else while the latter were kept in ledgers, files, card indexes, etc. Certain aspects of the non-geographic data have been "computerised" over the years and digital mapping in the 1970s and early 1980s went some way towards automating map production. There had been little link between the geographic and the non-geographic information and digital mapping did not do much to bridge the gap. Many of the earlier systems were derivatives of CAD which, although giving a reasonable graphic representation, lacked the functionality that good GIS should give.

Utilities are capital-intensive industries whose assets are absolutely central to their operation; those assets are geographically spread and knowledge of that geography is of critical importance to their effective running.

Few utilities will be fortunate enough to start with a clean sheet in computing terms, but when considering future strategy GIS should be at the heart, not on the edge. How many utilities run GIS on their corporate mainframe machines as fully integrated systems or at least with the right 'hooks' to allow full integration at some stage in the near future? The answer is not many, though a few such as Thames Water are going down that road and there is no doubt that such organisations will reap the benefit of this approach in years to come.

Undoubtedly, one of the main barriers in the past has been the cost. The integrated GIS does not come cheap and currently does not lend itself to the 'gently gently' approach. The result has been that companies have gone for the 'cheap option'—the standalone system, probably PC-based which may well have met some of their immediate needs. Soon they start enhancing those self same systems and before long they have invested substantial sums in a non-integrated system and there is a reluctance to change direction. The decision that should be made is avoided.

Why should GIS be integrated? First and foremost they allow a single data source with shared rather than duplicated data; the opportunities for inconsistent data are significantly reduced. Secondly GIS have the potential to be used as a very powerful tool for updating data on other parts of the business's systems with an ease that previously could only have been dreamt of. Take as an example a gazetteer used within a customer information system; typically it might hold a list of streets with their locations together with a whole range of other related information, for example local authority, health authority, inspector's name, pressure zone, etc. All this related information or attribute information is liable to change and keeping the gazetteer up to date using traditional methods is well nigh impossible. Sooner rather than later the records become out of date and the integrity of the information on the gazetteer is called into question. GIS and in particular polygon analysis techniques can render updating a relatively simple exercise which can therefore be carried out regularly. Integrity is improved and with it the users' confidence in the system; the process snowballs and the overall integrity of data can advance in leaps and bounds.

Data capture

One of the biggest tasks facing utilities is initial data capture or 'conversion'. Full benefit can only be obtained from a GIS once all data have been captured, though there may be substantial interim benefits, some of which may arise out of the very disciplines necessary to undertake the data capture.

The effort required for data capture is often underestimated, but the mix of techniques is also misunderstood. Manual digitising is often seen as the major element of data capture whereas preparation of the input records and checking of the output, if done properly, are likely to take more time than the digitising.

It is important that the quality of the source records is good in structural terms but this should not be confused with accuracy. The process should be to take existing manual records, check for and resolve any obvious conflicts, digitise the result, and then check the output. Some are tempted to carry out rigorous surveys before entering any data onto the system—such an approach is wasteful and misses one of the major advantages of GIS, the ability to update easily and be confident of the result.

It is well worth spending time taking a critical look at the various parts of the data capture process; even relatively small improvements in technique can bring large savings overall. A large data capture exercise can run into hundreds of man-years of effort—on this scale even a few percentage points mean big money.

Management of change

No matter how good the system, benefits will not be achieved without changes to practices and procedures. Developing a system to do wonderful things is relatively easy, getting users to make the most of it is far more difficult. Revised methods of working need to be carefully thought out, documented and put into practice. A detailed 'user training needs analysis' must be undertaken and followed up with a defined programme of training in all aspects relating to the system. Only in this way will one avoid the misunderstandings and misuse of new systems. Neglect this training and it will be the system that gets the blame when things go wrong.

Education

The need for retraining for those using GIS was highlighted in the last section, but there are many other areas where education is required. Many universities now include GIS as an option in some of their courses. In terms of general awareness this must be good news because anyone who has anything to do with spatially related data is likely to come in contact with GIS in the future. There are some words of warning however. In concentrating on the benefits of GIS there is probably insufficient attention paid to data capture aspects and the enormity of the task that has to be gone through before reaching the benefits. There is little point in having excellent analysis tools if there is no data to analyse.

Almost inevitably, most universities, if they have GIS available, currently have standalone systems; ones that are not 'integrated'. The danger here is that the benefits of integration may well not be emphasised sufficiently. To date there has been no 'low cost' entry into the integrated arena but that will hopefully be changing in the not-too-distant future.

Technology

Even after a short visit to an AGI exhibition one can hardly fail to be impressed at what is available. Clearly, a lot of work has gone in to developing products and there is no doubt that there are some very neat 'front ends'. Functionality has also improved over the years but a more detailed look is likely to reveal that many of the systems are still only 'skin deep'. They may be considered as good mapping systems but in years to come they will not bear comparison with the best GIS.

Until recently GIS have broadly fallen into two categories, by far the largest (by number) being PC-based with a few mainframe-based. The former, though becoming progressively more powerful, are still very limited; the latter can be fully integrated, are very powerful but are less flexible and are expensive to run. The increasing numbers of good UNIX platforms should provide an ideal way forward, allowing integration with the mainframe but having much of the flexibility of the current PC-based systems at a reasonable cost.

There has been much talk of distributed databases and no doubt there will be a progressive move in that direction over the next few years. However, most organisations have enough difficulty in managing centralised databases successfully; distribution will not provide a solution, but will make matters worse. Distributed processing is a different matter and again with the use of UNIX it becomes a practicable proposition.

Once past the data capture stage there will be the need for wide access to the data if maximum benefit is to be gained. Systems that require high-cost, specialist hardware for viewing purposes will find it difficult to succeed.

Suppliers have been working hard to try to meet the aspirations of users and, to be fair, they have probably been more successful in that than the users have been in taking on board the resultant technology. Future success in GIS may be less dependent on further advances in technology than on the ability of the end users to adapt and make use of the techniques already available.

48

GIS in a local authority

Kelvin Rufus
Information Technology Co-ordinator, West Somerset District Council

Local authorities are in a position to realise immense benefit from geographic information systems, probably more so than other public bodies. This opinion is expressed as a result of four years' work undertaken at West Somerset District Council in the development of its corporate geographic information system.

The use of information, both geographic, graphic and text, is critical to local authority administration. Information is stored in many places, on various media and in varied format. The combining of information in an acceptable and accurate format from the various sources is traditionally time-consuming, probably lacking in all pertinent information and fragmented.

The use of geographic information systems provides the vehicle to overcome the inadequacies of traditional information access and facilitates the reduction of duplication of effort, it rationalises data exchange between departments, supports analytical tools and provides an enabling effect for users hitherto unknown.

Individual items of information in their raw state, whilst having some value, do not present an adequate or full perception of the information; e.g. in the case of an address, it is purely an identifier, whilst the exact site location is unknown (other than loosely), the site area is unknown, and its relationship to other properties is unknown. However, a GIS can provide the necessary combination of information in a most acceptable form.

In consideration of the use of a GIS, important fundamental issues have to be appreciated before a structured approach to development can be achieved. The Council, in determining its development strategy, employed certain basic principles; of prime importance was the analysis carried out to determine what a local authority does. This was achieved by examining information usage, information flow patterns, operational requirements and information types (ie. map-based, text-based, etc.). As a result of the analysis it became clear that almost all functions carried out by a local authority had a relationship to location (land and property), whether it be traditional map-based records, textual records or data in computer-operational applications. The land parcel was recognised as the common feature relating to the Council's activities.

It is the recognition of the land parcel that is the basis of the Council's GIS and the interdependence of data flow/usage between departments, based on land parcel identification, highlights the need for a corporate system. The Council has progressed its corporate GIS through identifying land parcels to the GIS as a primary layer, by digitising a polygon around the curtilage of each parcel using the Ordnance

Survey digitised maps as a background (following site checking of boundaries). Against each polygon certain basic information is maintained relating to the address of the property and a Unique Property Reference Number (UPRN), the latter being autogenerated based on the grid co-ordinates of the centroid of the polygon.

The base (or primary) information is then available to all other departmental applications developed within the corporate GIS. Similarly the departmental applications developed recognise related data which are automatically passed to associated applications, e.g. planning conditions are automatically transferred into land charges.

Applications have been developed based on land parcel polygons relating to:

- Primary boundaries (including identification of land parcels, UPRN, plot amalgamation, plot dissection);
- Sewerage (including foul and SW sewers, pumping stations, sewer incidents, Section 17 & 18 sewers, drainage area plans, catchment studies);
- Land charge register (including identification of land parcels, entry facility of all charges, searches and outputs);
- Planning (including land parcel based identification, listed buildings, monuments, TPO's and grants);
- Housing (including land parcel based details of all Council property types, building fabric, facilities, tenant details);
- Private water supply (including land parcel attachment, source details, construction details);
- Highways and footpaths (permits entry of highway centre lines, public footpath location, text details, maintenance and amendments);
- Utilities (includes entry of PUSWA details, both received and issued, and attachment of details); and
- Land terrier (includes land parcel identification, land owned, land sold, land leased and agreements and wayleaves).

Further applications are, and will continue to be, developed.

Perhaps the most important feature of a GIS is the end-user's ability to access, analyse and display data without the involvement of programming skills. The Council's GIS, being IBM's Geographic Facilities Information System (GFIS), provides such a facility through menu-driven panels accessing the relational database thus providing access to information.

The development, introduction and use of the GIS has brought with it an enthusiasm for all staff involved. It has improved the accuracy of "map-based" work (the system is accurate to within 100 mm) and encourages staff positively to enhance their working operations.

The GIS is still in its infancy but the signs are clear that, as the system expands and further applications are developed, the regime of local authority administration will change for the better.

49

The Ordnance Survey: Spatial data and the future

Neil Smith
Ordnance Survey

The year 1991 will be the year in which Ordnance Survey (OS) celebrates its 200th year of operation. It will also mark the 20th year of map digitising.

OS strengths have been in adapting technologies to serve its own purposes. OS was a very early user of triangulation, tacheometry, photography (both aerial and in graphic production), lithography and scribing. In later years photomechanics allowed, for example to allow the rapid production of 1:50,000 scale mapping from the 1 inch series without degrading the cartographic quality, or the conversion of "County Series" large scales maps to "National Grid Series". Similarly, in marketing mapping OS has since the mid-1970s been selling copies of survey documents as a means of providing maps which are as up to date as possible to an adequate level of representation. This history of innovation has continued into the era of digital mapping. Experiments started in the late 1960s but by the early 1970s sufficient technological and marketing support was gained to start a pilot production line of large scale digital mapping. In 1974, under sponsorship of the Department of the Environment, a study was started to investigate the need for what would now be called a geographic information system. The project never reached fruition mainly because the technology was far behind the ideas—who nowadays would consider writing a GIS on a machine with 32k of available memory? The ideas, put forward by the consultants PMA and supported by OS, were for a system which would effectively "interpret" parcels, attaching names where appropriate. That large parts of this software worked at all in 1977–9 are remarkable because of the technological environment of the time; given today's technology of faster machines and better software, interactive graphics and high speed plotters, what was considered ambitious in 1974 should be simple today.

However, for large scale data there is no such software in place. Only on an experimental basis is there software which could match that of PMA. The reasons for this are not so much to do with technology but a reflection of the market environment. In 1990, OS map data is still digitised mainly to a specification designed to meet output devices of the early 1970s. There have been some changes in specification, but these have been concerned with the semantics, or descriptive elements, of the data. The specification for the geometric elements is still suitable primarily for driving a plotter. Up to now, this has reflected the market, which is to allow map production to be passed to the end user.

The levelling off of innovation aimed at exploiting map data in the early 1980s, after the abandonment of the PMA system, was counteracted by a growth in other areas. Techniques of capturing data by line-following and raster scanning were investigated and rejected, not on technical grounds but because the costs could not compete with the alternative of in-house digitising and later contracted work. The early digital mapping systems, which were all based around "blind" digitising, relied on very strict operator discipline to enable editing to take place. Update was very slow and cumbersome. As a spin-off from experiments in line-following, the benefits of interactive editing were identified and software written or purchased to exploit this. In the days of 24 MIP 15,000 workstations it is difficult to see how much agonising this caused, but it should be appreciated that the cost "per seat" of those early workstations' hardware and software was of the order of ten times the operators' annual salaries, whereas nowadays it is less than a factor of two. OS has at the end of 1990 about 100 such workstations involved in map data collection and maintenance, in both the headquarters and field offices, and is in the process of developing handheld, A4-sized graphics stations for use by surveyors in the field. Photogrammetry and terrestrial surveys are heavily automated in both capture and computation, allowing direct input and update of the map files. OS was among the first organisations in the world to sell map data, and to date has sold around 100,000 map units. It has facilities for licensing, linked to update, and has taken steps to protect intellectual rights in the data by copyright, in much the same way that suppliers of computer software protect their wares.

In the latter half of the 1980s a new phase of innovation in exploiting data started. It was realised that by reapplying the thoughts of ten years before about data structures, real savings could be made in map production, while serving the wider needs of users better. Whereas the production of digital maps was initially a semi-manual process, with all ornamentation applied by hand, software was written in the period 1987–89 to allow large scale map production to be fully automated. Plotting of these maps on electrostatic plotters has put map production at the point of sale, allowing the customer to specify, within limits, how the map should look and what it should contain.

In 1990 the first small scale *Motoring Atlas*, based on 1:250,000 scale data, was published, although the first digital map at small scales (1:625,000) was produced in 1986.

OS does not use any software marketed as a GIS except as a research tool. GIS has not therefore impacted upon the organisation directly to any degree. There are possibilities for use in, for example, data validation, repayment work and small scales data management. However, as more and more of the customers for OS data are using GIS for their own purposes, their needs for data which can be used successfully in GIS systems will cause changes in OS's own perception of the data, away from map production to being a source of spatial information. To assist this current projects are investigating data storage in the short, medium and longer term (to the end of the decade); new methods of marketing and distribution are being investigated and new "added value" products devised. The latter could alter OS's whole emphasis from mapmakers to suppliers of the data underlying navigation or land management system, used for applications as diverse as grass-cutting contract monitoring or traffic modelling.

This change in emphasis, within the constraints (and freedoms) of being a government department, has meant that OS personnel have to be flexible. They have had to move in the last ten years from relatively simple systems little changed from those of the early years of OS in many respects, over to complex computer-based systems using technologies never thought of.

Satellite positioning is now routine, as is the use of electrostatic plotting, desk-top publishing and electronic data transfer. The craft elements of the cartographer—the skill required to draw lines using pen or even scribing tools—are rapidly disappearing. The need to store large scale maps as paper documents is rapidly becoming superseded by production at point of sale. This has created human problems which have largely been overcome by the staff. There is still a tendency for ideas to be ahead of technology, but this can be argued as being no bad thing. The growth in technology in the past decade has been far faster than was thought possible ten years ago. Then a screen which drew light green lines on a dark green background was magic! A four megabyte memory minicomputer was enormous!

What will happen to OS in the next ten (or two hundred) years? One thing is certain, that people will have to adapt to use and exploit the technology available. Up to now, OS staff have been quite prepared to do so.

Part IX

Research trends

With the current spate of activity in GIS research choosing a selection of papers for this particular section of the Yearbook was difficult. The section commences with two papers, by Masser and Goodchild, which look at GIS research in the wider context. Masser's paper examines the products, potential and future of the Regional Research Laboratory initiative funded by the Economic and Social Research Council which has spearheaded a large amount of GIS research in the UK. The contribution by Goodchild extends the research debate even further and looks towards the future and examines if indeed there really can claim to be such a thing as a ''Science of Spatial Information''.

The remaining papers look at four research issues which are considered to be critical in the development of the technology; objected-oriented database design, raster-vector issues, error modelling and user interfaces.

50

Whither the Regional Research Laboratories?

Ian Masser
University of Sheffield

The end of 1991 marks the end of ESRC (Economic and Social Research Council) direct funding for the Regional Research Laboratory Initiative. It is useful, therefore, to reflect on the achievements of the Initiative as it moves into its last year of ESRC funding and to consider what lies ahead of the RRLs once full ESRC funding ends.

The Regional Research Laboratory (RRL) Initiative differs from most other research council initiatives in a number of important ways. The most important departure from standard practice is the emphasis that has been given in the initiative to institution-building. This places a very different kind of responsibility on each RRL by comparison with other research council contracts. The task of the grant-holders in this case is not only to carry out high quality research in the fields of data handling, data-base management, geographic analysis, software development and training and advice, but also to establish regional centres of excellence which are capable of sustaining their own activities.

The resources placed at the disposal of the RRL Initiative, over £2,000,0000, represent a considerable investment on the part of the ESRC. This has made it possible for each of the eight RRLs that are currently in operation to appoint up to three staff for the duration of the main phase of the Initiative from 1988–1991. In some cases the host institutions have provided funds for additional appointments to enable the RRLs to get off the ground. From the outset, however, the RRLs have been heavily dependent on the enthusiasm and energy of existing staff in post. An important indicator of the success of the initiative as a whole is the catalytic effect that RRL designation has had on the existing staff, to the extent that the total manpower resources at the disposal of the initiative during the main phase have generally been between 70 and 80 people.

By the time that this article is published the Regional Research Laboratory Initiative will have been the subject of an exhaustive evaluation by an International Panel chaired by Professor Peter Burrough from the University of Utrecht. Without pre-empting the findings of this panel, there can be little doubt that the initiative has been enormously productive from the standpoint of its research outputs. By the middle of 1990, just over 18 months from the start of the main phase, more than 180 technical papers had been produced by RRL staff and the products of RRL research were already featuring prominently not only in specialist academic journals, such as the International Journal of Geographical Information Systems but also in periodicals such as Mapping Awareness which have a very wide readership among practitioners in the geographic information systems field. The latter journal ran a series of features between December 1988 and April 1990 on the work of the initiative, which have now been reprinted as a separate volume entitled ''GIS and the Regional Research Laboratories'' (Masser, 1990).

Another notable product resulting from the initiative is the 17 minute video entitled ''Mapping the Future'' (ESRC, 1989) which features the work carried out at a number of Regional Research Laboratories. This has been shown at major conferences in Canada, Greece, the Netherlands, Malaysia, Switzerland and the USA as well as in Britain. Given the number of requests for sales and leases that have been received by ESRC it has also proved an invaluable medium for training and raising levels of awareness in public and private sector agencies as well as at educational institutions.

Since the beginning of the main phase of the initiative in the autumn of 1988 special sessions have been devoted to the work of the Regional Research Laboratories at conferences organised by the British Association for the Advancement of Science, the Institute of British Geographers, the Regional Science Association and the Royal Geographical Society. The RRLs themselves have also been very active in organising specialist meetings and workshops to provide a forum for debate at the regional as well as the national level. The proceedings of a number of these meetings are already in the course of publication in book or periodical form (see, for example, Masser and Blakemore, 1991).

Given the impetus that has been built up since the establishment of the four trial-phase Regional Research Laboratories at the beginning of 1987 it is unlikely that the RRLs will disappear from the scene when ESRC direct funding ends in December 1991. Over the last 12 months a great deal of effort has been devoted to agenda-setting activities with a view to identifying key research priorities for the post-1991 period. With this in mind a number of outreach papers have been commissioned to review the current state-of-the-art and define priority research areas in fields such as spatial analysis (Openshaw, 1990), environmental monitoring (Rhind, 1990), organisational databases (Lloyd, 1990), geodemographics (Brown, 1990), education (Burnhill, 1990), technological and natural hazards (Gatrell and Vincent, 1990) and the organisational impacts of geographic information systems (Campbell, 1991). The objective of these papers is to raise awareness regarding the potential opened up by recent technological developments in the data handling field and to stimulate debate within the wider academic community. The future of some RRLs has been assured by their success in obtaining further research grants from the research councils. The decision of the Advisory Board for Research Councils to allocate over £1,000,000 to the ESRC and the Natural Environment Research Council for a three-year research and training programme in Geographic Information Handling provided a welcome boost to some RRLs, particularly in the development of underpinning research projects which straddle the boundaries between socio-economic and environmental research. The RRLs have also been successful in obtaining support for specific projects through the response mode channels in the research councils and they will also be in a strong position when bidding for funds from future ESRC programmes such as the Global Environmental Change and the 1991 Census initiatives.

The Regional Research Laboratories have also enjoyed a considerable measure of success in their efforts to develop collaborative ventures with both public and private sector agencies. One of the most important issues facing the RRLs in this respect has been to build up their credibility with specific agencies and to create strategic alliances which satisfy academic as well as public sector or commercial needs. As a result of these efforts a number of RRLs are currently negotiating projects with private sector agencies under the Department of Trade and Industry supported Teaching Company Scheme. The outcomes of these efforts will not be fully evident until some time after 1991.

The immediate answer to the question posed in the title of this article is that the RRLs are now well-established centres which will continue to operate after direct funding ends in 1991. However, the future shape of the RRLs is likely to depend on a number of factors. Firstly, it will depend to a very large extent on the opportunities that are provided by the ESRC and the other research councils for the development of particular areas of research, including the extent to which long term programme support for such activities, rather than short term research contracts, is forthcoming. Secondly, it will depend on the nature of the strategic alliances that individual RRLs are able to develop with public and privat sector agencies. Given that geographic information systems research is very much an applications-led technology, this will have a profound influence on the direction of future research as well as on the financial viability of the RRL. Finally, the future shape of the RRLs will depend on developments in the institutions within which they are located and the motivations of the tenured staff involved. After 1991 the RRLs will require strong support from the former and a high level of commitment from the latter if they are to build upon their achievements during the main phase of the initiative and keep British research in the forefront of international efforts in the GIS field.

References

Brown, P.J.B. (1990). Geodemographics: a review of recent developments and emerging issues. RRL initiative disc paper 5, Department of Town and Regional Planning, University of Sheffield.

Burnhill, P. (1990). Research on Education: A Place and Time for Schooling. RRL initiative disc paper 6, Department of Town and Regional Planning, University of Sheffield.

Campbell, H. (1991). Organisational Issues and the Utilisation of Geographic Information Systems. RRL initiative disc paper 9, Department of Town and Regional Planning, University of Sheffield.

ESRC (1989). "Mapping the Future: the Regional Laboratory Initiative". Video film, ESRC, Swindon.

Gatrell, A. and Vincent, P. (1990). Managing Natural and Technological Hazards: The Role of GIS. RRL initiative disc paper 7, Department of Town and Regional Planning, University of Sheffield.

Masser, I. (Ed.) (1990). GIS and the Regional Research Laboratories. Department of Town and Regional Planning, University of Sheffield.

Masser, I. and Blakemore, M. (1991). Geographic Information Management: Methodology and Applications, London: Longman.

Openshaw, S. (1990). A spatial analysis research strategy for the Regional Research Laboratory Initiative. RRL initiative disc paper 3, Department of Town and Regional Planning, University of Sheffield.

Rhind, D. (1990). Environmental Monitoring and Prediction. RRL initiative disc paper 8, Department of Town and Regional Planning, University of Sheffield.

51

Towards a science of geographic information

Michael Goodchild
University of California, Santa Barbara

Introduction

It has often been said that GIS is technology-driven; that the interest we see expressed in conferences, software sales, textbooks and magazines is motivated not so much by the ability of GIS to solve real problems, as by a fascination with a field that combines the attraction of maps and geography with novel high technology. In this somewhat cynical view GIS is a technology in search of applications, and the ability to process geographic information and create colourful, impressive products may explain some of its current popularity.

At another level GIS is clearly application-driven, with an industry committed to defining and exploiting specific market niches. Many of the current range of GIS software products were developed for early applications in resource management, and later extended to meet the needs of utilities or municipal governments. Many of these applications emphasize the management of facilities in a spatial context, and rightly see GIS as a problem of providing geographical access to vast numbers of records on customers and facilities.

Neither of these views addresses the relationship between GIS and universities, or the nature of GIS research, or the role of GIS as a scientific tool. Yet the use of GIS raises a host of issues regarding the computerized handling and analysis of geographical data, and at the same time offers tremendous analytical power. In this article I explore these two roles of GIS in science—first as an enabling technology for a range of scientific disciplines, and second as a focus for research on a set of generic issues that together might constitute a science of geographic information.

GIS as a scientific tool

A useful starting point for a discussion of the value of GIS as a tool for science is to compare it to statistical packages, e.g. SAS, SPSS, S, or Systat. All were developed to satisfy a market in the scientific research community for integrated software to perform a variety of forms of statistical analysis. Statistics as a discipline has developed a vast array of techniques, and integration is desirable to avoid duplication of effort, in keeping track of data formats, variable names, etc., and to allow procedures to be linked together easily. Integration at this scale (many of these packages contain literally thousands of procedures) is possible because the techniques share a very simple, flat-file data model.

Initially packages like BMD and SAS were developed and marketed entirely within the scientific community, as general purpose scientific tools. Although the customer base has broadened considerably, there is no similarity to the dependence of the GIS industry on municipal, utility and government markets. The statistical packages are primarily tools for science, and very effective methods of delivering the theory and methodology developed in the parent discipline of statistics. From a marketing perspective, GIS by contrast is a tool for information management, with no obvious parent discipline and with only a minor interest in supporting scientific analysis and modelling. Only a small fraction of GIS installations currently support scientific research, and only a fraction of GIS vendors market actively to the research community.

Against this we must balance the fact that geographical relationships and the geographical perspective are enormously powerful devices in research. A map showing incidence of disease is invaluable in suggesting possible causal factors or processes of contagion. Geographical proximity and adjacency are powerful factors in processes as distinct as consumer shopping behaviour and migration. Maps are valuable ways of visualizing the results of modelling, for example in atmospheric science. Geographical space has a clear role in explanation not only in the earth sciences but also in social science research, in understanding distributions of artifacts or prehistoric settlement patterns in archaeology, or in regional economic modelling or sociology. Ultimately, GIS has a role as a display, analysis and modelling tool in all those disciplines that deal with geographically located information and processes on the earth's surface. Perhaps it may even be that the abundance and importance of geographic information is actually greater than the tabular, numeric information accommodated by the statistical packages—certainly the flat tabular file is a subset of the relational model so popular in current GIS.

What will it take to exploit the potential of GIS as a tool for science, rather than a system for information management? On the hardware side, many GIS vendors have already moved to the scientific workstation as the platform of choice, and UNIX offers enormous power for integrating GIS with more advanced modelling and analysis, and for networking. Workstation capabilities are expanding dramatically, and we can expect to see spectacular improvements in the next few years, particularly in 3D graphics.

However, the story on the software side is more complex, and the scientific market is not currently viewed as particularly significant by many vendors. The market is not large, and academics have an annoying habit of expecting major price reductions, if not outright gifts. GIS courses and programmes are proliferating, but many are oriented to training students for jobs in the GIS industry, rather than to educating future scientists in the use of an important tool. In this sense again there is little comparison with offerings of statistics courses.

To develop the role of GIS as a technology for science we must ensure that vendors see support of analysis and modelling as important. We need institutions, journals and texts that draw out this role for GIS, and courses that provide introductions to GIS and geographical analysis for scientists in the relevant disciplines. One of the critical factors in the early success of SAS was the ease with which specialized methods of analysis could be contributed by scientists and incorporated within the general framework, even though their applications might be limited to a few projects. We have yet to see the same sort of arrangement in GIS, although there is a similarly enormous range of forms of spatial analysis that might be contributed, and GIS could provide a very effective method of delivery.

A science of spatial information

The current range of GIS software and hardware products incorporates an impressive array of technological breakthroughs. Concepts such as the TIN and quadtree are the direct result of GIS research, and are only two among the many innovative ideas to have emerged over the past three decades. Any technologically based field must be constantly supplied with new ideas if it is to thrive, and needs to be supported by an active research and development community.

However, there is a strong feeling at the present time in the GIS community that the most important issues confronting the field are not necessarily technological. The GIS community seems to be converging not around a single, uniform software product (a standard GIS) or a single application, or around the technology itself, but around a set of generic issues that emerge from the technology. Whatever the application or EDP solution, every user of GIS faces the same set of problems in dealing effectively with digital geographical data, and these problems in turn form the agenda for discussion at GIS meetings—the true glue of the GIS community. Some of the more prominent are:

- data capture—how to convert data from raw to digital form in an efficient, cost-effective manner;
- data modelling—how to represent the infinite complexity of the real world in a discrete, digital machine—whether to use raster or vector, layers or objects, how to model complex objects;
- accuracy—how to cope with the uncertainty present to varying degrees in all geographical data;
- volume—how to deal with the fact that demands for geographical data will often exceed the space available for storage;
- access—how to design data structures, indexes and algorithms to provide rapid access to large volumes of geographical data;
- analysis—how to link GIS databases with advanced modelling capabilities;
- user interfaces—how to present the GIS database to the user in a friendly, comprehensible, readily used fashion;
- costs and benefits—how to measure the benefits of GIS information and compare them to the costs;
- impact on organizations—how to introduce GIS successfully into a complex organization.

All of these issues transcend the technology itself, and all of them in one way or another affect the technology's usefulness, whatever the application and whatever the platform. In recent years they have emerged in various guises as the basis of the research agendas of the NCGIA (NCGIA, 1989), URISA (Craig, 1989) and the Regional Research Laboratories (Masser, 1990; Maguire, 1990). Goodchild (1990) has argued that together they constitute a science of geographic information, and that the future of the GIS community lies in recognizing a common interest in geographic information science rather than in the technology of geographic information systems.

Once one begins to see the generic issues that underlie GIS, and transcend the particulars of its technology and its applications, then one can begin to understand how GIS can affect one's view of the world. Traditionally, information about places on the earth's surface has been stored and transmitted in the form of maps, images, text and to some degree sound. The focus of early GIS was on the digital database as a store of maps—maps were the input, the output and the metaphor of GIS applications. But increasingly GIS is seen as a means of access not to maps but to the real world that those maps represent. The purpose of the database must be to inform the user accurately about the contents of the real world, more than about the contents of a source document. A DEM (digital elevation model), for example, should be assessed on its ability to return the elevation of any point on the earth's surface, not the position of an abstract contour line.

GIS has also affected the role of geographic information within an organization. It encourages the notion that geographic information is a commodity that flows through the organization, and that has a value determined by its accuracy, currency, accessibility, etc. In fact it may be the central commodity in some organizations, for example forest resource management agencies. Geographical data needs careful planning and budgeting if it is to be collected and updated on a regular basis, and made accessible to the organization's analysts and decision-makers. Finally, if information is important, then it is rational to use different types of information as the basis for the organization of departments and systems.

In summary, GIS is a diverse collection of interests, software and hardware solutions, and applications. Two software products applied to the same problem (e.g. ESRI's ARC/INFO and IBM's GFIS applied to management of a utility company's facilities) can produce entirely different solutions; and the needs of forest resource management and school bus routing appear to have very little in common. But there is a growing sense that the issues that hold the GIS community together, and produce convergence rather than divergence, are the generic issues of dealing with geographic information—with representing it in a digital computer, and working effectively with it to produce answers to problems.

GIS as a discipline

The current growth of GIS shows no signs of abating and is likely to continue for some time into the 1990s. New magazines are appearing, and existing ones, such as *GIS World* and *Mapping Awareness*, are growing and increasing their circulation. Conferences are numerous and successful, offering workshops on increasingly specialized topics and access to the latest vendor products. New software vendors are entering the market with exciting and innovative products. GIS is finding new applications and strengthening its penetration into existing markets. GIS courses are proliferating at universities and colleges, and are finding increasing interest from students anxious to acquire useful skills.

On the other hand there are increasing signs of diversification, and this trend is likely to continue to strengthen in the next few years. GIS applications such as facilities management fall under the spatial information paradigm, whereas scientific and resource analysis applications fall under the spatial analysis paradigm. The former emphasizes the database and query aspects of GIS, whereas the latter tends to focus on modelling. The split is illustrated by the case of two Canadian companies—TYDAC and GeoVision—one marketing 'spatial analysis systems' with the very successful SPANS product, the other marketing ''geographic information systems'. Within the PC marketplace, there is increasing divergence between products aimed at GIS applications in resource management, facilities management, or market research (compare, for example, PAMAP, TYDAC's SPANS, Facility Mapping Systems' FMS/AC and Strategic Mapping's ATLAS*GIS).

This trend to diversification is appropriate and rational, as it matches software and platforms with different functions and applications. The complex modelling and analysis of resource management require a very different solution from intensive digitizing or the management of large facility inventories. In time, we can expect this trend to lead to more and more specialization within the GIS industry, as it becomes less and less possible to offer a single software solution for all platforms and all applications. One vendor may specialize in digitizing stations using PCs, another in database maintenance using large mainframes and terminals, another in spatial analysis using advanced personal workstations, and another in 3D applications.

Despite this sense of growing diversity in the GIS community, there is evidence of convergence. The past few years have seen the emergence of several series of conferences aimed at the full GIS community. In the US, the annual GIS/LIS series sponsored by a consortium of five societies (AAG, ASPRS, ACSM, AM/FM and URISA) has grown quickly to over 3,000 attendees. In Canada, the Ottawa meetings in early March have been similarly successful. The lone textbook of 1986 by Peter Burrough (*Principles of GIS for Land Resources Assessment*, Oxford) has now been joined by several others (e.g. Aronoff's *GIS: A Management Perspective*, WDL, Ottawa; Star and Estes' *Introduction to GIS*, Prentice-Hall), and many more are on the way. New organizations have appeared (the Association for Geographic Information seems to be a particularly successful example), and GIS now has its own journals.

All of these would be recognized in the sociology of science as symbols of an emerging scientific community—in short, a discipline. But unlike physics or biology, GIS has no fundamental problems to solve of the magnitude of the origins of the universe or the basis of life. One view holds that it is merely a tool, and that the GIS research community must wither away as the tool reaches perfection. Another, presented at some length above, holds that there are fundamental issues in GIS, not so much in the tools themselves as in the use of the tools. Or perhaps GIS is like statistics—a tool to most scientists, but a set of fundamental research problems to its parent discipline.

If GIS is a discipline, then it is clearly not 'owned' by any traditional one. Geography, cartography, surveying, photogrammetry and engineering have all been accused from time to time of trying to dominate GIS, but with little success, as GIS is fundamentally an interdisciplinary field. Whether it develops the institutional structures of a discipline in its own right, like statistics, or continues as an interdisciplinary consortium of interests like remote sensing remains to be seen.

Prospects for the future

There seem to be two contrasting views of the prospects for GIS in the coming decade. The first is negative and the second positive, and it seems more likely that the second will prevail. However there are actions that can be taken to strengthen the odds.

In the negative view, GIS will fragment and disappear, and by the end of the decade will be nothing but a memory. Geographers often draw a parallel between GIS and the introduction of quantitative methods to geography in the late 1960s (Taylor, 1990), and comment on the lack of interest in quantification, at least in human geography, in the 1980s. GIS will fragment because it is too loose to hold together, and because the glue is too weak and abstract. Users of IBM's GFIS, Intergraph's TIGRIS and Map/Info will cease to see any reason to attend the same conferences. The consortium of five organizations responsible for the North American GIS/LIS conference series will break up and each will concentrate on its own agenda. GIS will be seen as the Edsel of EDP, too awkward, complex and expensive except in some specialised applications.

In the positive view, the GIS consortium will continue to converge. A constant supply of better tools seems assured, particularly in computing speed, software integration, network communication, graphics and storage capacity. The infrastructure of the GIS community will continue to improve, with better magazines, organizations, textbooks, meetings, and all of the symbols of an emerging specialty. Less assured but essential is a constant supply of new players in the industry, since the pattern has been that they are the source of a disproportionate share of technological innovation. New players such as Prime/Wild with System/9, Small World Systems, or Caliper's GIS-PLUS bring new ideas to the industry.

In the positive view, the public agencies will promote and develop standards for data exchange formats, structures, models, and data quality. Training and education programmes will develop through cooperation between vendors and institutions, and lead to the emergence of a strong set of core concepts. Funds will be available through cooperative agreements to support the development of teaching facilities, and to ensure that these keep pace with developments in the technology.

The results of research currently under way will emerge in improved products. Of particular significance will be:

- data models to handle 3D and time dependence, and complex interactions between objects;
- support for complex analytical applications, including tracking of data lineage, tools for visual interaction with the stages in the analysis process, and propagation of uncertainty;
- support for quality assurance and quality control (QA/QC) especially in GIS applications where litigation is a constant problem;
- support for multiple media—unstructured images, both digital and NTSC, text and sound;
- integration of GIS with the capabilities of GPS for data collection and compilation;
- tools for visualizing 3D and time-dependent data;
- tools for data compilation, particularly in 3D;
- improved techniques for conducting functional requirements studies, evaluating costs and benefits, benchmarking and other aspects of the GIS acquisition and project management process.

Finally, the GIS community will converge around a common concern not only for the technology of GIS, but more importantly for the common issues that transcend the technology and pervade all of its applications. GIS can survive by constantly developing new and exciting capabilities, or by constantly finding new applications, but the really fundamental issues in GIS are those that are common to all users of geographic information—how to capture a complex and dynamic world in a digital database and provide access to it in a useful, accurate and cost-effective manner.

References

Craig, W.J. (1989). URISA's research agenda and the NCGIA. *Journal of the Urban and Regional Information Systems Association*, **1**(1): 7–16.

Goodchild, M.F. (1990). Keynote address: spatial information science. *Proceedings*, Fourth International Symposium on Spatial Data Handling, Zurich, *1*: 3–14.

Maguire, D.J. (1990). A research plan for GIS in the 1990s. In Foster, M. and Shand, P.J. (Eds) *AGI Yearbook* 1990, London: Taylor & Francis, 267–277.

Masser, I. (1990). The Regional Research Laboratory initiative: an update. In Foster, M. and Shand, P.J. (Eds) *AGI Yearbook* 1990, London: Taylor & Francis, 259–263.

NCGIA (1989). The research plan of the National Center for Geographic Information and Analysis. *International Journal of Geographical Information Systems*, **3**(2): 117–136.

Taylor, P.J. (1990). Editorial comment: GKS. *Political Geography Quarterly*, **9**(3): 211–212.

Based on the First Midlands GIS Lecture delivered at the University of Leicester, October 6th 1989, and sponsored by The Association for Geographic Information, The Midlands Regional Research Laboratory, and The Institute for Engineering, Surveying and Space Geodesy.

52

An object-oriented future

Hilary Hearnshaw
ESRC Midlands RRL, Leicester

The object-oriented approach

The last twelve months or so have seen a growth in interest in the applications of the object-oriented approach in the world of geographical information science. This interest reflects similar, developing, interest in the principles of object-orientation in computing science in general. "Object-oriented" is a descriptive label which has been applied to a range of activities, including:

- *Object-oriented data modelling* (OODM). This is the abstraction of the real world to a form appropriate for use by a digital system, which allows the properties of an object and its operators to be associated together (encapsulated), and for specified passing of properties between objects (inheritance). This allows a closer similarity in the model to the complexity of the real world than previous techniques for data modelling, such as entity-relationship modelling, are able to do. They are not able to include the required detail as easily as object-oriented data models can.
- *Object-oriented programming languages* (OOPLA). Languages such as Smalltalk, C+ +, Actor, are high level languages which allow the programmer, among other things, efficiently to represent objects which not only contain their own internal data model, but also contain their own operational functionality.
- *Object-oriented database management systems* (OODBMS). These are being developed in response to the need to handle highly complex data efficiently. Previous relational, or hierarchical, databases were failing to do this. OODBMS are more suitable than previous DBMS in such fields as multi-media databases, complex engineering databases or geographical information systems (GIS). Attempts at implementing successful object-oriented data modelling techniques also motivated the development of appropriate database management systems.

As the development, and evolution, of object-oriented techniques continue, and there is no question that they will, so the application to geographical information processing will grow. It is, of course, realised that neither object-oriented data modelling, nor object-oriented database management systems, will solve all, or even most of, the problems in geographical information science. However, object-oriented techniques seem to provide more appropriate solutions than previous attempts, in the requirement for handling, and using, large amounts of spatially referenced data which require complex data models (Egenhofer and Frank, 1987; Maguire *et al.*, 1990). We consider next just what are the distinguishing features of an object-oriented database.

Requirements of an object-oriented database

Atkinson and his colleagues, from the ALTAIR project (Atkinson *et al.*, 1989), have drafted a manifesto for the required characteristics of an object-oriented database management system, which they hope will lead to a formal definition. This is something which, so far, is lacking.

Mandatory features of an OODBMS should include the following:

Support for	complex objects	—built from simpler objects by set, list and tuple;
	object identity	—the object's existence is independent of its value;
	encapsulation	—both data and operations are in the object;
	types or classes	—either concept to be available;
	hierarchies	—to allow inheritance between objects.
Presence of	overriding	—redefinition of operators according to type;
	overloading	—multiple application to types of operator name;
	late binding	—run-time translation of operators, computational;
	completeness	—any computable function can be expressed.
Provision of	extensibility	—both system- and user-defined types;
	persistence	—survival of data over process execution;
	concurrency	—harmonious coexistence of processes;
	recovery	—return after hardware or software failure to secondary storage;
	management	—for invisible handling of very large databases.
	A simple way of querying data	—high level, efficient and independent of application.

Relational DBMS provide only the last five of these.

Optional features of an OODBMS can include the following:

Support for	multiple inheritance	—from more than one object;
	type checking	—at compile time;
	distributed systems	—over multiple nodes;
	design transactions	—for customised use;
	versioning	—managing alternative versions of data.

These are not necessarily a minimal set of characteristics, and other authors have provided more succinct descriptors of object-oriented systems. For example, Yokota (1989) writes "an OODBS is just a database, based on an object-oriented paradigm, the intention of which is natural modelling of entities in the real world, including behaviours. The paradigm is independent of specific problem domains and is characterised by encapsulation and autonomy".

Available object-oriented databases

Implementation of truly object-oriented DBMS has been slow getting started. However, there are many attempts being made to develop them, either by extending and modifying an existing relational DBMS, or by starting from scratch.

One attempt at *ab initio* development is by the ALTAIR consortium's project, in France. An OODBMS, called O2, has been designed, prototyped, redesigned and re-prototyped by this team (O. Deux, 1990). It is claimed that O2 has DBMS properties: persistence, disk management, sharing and query language; as well as OODBMS properties: complex objects, object identity, encapsulation, typing, inheritance, overriding, extensibility and completeness. Thus it fits the mandatory requirements above, to be a true OODBMS. It should be noted that the manifesto for OODBMS is also published from the ALTAIR project, and so the team, using the pseudonym "O.Deux" is at least satisfying its own suggested criteria. In O2 the user can define both objects, which have identity and encapsulate data and behaviour, and values which are assigned to objects such that objects are (identifier, value) pairs.

Also starting from scratch on the database side, Servio Logic Development Corporation have produced the GEMSTONE OODBMS. It is based around an Object Storage Manager and supports a database language OPAL, which is an object-oriented language, developed from Smalltalk. OPAL provides for concurrency control and data integrity maintenance, in this multi-user system. All the mandatory aspects of an OODBMS seem to be provided by GEMSTONE.

An example of an OODBMS developed by extending an RDBMS is POSTGRES, which is an extension of the familiar RDBMS, INGRES, so that objects can be handled. The POSTGRES query language is an extension of QUEL (the INGRES query language). POSTGRES satisfies some, but not all, of the ALTAIR project's requirements.

Object-oriented GIS

Morehouse (1990) claims that there are as yet no truly object-oriented GIS. He says that "To be truly object-oriented, a system must support the development of a specific geographic data model from a generic one by means of inheritance and encapsulation". However, there are manufacturers who would claim that their product is a truly object-oriented GIS, for example Intergraph's TIGRIS (Topologically Integrated Geographic Resource Information System; Herring, 1990). It is claimed that the TIGRIS system is based on a data model which is object-oriented and an extension of the standard relational model. The extensions include: the existence and persistence of the object beyond the value of its attributes; relations based upon objects; and extensibility of the query language. The Object

Manager is the element of the system which is responsible for the creation, deletion and linking together of the objects defined. TIGRIS contains a two-dimensional topological model comprising nodes, edges and faces. All topological relations between these can be derived from explicitly stored primitive relationships. Thus, TIGRIS provides handling of geographic data, based on topological models, as well as non-spatial data, using the characteristics of an OODBMS and providing the functionality of a GIS.

A successful compromise approach has been taken by Smallworld Systems (Chance *et al.*, 1990) who have created a system which marries a relational (tabular) database with an object-oriented language, thus getting the best from both worlds. This maintains the interactive features of a relational database while also allowing interactive program development and customisation. While not claiming to be a fully object-oriented GIS, the creators feel that their object-oriented front end to a relational database can bring the advantages of object-orientation to previously existing data stores, thus not losing the resources already committed to such data storage.

Areas for research

As can be observed from this chapter, there is no general agreement yet about how object-orientation can be defined. Furthermore, the term is sometimes used in an inappropriate manner by manufacturers, in order to promote the sale of their systems. The computer scientist Roger King once remarked that the way to sell your cat was to advertise it, not as cuddly or nice to humans, but as object-oriented. This kind of sales hype has done the object-oriented community some harm, in encouraging misunderstanding, and hence misuse, of the term.

One of the nice features of relational systems is the elegant formal model which stands alongside them. This is probably the one single feature which has led to their growth and widespread use. Unfortunately, there is no such formal model given in the field of object-orientation. Thus, a first priority research problem is the provision of such a formal model of what an object-oriented system is (and is not). Moving from the general field of object-orientation in all its aspects, to the specialist area of spatial systems, a connected further research area is the provision of a generic model of spatial information, for example Egenhofer *et al.* (1990) or Worboys (1990). Using an underlying structure taken from the field of algebraic topology, it is possible rigorously to define the complete set of primitive spatial objects (point, line, area, etc.), but a complete set of methods associated with these classes and describing spatial relationships (nearness, containment, etc.) has yet to be given.

Another research area which is already being tackled relates to the user interface to an object-oriented system when spatial information is involved. It appears that a more fundamental approach is required for spatial systems than just adapting general purpose interfaces to other object-oriented systems. This is, of course, also true for non-object-oriented spatial systems. Both iconic interfaces and query language interfaces need to be investigated. The SQL (Structured Query Language) Working Party, set up by the AGI, has worked on recommendations on how SQL can be extended in a sufficiently meaningful way to handle spatial data.

In trying to look further ahead into the future, we see the possibility of adding temporality to an object-oriented spatial system. There is a distinction between 'system time' and 'real world time', and it has been recognised for a while that there are special problems for GIS in handling system time. Smallworld Systems have tackled the question of handling long duration transactions by using alternative versions (Easterfield *et al.*, 1990). Work is in progress to provide a formal model of objects which include both spatial and temporal aspects (Langran, 1989; Worboys 1990). The inclusion of time features in an object-oriented spatial system is thus a current research issue.

Conclusion

There is clearly a future for object-oriented techniques in many areas of computing. OODBMS is one of those areas, where development of systems which can be successfully implemented will be of major benefit. In particular the world of GIS will benefit from further research and development of object-oriented systems, as this seems to offer the most appropriate vehicle for handling the complexities of spatial information. Until superseded, OODBMS and object-oriented GIS may be supplying the needs of many users and developers. The urgent needs are, on the theoretical side, for both a formal model of object-orientation and an agreed definition of an OODBMS, while on the application side the need is for a successful implementation of an OODBMS and a fully object-oriented GIS.

References

Atkinson, M., Bancilhon, F., DeWitt, D., Dittrich, K., Maier, D., Zdonik, S. (1989). The Object-Oriented Database System Manifesto. ALTAIR Technical Report 30–89.

Chance, A., Newell, R., and Theriault, D. (1990). An Object-Oriented GIS—Issues and Solutions. *Proceedings*, EGIS'90, Amsterdam, **1**, 179–188.

Deux, O. (1990). The Story of O2. *IEEE Transactions on Knowledge and Data Engineering*, **2**(1), 91–108.

Easterfield, M., Newell, R., and Theriault D. (1990). Version Management in GIS—Applications and Techniques. Proceedings, *EGIS'90*, Amsterdam, **1**, 288–297.

Egenhofer, M. and Frank, A. (1987). Object-Oriented Databases: Database Requirements for GIS. *Proceedings of International GIS Symposium: The Research Agenda*, Washington, 2, 189–211.

Egenhofer, M., Frank, A., and Jackson, J. (1989). A Topological Data Model for Spatial Databases. *Proceedings, 1st Symposium S.S.D., "Design and Implementation of Large Spatial Databases"*. Santa Barbara, 271–286.

Herring, J.R. (1990). TIGRIS: A Data Model for an Object-Oriented Geographic Information System.

Kim, W., and Lochovsky, F.H. (1989). *Object-Oriented Concepts, Databases and Applications*. New York: ACSM Press.

Langran, G., (1989). A Review of Temporal Database Research and its Use in GIS. *International Journal of Geographical Information Systems*, 3(3), 215–232.

Maguire, D.J., Worboys, M.F., and Hearnshaw, H.M. (1990). An Introduction to Object-Oriented Geographical Information Systems. *Mapping Awareness*, **4**(2) 36–39.

Morehouse, S. (1990). The Role of Semantics in Geographic Data Modelling. *Proceedings, 4th International Symposium on Spatial Data Handling*, Zurich,

2, 689–697.
Worboys, M.F. (1990). The Role of Modal Logics in the Description of Knowledge in a GIS. NATO Advanced Study Institute, Spain.
Yokota, K. (1989). What is Expected of an Object-Oriented Data Model? In Ritter, G.X. (Ed) *Information Processing 89*, Amsterdam: Elsevier.

53

The raster–vector issue revisited

David Mitchell and Neil Stuart
Department of Geography, Edinburgh University

Introduction

Within GIS few debates have been as long running as the debate over whether raster or vector representations of geographic phenomena are the most appropriate. Today the debate appears in various forms, from the academics writing in journals through to the vendors on the sales floor, each extolling the relative merits of their own implementations.

In this article, we look briefly at the raster and vector models, showing how they are in essence dual representations of space. From this premise we argue that, fundamentally, any operation in a GIS may be performed using either model. Turning then to the present state of the art in GIS, we draw conclusions as to the motivation and merit in attempting to bridge the 'raster–vector divide'.

What geographic entities do we need to model?

Essentially the geographic world, as represented in a geographic information system, consists of a series of spatial data and associated attribute data. The spatial data are either points, lines or polygons. For example, a GIS which was designed to monitor the impact of pollution in a hydrological system might have sources of pollution represented as points, rivers as lines and lakes as polygons. The attribute data are items such as cadmium concentration, flow rate, or total volume of water, which are associated with spatial phenomena. The raster–vector debate concerns whether raster or vector data models are better at representing the spatial data in which we are interested.

However, it is not only the capability of each model to represent geographic entities which is in question; it is also the suitability of each of the models for the variety of spatial operations that we would wish to carry out. Both raster and vector data models are able to handle these basic geographic entities and permit analysis of the relationships between them. For example, the operation of polygon overlay, one of the most common GIS operations, can be carried out using either raster or vector data, although polygon overlay is inherently easier with raster data.

What are rasters and vectors?

Before progressing further we should define what we mean by the terms raster and vector. Both raster and vector are best thought of as spatial data models; ie. they are both ways of conceptualising the real world and enabling us to represent that real world within a GIS.

The raster model is usually presented as being simpler than the vector model (see for example, NCGIA, 1990). The grid square representation is the most common presentation used to explain raster data (Aronoff, 1990). A value is associated with each of the grid squares, which denotes the spatial entity to which that grid square belongs.

It should be pointed out that the raster model is by no means tied to the geometric grid. Raster data can be equally well represented by other shapes which form part of a regular tesselation. For example, the equilateral triangle could and has been used. However, grid data are the most common raster representations, primarily through their association with data capture mechanisms such as scanners or satellite recorders.

The vector model positions each geographic feature precisely within continuous space. Each coordinate is assumed to be mathematically exact and space is conceived of as infinitely divisible. The vector model is often contended to be more elegant, in that vector graphical output more closely approximates to conventional cartographic work than its raster counterpart. Vector data represents geographic reality through features which are either points, lines or polygons. A point is recorded as a single XY paring, a line is a stream of such pairs and a polygon is a closed geometric object the boundary of which is a series of XY coordinate pairs.

Vector data, unlike its raster equivalent, has a much better inherent ability to represent topological relationships between geographic entities. As well as representing the basic geometric data the vector model is often extended to include topology. The structuring of data into points and lines enables topology to be explicitly stored, or to be implicitly calculated where necessary.

Raster and vector data models thus have radically different ways of representing the same geographical reality. As Aronoff (1990) explains.

> "In both models, the spatial information is represented using homogeneous units. In the raster approach, the homogeneous units are the cells [while] in the vector approach, the homogeneous units are the points, lines and polygons" (p. 165).

In other words the models are logical duals of each other and each has its own advantages and disadvantages: neither is inherently superior *per se*.

Spatial analysis in raster or vector mode

The fact that the raster-vector issue continues to arouse such interest is witnessed by a recent conference dedicated to the issues of "GIS Design Models and Functionality" (Anon, 1990). The meeting attempted to demonstrate, at the generic (non-proprietary) level, certain classes of functionality that are inherently easier for a GIS using one type of model than the other.

For example, the vector model lends itself to analysis that uses the explicit description of topology through node and arc linkages, such as linear network analysis. The raster model, where topology is implicit and if needed must be recalculated, finds network analysis very difficult. On the other hand, the vector model requires a large number of floating point operations to determine intersections of arcs which have been overlain. Overlay analysis in raster mode is a trivial operation, since it reduces to a logical comparison of the values (colours) present in each theme, on a pixel-by-pixel basis. The above example illustrates the vector–raster divide at one of its widest points. Note, however, that it is possible to do either overlay in vector mode or network analysis in raster mode, should one wish to. In fact, any spatial operation can conceivably be carried out in either mode. This stems from the models being essentially logical duals of each other. For example, rotation, translation, reclassification, reselection, buffering and neighbourhood analysis are equally possible in either raster or vector GIS. Each system scores over the other for some of the time.

Into the real world

At present the systems provided do not have the capability to offer us full functionality in either model. Partial solutions are provided which exploit the particular benefits of the hardware to gain a competitive advantage. This means that GIS vendors are optimising their systems for a subset of all possible queries. Purchasers are learning by experience to define carefully their requirements in order to avoid selecting an inappropriate solution. Consideration needs to be given to.

(1) What is the source of majority of input data?
(2) What will be the most common types of analytical operation?
(3) What types of output are required?

We believe that, just as (3) is now becoming less of a constraint with improvements in raster output devices, so (1) and (2) in future will become less vexed issues as system software matures and overall power rises. For example, some vendors are now claiming that the vector–raster divide is irrelevant. They argue that high powered vector graphics workstations can optimally deal with the floating point operations, whilst throwing raw computer power to 'crunch' raster arrays.

From a different perspective, Maffini (1988), speaking as a designer of raster-based systems, argued that he expected to see the traditional dominance of vector GIS eroded in certain areas where the raster model can offer significant advantages for the majority of operations.

One may infer from these two viewpoints that the vector–raster issue may not be laid to rest in the near future, at least not by the vendors.

Conclusions

Our argument is that the raster–vector debate is one which should be a side issue for most GIS practitioners, who should focus upon the analytical functions offered by various GIS rather than upon the technicalities of their underlying data models.

If we were to encapsulate our argument in a single phrase it would be "A plague on all your houses". The implication is that we, as GIS professionals, should be able to carry out the analyses which we require irrespective of the underlying architecture of the system or package that we are using.

References

Anon (1990). "GIS Design Models and Functionality", Proceedings of a Conference held at the Midlands Regional Research Laboratory, Leicester, 21–22 March 1990.

Aronoff, S (1990) *GIS: A Management Perspective*, Ottawa: WDL Publications.

Peuquet, D. (1984). A Conceptual Framework and Comparison of Spatial Data Models. *Cartographica*, **21**(4), 66–113. (Reprinted in Peuquet, D. and Marble, D. (Eds), Introductory Readings in GIS, London: Taylor & Francis.)

Maffini, J. (1988) Raster and vector encoding: A commentary. *Photogrammetric Engineering & Remote Sensing*.

National Centers for Geographic Information and Analysis (NCGIA) (1990). *Core Curriculum for GIS: Introduction* (Vol. 1).

54

Error modelling in GIS: Who cares?

Steve Carver
University of Newcastle-Upon-Tyne

The current situation regarding errors in GIS

At present no proprietary GIS software provides any procedures for handling input data uncertainty, nor does it allow the effects of error to be tracked as it propagates through a sequence of GIS operations. To put it bluntly, a key section of the GIS tool kit is missing. Worse still, neither users nor vendors seem to be particularly concerned about it. Yet this problem affects all GIS applications since the presence of some level of uncertainty is a feature of all geographic information. The problem facing the concerned user is that currently it is extremely difficult to assess its magnitude and whether it really matters (a fact compounded by the data and application-specific nature of the error problem). Meanwhile, the ordinary users of GIS continue to ignore the problem mainly because of the complete absence of any error handling and/or analysis functionality.

Many researchers working in the GIS field, including those in the North East Regional Research Laboratory, view the problem of error very differently. Error propagation may be seen as the 'Achilles heel' of GIS and it is imperative that mechanisms are found for handling rather than ignoring the problem. Without error handling functionality, GIS will remain a fragile and potentially vulnerable technology which might not always yield reliable results. This provides no confident basis for progressing GIS through the 1990's. It is important to identify those particular circumstances under which error propagation becomes a problem and devise methods for coping with it. Raising the level of end-user awareness is another aspect of the problem that has to be tackled but cannot be approached satisfactorily until there is a solution in sight.

The problem of error propagation is not conceptually difficult to understand. It is well known that there are many sources of error in digital map data. These are well documented by Burrough (1986) and are listed in Table 1. There is nothing new about many of these. Error and uncertainty have always been a feature of cartographic information. Previously, however, the worst effects were largely avoided by a combination of the expertise of the cartographer, who knew about the inherent generalisation implicit in analogue maps, and the difficulty of manipulating maps by manual means which precluded most forms of analysis likely to be sensitive to the effects of error. This has, in the pre-GIS era, reduced the impact of pushing geographic information beyond the limits which it could reasonably sustain. The advent of practical GIS has provided the user of geographic information with virtually unlimited freedom for map manipulation. GIS output resulting from the mixture of data taken from maps of differing scales can possess considerable artistic appeal, but may be considered lacking from a more rigorous scientific viewpoint. In extreme cases the dangers are easily identified but what happens when GIS are used in a careful and considered manner? What impacts will data uncertainty have on the results and on the performance of decision-support systems? Alas, there are no tools available for answering such questions. It is, however, reasonable to suggest that the effects of error propagation will not be negligible.

Table 1. Typology of errors in GIS

Obvious sources of error:	age of data areal coverage map scale density of observations
Errors resulting from natural variations or from ordinal measurements:	positional accuracy attribute uncertainty generalisation
Errors arising through processing:	numerical computing errors faulty topological analyses errors in interpolation

Not surprisingly, various aspects of the error propagation issue have been the subject of academic research. However, it is suggested here that whilst it is important to seek to develop an improved level of theoretical and empirical understanding of the nature of the various types of error affecting GIS, it is envisaged that such a task may take several years. There is also some risk that a full and clear understanding of all aspects of this complex and multifaceted problem may never actually materialise, and even if it does it may well not be of much practical use to the end-user. It is equally important to provide users with a practical error handling/analysis technology now so that some, however crude, estimates can be made of the effects of error on GIS output. At the very least it should offer a measure of result robustness, at best something analogous to confidence intervals or credibility regions (although it is recognised that high degrees of accuracy are likely to be unattainable at present).

A Monte Carlo simulation approach?

Openshaw *et al.* (1989), Charlton *et al.* (1989) and Brunsdon *et al.* (1990) describe a Monte Carlo simulation approach to handling error propagation in GIS. The procedure adopted is conceptually simple, but presents a number of operational difficulties. Briefly, the Monte Carlo approach simulates the effect of input data uncertainty by the addition of random 'noise' to the map data. Each input data source for an arbitrary sequence of GIS operations is assumed to be characterised by an error model which represents reasonable estimates of the levels and nature of the data uncertainty thought to be present. This allows the map data to be 'perturbed' using probability distributions of known form and parameters. A single simulation is made by generating random numbers from these distributions and adding them to the observed geographic coordinates of the input data. The randomised or 'wobbled' input maps are then subject to the required sequence of

GIS operations and the results saved. This entire process is repeated *n* times (where *n* may be about 100). If the output is entirely numeric, then the distribution of *n* results gives some indication of the effects of input data uncertainty. If the result is a map, then the total set of *n* output maps can be used to draw confidence intervals or 'credibility regions' which can be overlaid on the deterministic results as a visual indication of the effects of uncertainty. The method described above has been tested using a vector GIS (ARC/INFO), but is equally applicable to raster-based systems.

There are four areas of difficulty associated with the Monte Carlo approach. These are:

(1) knowing what error model to apply and how to parameterise it;
(2) the 'wobbling' of line segment data either has to be constrained to avoid altering the map topology or else a new topology created without the need for manual intervention;
(3) the two magnitude increase in processing time required; and
(4) the difficulty of tagging outputs with error information if they are to be used in further GIS operations.

None of these problems is insurmountable and if overcome this approach may represent a universally applicable procedure for assessing some of the uncertainty in GIS outputs.

The epsilon alternative

The Monte Carlo approach described above is considered to present some difficult problems and could only be justified if there were no alternatives that would be easier to operationalise within existing GIS environments. With this in mind, Brunsdon *et al.* (1990) and Carver and Openshaw (1990) define a probabilistic epsilon error approach (PEE). The epsilon model defined by Perkal (1956, 1966) and Blakemore (1984) is well known as a means of generalisation and conversely describing the effects of cartographic generalisation on the accuracy of GIS overlay operations.

Perkal used a buffer distance (epsilon) about a cartographic line as a means of objective generalisation, whereas Blakemore reverses the concept to indicate an error band about a digitised line. The latter can be applied to polygon overlay operations to ascribe descriptive levels of uncertainty in the resulting map. Recently this method has been placed in a formal probabilistic framework by Brunsdon *et al.* (1990). The basic Perkal-Blakemore epsilon model is enhanced by specifying a function describing the probability of containment within the epsilon buffer distance. The true shape of the probability function remains to be determined, but one possible suggestion is that it approximates to a piecewise quartic function.

The probabilistic epsilon error approach or PEE has a number of advantages. These are:

(1) it does not involve excessive amounts of processing time;
(2) it uses only a combination of existing GIS tools;
(3) each line segment can have a different error model associated with it depending on its individual characteristics; and
(4) the probabilistic framework can be applied to all GIS operations likely to be performed on uncertain data.

The PEE approach seems to offer a useful way forward that can be implemented within both vector and raster GIS and without having to use external routines, only existing parts of the GIS tool box.

Enhancing the ARC/INFO tool kit

The following discussion outlines how the PEE approach might be operationalised within the ARC/INFO GIS and thus provide a viable technology for handling error propagation within a proprietary system. It is suggested that ARC/INFO is the most suitable development platform given its extensive range of functionality and its wide distribution in both the academic and commercial sectors. Experiments have already shown that a PEE approach can be programmed using the ARC/INFO macro language (AML) and existing ARC/INFO commands with no need for additional external routines. Indeed it is possible to modify all the required point-line-polygon overlay combinations in GIS using this approach. So far, four new ARC/INFO commands have been created by applying the PEE approach to the boolean logical operators AND, NOT, OR and XOR. Input, dialogue and output from these new commands (PAND, PNOT, POR, PXOR) follows the ARC/INFO model. A novel feature of these particular commands is the ability to specify the directional bias of the error model for each digital map input. This is illustrated in Figure 1 for a simple AND overlay using venn diagrams. Figure 2 shows a more complex example describing the results from an area screening exercise for radioactive waste disposal sites in Britain (Openshaw *et al.*, 1989). Both the PAND and POR commands are utilised here to identify areas of the country which simultaneously satisfy all the siting criteria based on geology, population distribution, accessibility and conservation issues. Overlays are performed to a 95 per cent probability level with exclusive bias. The feasible areas shown in Figure 2(b) have a 95 per cent probability of

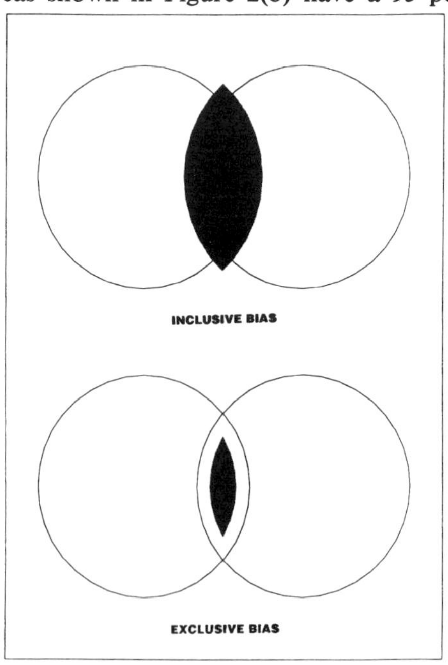

Figure 1. Venn diagrams showing effect of bias on boolean AND overlay.

Figure 2. Identifying potentially suitable areas for radioactive waste disposal: (a) results from normal sequence of ARC/INFO overlay commands; (b)results from PEE-based ARC/INFO commands; (c) areas 'eroded' from (a) by applying the PEE approach in (b).

being within the correct area assuming the error model and the epsilon distance are correct. These can be compared to the results from a normal overlay sequence using the same data and standard ARC/INFO commands in Figure 2(a). Figure 2(c) shows the areas where the two sets of results differ.

Other ARC/INFO commands (CLIP, ERASE, IDENTITY, INTERSECT, NEAR, POINTDISTANCE, UNIO and UPDATE) could be modified in a similar manner using the PEE approach. This is outlined in more detail by Carver and Openshaw (1990).

The way forward?

As regards error in GIS, the new decade is likely to be the beginning of a new era of research, user awareness and the availability of practical error handling/analysis tools. Research at Newcastle University and elsewhere has already gone some way towards providing practical, operational, relevant and generally applicable error handling technology for GIS. Given suitable encouragement and the cooperation of vendors and users, first generation tools could soon be available for general use and a key deficiency in the GIS tool kit remedied. It is important that developments in providing error handling routines for GIS users and developments in our understanding of the theoretical and empirical nature of error propagation in GIS go hand in hand. Whilst certain error handling functionality can be made available within the year (given vendor cooperation), it is likely that the error models initially used will be rather crude and an over-simplification of reality. As our understanding of the error propagation process increases, then error handling routines can be improved, thereby providing the end-user with steadily more and more accurate error estimates for their GIS outputs. As a result, confidence in GIS solutions and their acceptance in all sectors of the economy, government and academia can only increase.

Bibliography

Blakemore, M. (1984). Generalisation and error in spatial data bases. *Cartographica* 21, 131–139.

Brunsdon, C., Carver, S., Charlton, M. and Openshaw, S. (1990). A review of methods for handling error propagation in GIS. Proceedings, *1st European Conference on GIS*, Amsterdam, 106–116.

Burrough, P.A. (1986). *Principles of Geographic Information Systems for Land Resource Assessment*. Oxford: Clarendon Press.

Carver, S. and Openshaw, S. (1990). Handling error propagation in GIS: a proposal for enhancing the ARC/INFO tool kit. NE.RRL Report, University of Newcastle upon Tyne.

Charlton, M., Openshaw, S. and Carver, S. (1989). Monte Carlo simulation of error propagation in GIS. Paper presented at *6th European Colloquium of Theoretical and Quantitative Geography*, Chantilly, France.

Openshaw, S., Carver, S. and Fernie, J. (1989). *Britain's Nuclear Waste: Safety and siting*, London: Belhaven Press.

Perkal, J. (1956). On the epsilon length. *Bulletin de L'Academie Polonaise des Sciences*, **4**(7), 399–403.

Perkal, J. (1966). On the length of empirical curves. Discussion paper 10, Ann Arbor, Michigan Inter-University Community of Mathematical Geographers.

55

"What have I done now?"

David Medyckyj-Scott
ESRC Midlands RRL, Loughborough

GIS user interfaces

With increasing frequency one hears the assertion that user interfaces are an important component of GIS while vendors talk about their systems being 'user friendly'. The reality is that while GIS have come a long way in the last decade they are still difficult to use. True, the techniques of user interaction have progressed considerably, from command line input to graphic user interfaces with direct object manipulation, so that today the user has greater control over the computer environment than ever before. However, GIS still make considerable demands on users: they are conceptually complex to understand; the range of functions on offer can be daunting and interaction (particularly the query languages) can be non-intuitive and fail to map onto the types of tasks the user wants to perform and the way he thinks. Current systems are also very primitive with regard to the questions that users can ask of them. Some are very user unfriendly with users spending hours learning the use of the system rather than producing useful output. Such difficulties can actually limit the use of the GIS with users failing to explore the functionality on offer in the rest of the system. In addition, we are beginning to see a different type of GIS user appear, one who is less technically orientated and has neither the time or the inclination to learn what appear such complex systems.

A user interface has at least two roles within a GIS (Healey and Gittings, 1989). Firstly, it is the interface a particular software package presents to a user, so the user is made aware of the operations provided and can employ them. Secondly, it is a facilitator by which a user can analyse the spatial databases available. Most GIS user interfaces support these two roles by the use of a combination of interaction styles: command driven; Q&A (question and answers); menus; form filling and maps. Which of these is used and when depends upon the task in hand.

The specific requirements for GIS interfaces are (after Healey and Gittings, 1989):

- separation of lexical and graphical interaction;
- window management;
- support for simultaneous manipulation of digital cartographic and attribute data;
- sophisticated graphics display capabilities;
- effective context management;
- customisable;

- modifiable;
- uniformity;
- portability.

To achieve some of these requirements many vendors are migrating their GIS to the X-windows environment. X-windows offers a number of advantages to the GIS developer: the software is portable across different hardware platforms running X; a consistent user interface can be maintained across a range of platforms thus permitting the user to move easily between platforms with little re-learning and, being a network-based protocol, users can directly access data residing at distributed sites and display it locally.

Over and above the requirements, does a good user interface matter and what exactly is a 'good' user interface anyway? As yet there is no empirical evidence on the difference a good GIS user interface makes. Evidence from other computer application areas suggests that interfaces are crucial for system efficiency, effectiveness and acceptability. Keister and Gallaway (1983) report on a study where they took a reasonably successful software product and made changes to elements such as screen format and content, consistency, error correction and feedback. Comparing the original and modified versions they found a reduction in time correcting errors of 32 per cent and 75 per cent fewer commands cancelled because of the wrong command being issued. The changes made equated to savings of 13.5 person weeks in improved throughput or $20–30,000 / year for a typical customer site.

The speed at which researchers have started investigating various aspects of GIS user interfaces would lead one to think that all the problems have been isolated and catalogued. In reality this is far from true. In fact there has been no systematic empirical evaluation of what the usability issues relating to GIS are, most commentary on the subject being based on the anecdotal comments of GIS users. We are currently no nearer to isolating formally the unique aspects of GIS interfaces from those which are important to any software user interface. As a result it is quite possible that considerable effort is being expended researching aspects of user interaction with which users have no problems. It also needs to be remembered that the user interface is a means to an end (i.e. making the system usable) and not the end itself. System usability is concerned with ensuring that a system can be used easily and effectively by a user who has been given specified training and support in order to fulfil a specified range of tasks within a specified range of scenarios. To be usable, a system must have the utility to provide a benefit to the user but minimise costs in terms of user effort. Usability depends upon the characteristics of the user (e.g. previous experience with similar systems and experience of the task domain) and the user's tasks. GIS usability is thus a crucial issue with the user interface being just one, albeit important, aspect of this (Medyckyj-Scott *et al.*, 1990). Improvements in usability should come about through the application of ergonomic principles in the design of GIS but very few commercial GIS show the application of such principles.

In the past the user interface has often been considered as an afterthought by vendors, produced only when the rest of the system is nearly complete. This has led to poorly designed interfaces that are difficult to change. This seems to reflect the fact that devoting human, financial and computer resources to user interface design as an element of product development has been, and still is, considered a luxury by many vendors. Generally vendors see user interface issues in terms of isolated problems, e.g. misunderstanding a command. There is little attempt to see if the problems are the result of more fundamental problems; for example, the lack of a overall conceptual model of user interaction. Additionally the interaction of user interface developers with users often seems intentionally limited with their primary source of information being through the marketing section who, in turn, base their information on contacts with managers in customer organisations and not the end-users themselves. If vendors have discovered issues of importance these have not been published so that basic research can take place to try to solve them.

As GIS begin to offer similar levels of functionality and the need for product differentiation in the market place becomes apparent, a 'good' user interface is being seen as a necessity. Given the variety of potential GIS users that now exist, however, and the large number of application domains, many vendors have realised that building a generic interface that can be used by all types of users is not possible. Instead they now supply tools by which an organisation can customise the user interfaces to their own needs, thereby passing the problems of designing the user interface onto the shoulders of the users! While the users may be able to decide the functions that are important to them and even the terms they want to use, they have no experience of user interface building. Moreover building an interface is a time consuming business. While any interface may require some customisation to a user's requirements this should be tinkering rather than major construction. Vendors should provide a number of generic interfaces tailored to different types of user (systems manager, analyst, decision-maker, 'browser') which can be finely tuned. Just as importantly though, GIS purchasers should be more particular about the general usability of the systems they are considering, looking at the usability of alternative systems during benchmarking as well as functionality and cost.

A more immediate problem is the profusion of different GIS/spatial terms used by vendors. Terminology for GIS operations etc., is used differently between vendors or, alternatively, terminology is the same but to perform slightly different operations. Agreement on terms and their meaning is urgently needed. Standardisation would lead to shorter learning curves for those training on new systems and fewer errors (e.g. initiating the wrong operation) for those who use more than one GIS.

Academic research has so far concentrated on what has interested researchers and/or what they could obtain research grants for. The initial focus was on the design of WYSIWYG based interfaces for spatial data handling (Egenhofer and Frank, 1988) including the use of icons to represent classes of data (Cassel and Parker, 1989) or operations (Phillips, 1990). A recent extension of this approach can be seen in the work of Campari *et al.* (1990) and Kirby and Pazner (1990). Noting the problems non-programmers have in constructing macros (e.g. ARC/INFO's AML), Campari *et al.* (1990) suggest the idea of a visual language based on icons by which a user can construct a procedural flow made up of the GIS operations required to achieve a particular goal.

As well as spatial query operations such as overlay, distance and proximity, users need to query the attributes of spatial databases and this interaction has been the subject of recent research. Because query languages are often linked to the data model employed by the GIS, many queries must be expressed in terms of software-specific query languages. This has led to a proliferation of languages, most of which are non-intuitive to the user. Systems such as Prime's System 9, and Intergraph's TIGRIS have standardised on SQL for attribute queries, taking the lead from the standardisation on SQL by the database field in general, and many systems allow the specification of a spatial query through the definition of spatial search areas in conjunction with SQL. While SQL can be used for expressing spatial queries across many data models there is a recognised need for a more user friendly language which also incorporates spatial extensions.

Maps form an important aspect of GIS user interfaces and have two roles to play. Firstly they are one way by which the user interacts with the system. In this way a map is not unlike any graphic user interface onto a large database but the importance of maps to user interaction seems to have received little consideration to date. Secondly, maps serve as a decision-making tool and as such the way information is portrayed is important. Currently most GIS display maps look very much like conventional paper maps. While these are familiar to the user, the information they contain may be inappropriate and become difficult to see and interact with when overlaid with other information.

Poorly designed maps may also convey incorrect ideas about spatial phenomena. Yet many GIS do not provide the necessary technical capabilities or flexibility to allow the user to produce effective maps. At a minimum, default values for the display of map features, etc., based on map design principles, should be provided by GIS vendors. The difficulty is establishing what these default values should be. Research into the design of paper-based maps is substantial but the guidelines derived from such studies may not be applicable to CRT displays. Even where guidelines based on empirical testing of CRT-displayed maps exist they are rarely incorporated into GIS software. Two solutions seem to exist, either the user must be provided with knowledge about map design or GIS software must incorporate knowledge about proper design principles through the development of expert systems or intelligent help facilities (Weibel and Buttenfield, 1988).

While the amount of research into GIS user interfaces is growing there are some who argue that there needs to be a completely new approach to GIS design (e.g. Gould, 1990). They point out that people do not think about space in terms of nodes, areas and arcs, or in terms of cells. Therefore, rather than teach users to adapt to the GIS they use, attempts should be made to incorporate the ideas of how people represent and think about space into the design of GIS and thus reduce the amount of cognitive and behavioural modification required by the user (see Medyckyj-Scott and Blades (1990) for a review of this). Using cognitive science and linguistics we should be able to design GIS that are more usable and flexible, through more effective:

- user interfaces e.g. natural language;
- spatial query languages;
- spatial metaphors (much like the 'desktop' metaphor) allowing the GIS designer to tap knowledge already held by the user group to aid the user as he/she learns, understands and interacts with a GIS;
- conceptual designs (Mark *et al.*, 1990).

By doing so we can also make these systems generally accessible to a larger number of ordinary people instead of leaving them to trained technical specialists. Trying to implement GIS based on users' conceptualization of space is difficult however. Currently knowledge of the user is either not formal enough or is formally defined but due to technical limitations cannot be implemented (Frank, 1990). It will therefore be some time before results from this area feed into mainstream GIS design.

Many other aspects of the user interface are yet to be investigated; for example, adaptive user interfaces and user models or interfaces for GIS as decision-support tools or command and control systems. Even where research has begun our knowledge is still limited.

The future

The usability of GIS is going to be one of the issues of the 1990's. Specifically this will be seen through improvements to the user interface and user interaction. In the short term we are likely to see:

- an increase in the number of GIS with direct manipulation or graphic user interfaces (GUI's);
- more vendors migrating to the X-windows environment;
- the re-emergence of other input devices such as touch screens and pens following improvements in the technology;
- the integration of GIS with Hypermedia, this providing users with multi-media interfaces, browsing, and simpler interfaces;
- more GIS vendors providing tools by which the user can customise their interfaces;
- easy-to-use front ends built for the casual end user in specific application domains.

We are also likely to see GIS becoming viewed as visualisation tools that allow users to produce 'pictures' of complex phenomena as a means of exploring data in order to gain an insight into problems. To a certain extent this view of GIS can be seen to have existed implicitly in the earliest GIS. Even so, viewing GIS as visualisation tools offers the prospect of exciting additions to the functionality offered and will be of use to the user in analysis, illustration and decision-making (Buttenfield and Ganter 1990). Becker *et al.*'s (1990) paper on dynamic graphical techniques for displaying network data is an early example of things to come.

In the more medium term we are likely to see:

- natural language interaction and voice recognition based systems;
- intelligent user interfaces—an interface that is able to identify the processes and data most appropriate to a query and explain to the user the reasoning behind the answer the system produces;
- adaptive user interfaces—systems containing models of each user which are used to tailor the user interface and user interaction to the characteristics of each user;
- studies that are now taking place into spatial cognition having some practical impact, e.g. systems that can handle spatial propositions and locative expressions.

In the far future (10–20 years) virtual world technology may have a dramatic effect on the nature of GIS. Such technology permits the simulation of a user's virtual sense by immersing the user in deliberately designed, external virtual worlds (a three-dimensional model programmed in a computer's memory which is experienced by its participants as an all encompassing environment of sights, sounds and tactile cues (Jacobson, 1990)). Such technology is developing at a fast rate and within 20 years may allow us to create GIS that more accurately reflect the way users conceptualise the physical and cognitive worlds they inhabit. The user of a GIS in the future may don special headgear and manipulative devices e.g. earphones, 'eyephones' and a hand glove, and through these view and interact with a virtual world that represents their geographic problem space. Will such technology resolve the problem of GIS being difficult to use? The answer has to be yes and no. While solving some aspects of user interface design it will throw up a different set of problems to be solved.

The successful development of user interfaces for GIS in the future is likely to require an inter-disciplinary team made up of computer specialists, cognitive scientists and spatial scientists. Such a team will focus on usability of the GIS rather than just the design of a user interface. This is likely to involve as much effort (through prototyping and iterative testing with real users, Medyckyj-Scott *et al.*, 1990) as is now currently given to the design and development of the underlying GIS functionality.

References

Becker, R.A., Eick, S.G., Miller, E.O. and Wilks, A.R. (1990). Network visualisation. *Proceedings of the International Symposium on Spatial Data Handling*, Zurich, 285–294.

Buttenfield, B.P. and Ganter, J.H. (1990). Visualisation and GIS: What should we see? What might we miss? *Proceedings of the International Symposium on Spatial Data Handling, Zurich*, 307–316.

Campari, I., Scopigno, R. and C. Magnarapa (1990). Visual language approach: a proposal for improving user friendliness and effective use of GIS. *Proceedings, First European GIS Conference*, April 10–13th, Amsterdam, Netherlands, 1024–1034.

Cassel, R. and Parker, H.D. (1989). Design and user testing of a GIS user interface. *Proceedings, GIS/LIS'89*, Nov. 26th-30th, Orlando, Florida, Vol. 2, 538–541.

Egenhofer, M. and Frank, A. (1988). Designing object-orientated query languages for GIS: human interface aspects. *Proceedings of the 3rd International Symposium on Spatial Data Handling*, Sydney, Australia, 79–96.

Frank, A. (1990). Spatial concepts, geometric data models and data structures. *Proceedings of the workshop on 'GIS design models and functionality' organised by the Midlands Regional Research Laboratory*, Leicester University, 21–22 March 1990.

Gould, M. (1990). Integrating spatial cognition and Geographic Information Systems. In Mark, D.M. and Frank, A. (eds.) *Cognitive and linguistic aspects of geographic space*. The Hague: Kluwer Academic Press, (in press).

Healey, R.G. and Gittings, B.M. (1989). Approaches to user interfacing in geographical information systems. Paper read at Regional Science Association Conference, Cambridge University, United Kingdom, August.

Jacobson, R. (1990). Virtual Worlds, inside and out. In Mark, D.M. and Frank, A. (eds.) *Cognitive and linguistic aspects of geographic space*. The Hague: Kluwer Academic Press, (in press).

Keister, R.S. and Gallaway, G.R. (1983). Making software user friendly: an assessment of data entry performance. *Proceedings of the Human Factors Society 27th Annual Meeting*, Santa Monica CA, 1031–1034.

Kirby, C. and Pazner, M. (1990). Graphic map algebra. *Proceedings of the 4th International Symposium on Spatial Data Handling*, Zurich, Switzerland, 413–422.

Mark, D., Gould, D.G. and Nunes, J. (1990). "Spatial language and Geographic Information Systems". National Center for Geographic Information and Analysis, State University of New York at Buffalo, Technical Paper 90–2.

Medyckyj-Scott, D. J., Newman, I. A., Ruggles, C. and Walker, D. (1990). Usability testing in user interface design for GIS and the role of rapid prototyping. *Proceedings, First European GIS Conference*, April 10–13th, Amsterdam, Netherlands, 737–746.

Medyckyj-Scott, D.J. and Blades, M. (1990). 'Users' Cognitive Representations of Space: Their relevance to the design and use of GIS". Research Report No. 13, Midlands Regional Research Laboratory, University of Leicester and Loughborough University, Leicester, UK.

Phillips, S. (1990). "An investigation into the design of icons for use within Geographical Information Systems". Unpublished M.Sc Thesis, Department of Geography, Leicester University.

Weibel, R. and Buttenfield, B.P. (1988). Map design for Geographic Information Systems. *Proceedings, GIS/LIS '88*, San Antonio, USA, Vol. 1, 350–359.

PART X

“What people said”

"What people said"

Introduction

With the growing problem of keeping track of all that is said and published on GIS in the UK in a given year the structure of the "What People Said" section of the *Yearbook* has been changed substantially from its format in preceding years. The emphasis now is on providing a pointer to new sources of published GIS-related material rather than providing an in-depth review. This change was felt necessary since the reviewing task had become a somewhat overwhelming objective. The only exception is for the AGI National Conference and sponsored seminars where the reviews presented in the *AGI News* have been reprinted with kind permission of the editor.

The new structure for this part of the *Yearbook* is as follows:

- New/forthcoming books
- Journals
- Research reports
- AGI National Conference
- Seminar reviews.

We appreciate that there will inevitably be some omissions. Once again we hope that we have kept this to a minimum and would ask that those who have been omitted contact us and keep us informed so that we can amend the situation in future additions.

New and forthcoming book titles

The following is a list of new or revised editions of GIS texts which have come out in 1990/91 or are due to emerge in 1991. This section has been organised alphabetically by publisher; the data and descriptive comments are based upon information provided by the respective publishers and do not constitute independent review or endorsement.

Longman Scientific and Technical

Geographic Information Systems: Principles and Applications
Editors: D.J. Maguire, M.F. Goodchild and D.W. Rhind
Size: c. 1200pp
Publication due: autumn 1991
Cost: c. £80.00.

This major new book has over 50 chapters written by an internationally renowned team and covers the basic principles and practice of GIS world-wide. The book has four main sections. The overview section provides the context discussing the main intellectual, technical and organisational forces which continue to influence the development of GIS. The principles section deals with the nature of spatial data, technical, functional, organisational and display issues, while the third section covers the applications of GIS through a series of general chapters which review developments in the different geographical and thematic areas. Detailed case studies are provided for carefully selected applications, and the Editors close with a summary of the main themes of the book and reflect on the current and future trends in GIS.

Map Generalisation: Making Rules for Knowledge Representation
Editors: B.P. Buttenfield and R.B. McMaster
Size: 264pp
Publication due: late 1991
Cost: c. £35.00

This book focuses on the development of a set of rules for the generalisation of digital maps. The National Center for Geographic Information and Analysis set up an initiative to formulate rules that will improve efficiency, preserve consistency and incorporate sound principles into digital mapping. This book is the first work to address the problems of map generalisation. Professionals from the disciplines of cartography, geography, computer science and surveying engineering have been invited to contribute chapters which provide a broad synthesis of ideas and knowledge from which guidelines for a set of rules can be formulated.

Geographical Information Management: Methodology and Application
Editors: I. Masser and M. Blakemore
Size: 336pp
Publication due: Mid 1991
Cost: c. £25.00.

This book contains 12 specially commissioned contributions which examine the methodological implications of recent developments in GIS technology and which explore potential applications in various fields. The discussion of methodological issues centres on five key themes: spatial analytical techniques; data integration; area transformation; error propagation; and the design of user interfaces. The applications reviewed include environmental monitoring, natural and technical hazard management, settlement and infrastructure planning, geodemographics and market analysis, and the evaluation of educational provision. The contributions draw heavily of the findings of research undertaken as part of the Regional Research Laboratory initiative which is funded by Britain's Economic and Social Research Council.

Oxford University Press

An Introduction to Urban Geographic Information Systems
Author: W.E. Huxhold
Length: 304pp
Publication due: April 1991
Price: £36.00.

This book is the second in the new series *Spatial Information Systems*, and describes the basic concepts of automated mapping, database structures, topological structures, land records information, and spatial analysis techniques with specific reference to US models. There are numerous black and white maps and a section of four-colour maps.

Statistical Methods in Soil and Land Resource Survey
Authors: R. Webster and M.A. Oliver
Length: 328pp
Publication date: November 1990
Price: £40.00 (hardback), £20.00 (softback)

This book describes suitable statistical methods for analysing variation in soil and for relating the soil to its environment. Sound land planning increasingly needs such quantitative information and decisions have to be made against a background of variation in soil and other land resources. This is essentially a second edition of Webster's book *Quantitative Numerical Methods in Soil Classification and Survey*. It tackles the topic of regression from the viewpoints of both statistics and natural science. Three new chapters are devoted to geostatistics and introduce regionalized variable theory. The accent throughout is on fully quantitative survey of land resources, measurement and estimation. The more advanced techniques are liberally illustrated with simple diagrams.

Pergamon Press

Geographic Information Systems: The Microcomputer and Modern Cartography
Editor: D.R.F. Taylor
Size: 296pp
Publication date: February 1991
Price: £50.00 (hardback), £19.50 (softback)

This book is an introduction to cartographic modelling, emphasising environmental decision-making applications, that develops a high-level computing language in presenting cartographic data processing conventions, capabilities and techniques.

Taylor & Francis

Introductory Readings in Geographic Information Systems
Editors: D.J. Peuquet and D.F. Marble
Length: 371pp
Publication: September 1990
Price: £40.00 (hardback), £18.00 (softback)

The widely varying application fields of GIS have meant that much of the GIS literature is scattered throughout journals and technical reports of numerous disciplines, not all of which are readily accessible to the newcomer to the field. This book assembles a balanced sample of written works covering important aspects of the basic principles of GIS and selected examples of current applications. The principles and methods of analysis illustrated have lasting value. The book is intended for students and others training in GIS, and practising professionals wishing to deepen their understanding of GIS. The five sections include: What is a GIS?; Examples of practical applications; Operations and problems of building a database; GIS internals: data representation and analysis techniques; Design and evaluation of GIS.

Interpreting Space: GIS and Archaeology
Editors: K.M.S. Allen, S.W. Green and E.B.W. Zubrow
Length: 398pp
Publication: November 1990
Price: £42.00 (hardback only)

This is the first volume in a series *Applications of GIS* and the first book to address the use of GIS in archaeology and anthropology. Offering original theoretical, methodological and practical guidance, it seeks to inform the scientist on the evaluation, purchase and application of GIS in the study of human development (along with how relevant data can be acquired). The contributors consider the analytical and technical capabilities of GIS as a fundamental method for coping with multidimensional data; from their use of the technology as a method to answer questions about spatial changes, readers in wider fields will find much of relevance to their own needs for analysing and modelling spatially orientated data.

Time in the Geographic Information System
Author: G. Langran
Length: c. 200pp
Publication due: summer 1991
Price: c. £18.00

Time in the GIS will be the first book to examine in detail the conceptual, logical and physical basis for developing a temporal GIS. It provides a broad survey of research on spatiotemporality, the philosophy of time, temporal databases and spatial data structuring. It includes much new and unpublished research concerning the extension of existing spatial data structuring techniques to a three-dimensional space-time application. Discussion of multidimensional access methods is then extended to assess issues such as clustering, quality and volume control. The book forms part of the Taylor & Francis series on *Technical Issues in GIS.*

Understanding GIS
Authors: D.W. Rhind, H.M. Mounsey and J.R. Raper
Length: c. 280pp
Publication due: late 1991
Price: c. £35.00 (hardback), c. £15.00 (softback)

Understanding GIS is the eagerly awaited, first new-generation text for the study of GIS in the 1990s. Its authors write with considerable experience of teaching in the academic and commercial sectors, research and implementations. The book is aimed at students in higher education but equally those encountering GIS in the professional community. Readers are led through the fundamentals of the data, the workings of the GIS, and the management issues behind successful GIS endeavour. Throughout there is repeated reference to operational application examples and the use of GIS within a commercial world.

Journals

There are now three major journals readily available in the UK which report on GIS activities.

International Journal of Geographical Information Systems, Taylor & Francis, Rankine Road, Basingstoke, Hampshire, RG24 0PR.
Mapping Awareness, Miles Arnold, High Winds, Cassington, Oxford, OX8 1DL.
GIS World, GIS World Inc, 2629 Redwing Road, Suite 280, Fort Collins, CO 80526, USA.

In addition GIS activities are being reported in an ever-increasing number of journals. The following list is by no means exhaustive but should at least provide the newcomer to GIS with a starting point from which to start searching.

AREA
Cartography and GIS (formerly *The American Cartographer)*
BURISA (British Urban and Regional Information Systems Association)
Cartographical Journal
Computer Journal
Computers in Geo-Science
Computer Vision, Graphics and Image Processing
Environment and Planning
Geographical Journal
Geo-Processing
International Journal of Remote Sensing
LARIA (Local Authorities Research and Intelligence Association)
Photogrammetric Engineering & Remote Sensing
The Planner.

Research reports

As the second phase of the Economic and Social Research Council's Regional Research Laboratory Initiative draws to a close we felt it would be useful to produce a summary list of the GIS-related research reports produced by the initiative.

Regional Research Laboratory Initiative

Discussion Paper Series: ISSN 0959–7158

1 "The Regional Research Laboratory initiative: an overview" Ian Masser (March 1990).
2 "The ESRC Regional Research Laboratories: A Technical Profile" Steve Plummer (March 1990).
3 "Towards a spatial analysis research strategy for the Regional Research Laboratory initiative" Stan Openshaw (May 1990).
4 "Organisational Databases: servicing the data needs of an expanding and under-privileged research community" Peter Lloyd (May 1990).

5 "Geodemographics: a review of recent developments and emerging issues—towards an RRL Research Agenda" Peter Brown (May 1990).
6 "Evaluating educational provision: a place and time for schooling" Peter Burnhill (May 1990).
7 "Managing natural and technological hazards: The role of GIS" Anthony C Gatrell (August 1990).
8 Environmental monitoring and prediction" David Rhind (August 1990).

All reports are priced at £4.00 (or US $10) including post and packing. They are available from Ms Christine Hobson, Department of Town and Regional Planning, The University of Sheffield, Sheffield S10 2TN; tel: 0742 768555 ext 6185.

Midlands Regional Research Laboratory

Research Report Series

1 "An Introduction to Area-Based Spatial Units: A Case Study of Leicestershire" Hilary M Hearnshaw, David J Maguire, Michael F Worboys. (December 1989). £3.00.
2 "A Syllabus for Teaching Geographical Information Systems" David J Unwin. (December 1989). £1.50.
3 "Modelling Population Distribution using Remote Sensing and GIS" Mitchell Langford, David J Maguire, David J Unwin. (December 1989). £3.00.
4 "The Problems of User and Organisational Acceptance of Geographical Information Systems" David J Medyckyj-Scott (December 1989). £3.00.
5 "Structuring Spatially Related Information for Effective Entry and Retrieval" Clive L N Ruggles, Ian A Newman, David J Medyckyj-Scott, David R F Walker (December 1989). £3.50.
6 "Towards a Data Directory for Leicestershire: Developing a Methodology for Collecting Information about Data" Sarah C Cornelius, Alan J Strachan (December 1989). £3.00.
7 "A Geographical Information System for Waste Disposal Site Selection" David J Maguire, Trajco V Mesev (December 1989). £2.00.
8 "Mapping the Density of Population: Continuous Surface Representations as an Alternative to Choroplethic and Dasymetric Maps" Mitchell Langford, David J Unwin, David J Maguire (February 1990) £3.50.
9 "The Use of Hypercard™ as a Rapid Prototyping Tool: A Case Study of the Development of Two Geographical Information Retrieval Systems" David J Medyckyj-Scott, Ian A Newman, David R F Walker, Clive L N Ruggles (March 1990) £4.00.
10 "Structuring Spatially Related Information for Effective Entry and Retrieval; Part 2: The Interaction Handler" Clive L N Ruggles, Ian A Newman, David J Medyckyj-Scott and David R F Walker (November 1990) £4.00.
11 "An Introduction to Object-Oriented Geographical Information Systems" David J Maguire, Michael F Worboys, and Hilary M Hearnshaw (August 1990) £3.50.
12 "The Development of a User-Interface for the Handling of Spatial Language Queries in a Geographic Information Retrieval System; Part 1: Theoretical and Conceptual Issues" Rob Angel, David J Medyckyj-Scott and Ian A Newman (October 1990) £4.00.

13 "Users' Cognitive Representations of Space: Their Relevance to the Design and Use of GIS" David J Medyckyj-Scott and Mark Blades (November 1990) £4.00.
14 "An Object-Oriented Future" Hilary M Hearnshaw and Michael F Warboys (October 1990) £4.00.
15 "Object-Oriented Data Modelling for Spatial Databases" Michael F Worboys, Hilary M Hearnshaw and David J Maguire (October 1990) £4.00.
16 "Object-Oriented Data and Query Modelling for Geographical Information Systems" Michael F Worboys, Hilary M Hearnshaw and David J Maguire (October 1990) £4.00.
17 "The Role of Modal Logics in the Description of a Geographical Information System" Michael F Warboys (October 1990) £4.00.
18 Reasoning About GIS Using Temporal and Dynamic Logics" Michael F Worboys (October 1990) £4.00.

For further information regarding the above, including the costs of postage and packing, please contact Mrs Elzbieta Radny, Midlands Regional Research Laboratory, Bennett Building, University of Leicester, Leicester LE1 7RH; tel: 0533 523849.

North East Regional Research Laboratory

Research Reports: ISSN 0957–4436

RR87/0 "A review of the problems of transferring data from one set of areal units to another incompatible set" Robin Flowerdew, Stan Openshaw
RR87/1 "Integrating geographic data for policy purposes: some recent UK experiences" Stan Openshaw, John Goddard, Mike Coombes
RR87/2 "Analysing and exploiting client data by creating a geographic marketing information and modelling system" Stan Openshaw
RR87/3 "A proposal and specification for a geographical subroutine library" John Dixon, Stan Openshaw, Colin Wymer
RR87/4 "Integrated local databases: progress in the RRL for the North" Mike Coombes
RR87/5 "An evaluation of the BBC Domesday interactive videodisc as a GIS for planners" Stan Openshaw
RR87/6 "An outline specification for an interactive video disk based target marketing and site evaluation system" Stan Openshaw
RR87/7 "A review of the opportunities for using GIS to find sites for Britain's future nuclear power stations" Stan Openshaw
RR87/8 "The use of gazetteers as research tools in geography: an example from epidemiology" Simon Raybould
RR87/9 "A study of visualisation of crime incidence mapping by police officers" Chris Brunsdon
RR88/0 "ODYSSEY: a low cost package for GIS research" Tony Gatrell, Martin Charlton
RR88/1 "Environmental use of Geographical Information Systems at the regional scale in Britain: some methodological and institutional constraints" Malcolm Newson
RR88/2 "Siting issues in the disposal of low and intermediate level radio-active wastes: the role for Geographical Information Systems" Steve Carver, Stan Openshaw, John Fernie

RR88/3 "Making sense of large flow datasets for marketing and other purposes" Stan Openshaw, Colin Wymer, Mike Coombes
RR88/4 "1991 Census geography and the case for some new census outputs" Stan Openshaw, Mike Coombes
RR88/5 "GIS and the social sciences: a binational review" John Goddard
RR88/6 "Restructuring Italy's local labour market areas: an evaluation of some alternatives" Stan Openshaw, Colin Wymer, Fabio Sforzi
RR88/7 "Managing environmental radioactivity monitoring data: a GIS approach" Ian Heywood, Sarah Cornelius
RR88/8 "Modelling the North-South divide" Stan Openshaw, Martin Charlton, Chris Brunsdon, Mike Coombes
RR88/9 "North East Regional Research Laboratory: proposed strategy" Stan Openshaw, John Goddard, Mike Coombes
RR89/0 "A review of methods for the short-term forecasting of crime incidence in small areas" Chris Brunsdon, Stan Openshaw, Jim Lillee
RR89/1 "Is there a role for spatial information systems in formulating multi-agency crime prevention strategies?" Ian Heywood, Neil Hall, Peter Redhead
RR89/2 "A guide to industrial databases: uses and pitfalls" Anna Cross
RR89/3 "Local 'Enterprise Activity' Potential (LEAP): variability in new small firm growth" Mike Coombes, Simon Raybould
RR89/4 "Adaptive data structures for multidimensional point data" Colin Wymer, Stan Openshaw, Martin Charlton
RR89/5 "New measures of house price inflation at the local scale" Mike Coombes, Tony Champion, Chris Brunsdon
RR89/6 "Putting some analytical functionality into police computer systems: the case for geographic crime incident analysis systems" Chris Brunsdon, Stan Openshaw
RR89/7 "The Domesday Data Access System: description and user guide" Sheila Barton
RR89/8 "Spatial analysis by computer movie" Martin Charlton, Stan Openshaw, Mike Rainsbury, Chris Osland
RR89/9 "A proposal for a centre for advanced research computing in the social sciences" Stan Openshaw, John Goddard
RR90/0 "Geographical analysis of child health measures: a pilot study in Alnwick, Northumberland" Richard Reading
RR90/1 "Profiling neighbourhoods: social and environmental statistics for Tyne and Wear" Mike Coombes, Simon Raybould
RR90/2 "Northern Region child health database: a pilot study" Sheila Barton, Mike Coombes, Steve Jarvis

The price for each Research Report is £5.00 including post and packing, with a 50% discount for British academics. They are available from Mike Coombes, NE.RRL Research Co-ordinator, CURDS, The University, Newcastle upon Tyne NE1 7RU; tel: 091 222 6000 ext 8014. This list includes a number of reports which were originally published jointly by NE.RRL and NWRRL as Northern RRL Research Reports.

Two other publications still available at present, although they are no longer published in the above series, are:

1 "Building a Mark 1 Geographical Analysis Machine for the automated analysis of point pattern cancer and other spatial data" Stan Openshaw, Martin Charlton, Colin Wymer and Alan Craft (Nov 1987).
2 Handling Geographic Information for Health Studies" Anthony C Gatrell (Jan 1988).

North West Regional Research Laboratory

Research Report Series

1 "Statistical Methods for Inference between Incompatible Zonal Systems" R Flowerdew, M Green (December 1989) £2.50.
2 "Evaluating the Potential Role of GIS for a Market Analysis Company" R Flowerdew, D Banting (December 1989) £2.50.
3 "Statistical Methods for Areal Interpolation: the EM Algorithm for Count Data" M Green (December 1989) £2.50.
4 "Notes on Shortest Path Algorithms for GIS" C E Dunn, D Newton (December 1989) £2.50.
5 "Transferring Data Files between ARC/INFO and GLIM" E Kehris (December 1989) £2.50.
6 "Inference between Incompatible Zonal Systems using the EM Algorithm" R Flowerdew, M Green (December 1989) £2.50.
7 "Redefining Enumeration District Centroids and Boundaries" P Boyle, C E Dunn (January 1990) £2.50.
8 "Burning Questions: Incineration of Wastes and Implications for Human Health" A C Gatrell, A A Lovett (December 1989) £2.50.
9 "On Modelling Spatial Point Patterns in Epidemiology: Cancer of the Larynx in Lancashire" A C Gatrell (December 1989) £2.50.
10 "Map Generalisation using ARC/INFO" R Daly (December 1989) £2.50.
11 "Modelling Binary Maps using ARC/INFO and GLIM" P.J. Vincent (August 1990) £2.50.
12 "GIS in Epidemiological Research: Analysing Cancer of the Larynx in North West England" A.C. Gatrell, C.E. Dunn (August 1990) £2.50.
13 "A Geographical Modelling Environment built around ARC/INFO" E Kehris (August 1990) £2.50.
14 "Using Geographical Information Systems for Emergency Planning" C.E. Dunn (August 1990) £2.50.
15 "The relative utility of the Central Postcode Directory and PinPoint Address Code in Applications of Geographical Information Systems" A.C. Gatrell, C.E.Dunn, P.J. Boyle (August 1990) £2.50.
16 "Spatial Autocorrelation Statistics in ARC/INFO" E. Kehris (August 1990) £2.50.

For further information regarding the above, including the costs of postage and packing, please contact Dr A C Gatrell, Co-Director, NWRRL, Department of Geography, University of Lancaster, Lancaster LA1 4YB; tel: 0524 65201 ext 3754.

Regional Research Laboratory for Scotland

Working Paper Series

1 "Summary of Enquiry Chaired by Lord Chorley on Geographical Information Systems" A C Messer (August 1987).
2 "User Interfaces—Views of the Past, Present and Future" B M Gittings, A Stacey (May 1988).
3 "The RRL Browser—First Stage Design and Prototype Implementation" M L Chu, R G Healey, A C Messer, C J Nailer (May 1988).
4 "A GRIDMAP user guide" A Bayley (May 1988).
5 "Deprivation Indices and Multilevel Models" C Garner, I Diamond (September 1988).
6 "Bibliographic Control of Research Data" P M Burnhill, A Carruthers, A Messer (September 1988).
7 "The RRL Browser—Issues in the Design and Implementation of a SASPAC Interface" R G Healey, A Messer (October 1988).
8 "CATAN—GIS for the Spatial Analysis of Educational Provision, Based on School Catchment Areas" K Tolgu (October 1988).
9 "Monitoring the Social and Spatial Impact of Educational Reform" C L Garner, B M Gittings, K Tolgu (January 1990).
10 "The Potential of the PostScript graphics model for GIS Applications" D Mitchell (January 1990).
11 "Land Use and Agricultural Activity: an Areal Approach for Harnessing the Agricultural Census for Scotland" J McG Hotson (November 1988).
12 "A Place and Time for Schooling: Prospects for RRL Activity" P M Burnhill (April 1990).
13 "GIS Training Course, Part One" P J Dowie, H Ewington, B M Gittings, R G Healey. (March 1990).
14 "GIS Training Course, Part Two" P J Dowie, B M Gittings, R Healey (Forthcoming).
15 "An Evaluation of Quadtree Data Structures in GIS for Rural Planning" J Selwood, N Stuart (Forthcoming).
16 "GIS and Rural Land Use Strategies: A Case Study of Islay, Part 1" P J Dowie, R G Healey, J Holbrook, A Mowle, T Rideout. (Forthcoming).
17 "User Requirements for the Integration of GIS and Locational Analysis Techniques" N Anderson, S Finlayson, R Hodgart (Forthcoming).
18 "Parallel Polygon Formation: An Investigation of the Effectiveness of Parallelism in a GIS" A Munro (Forthcoming).
19 "A Parallel Algorithm for Polygon Linkage" S Hopkins, T Waugh (Forthcoming).
20 "GIS User Interfaces" B M Gittings, R G Healey, D Mitchell (Forthcoming).
21 "The 1991 Census: Spatial Referencing Considerations" J Forbes, A McCleery (Forthcoming).
22 "GIS Management Issues" B M Gittings (Forthcoming).
23 "Dataset catalogue of Edinburgh University Data Library: a full statement of descriptive fields" P M Burnhill, H Ewington (October 1990).
24 "Common Service Boundaries: from conflicting to conterminous boundaries" P M Burnhill, P J Dowie, R G Healey (August 1990, revised January 1991).
25 "A parallel implementation of Franklin's uniform grid technique for line

intersection detection on a large transputer array" R G Healey, S Hopkins (October 1990).

26 "Implementation of a Parallel Architecture within the Digital VMS Operating System: an overview of symmetric multi-processing" B M Gittings (October 1990).

27 "Postgraduate Research in Geographical Information Systems at the University of Edinburgh 1986–1988" B M Gittings (October 1990).

28 "An approach to the design of parallel algorithms for 3-D GIS" G B Desa (October 1990).

For further information regarding the above, including the costs of postage and packing, please contact Heather Ewington, RRL Scotland, University of Edinburgh, 57 George Square, Edinburgh EH8 9JU; tel: 031 667 1011 ext 6383.

South East Regional Research Laboratory

Working Paper Series

1 "Recent Developments in Geographical Information Systems in the UK" D W Rhind £3.00.

2 "Database Design and Implementation" N P A Green £3.00.

3 "UAGIS, A Geographical Information System in Action" J W Shepherd, N P A Green £3.00.

4 "OS Digital Data in the National Transfer Format (NTF): A User Perspective" N P A Green, N Mackay, J W Shepherd £10.00.

5 "The Chorley Committee and 'Handling Geographic Information'" D W Rhind, H M Mounsey £3.00.

6 "A GIS Research Agenda" D W Rhind £3.00.

7 "The Design and Construction of the DoE Demonstrators: Objectives, Execution and Software Listings" N P A Green, J W Shepherd £10.00.

8 "GIST: an Object-Orientated Approach to a GIS Tutor" J F Raper, N P A Green £5.00.

9 "London 2001" P Hall £3.00.

10 "Integrating Remotely Sensed Imagery into a GIS for Town Planning Purposes" M Barnsley, J W Shepherd, G Sadler 10.00.

11 "Analysis of Census Data Available for Measuring Local Level Demographic and Socio-Economic Change in the South East of England between 1971 and 1981" C McKee £5.00.

12 Manpower Statistics and Social Survey Data, their Availability and Suitability for Measuring Local Scale Change in South England" C McKee £5.00.

13 "The 1971/1981 Census Change Files: A User's View and Projection to 1991" C McKee £5.00.

14 "An On-Line, Secure and Infinitely Flexible Database System for the National Population Census" D W Rhind £5.00.

15 "First Unix, then UGIX" D W Rhind, J F Raper, N P A Green £5.00.

16 "Environmental Monitoring and Predicting" D W Rhind £5.00.

17 "Generalisation and GIS Databases" E Joao, D W Rhind, S Openshaw, B Kelk £5.00.

18 "A scale theory of urban form and growth" P A Longley, M Batty, J W Shepherd £5.00.

19 "The Measurement and Control of Generalisation Effects" E Joao, G Herbet, D Rhind £5.00.
20 "The Utility and Value of Geographic Information" by M Didier. An edited translation by J Shepherd £5.00.
21 "UGIX: A GIS Independent User Interface Environment" J F Raper, M Bundock £5.00.
22 "Use of population density data to improve classification accuracies in remotely sensed images of urban areas" G J Sadler, M J Barnsley £3.00.
23 "The Application of GIS in the Development of a National Sample Survey of Vacant Land in Urban Areas" P Brignall, G J Sadler, J W Shepherd £3.00.
24 "Hyperarc—A Task Orientated Interface to ARC/INFO" T K Linsey, J F Raper, T Connolly £3.00.
25 "The Use of Knowledge-Based Techniques for Map Design and Generalisation" G Herbert £5.00.
26 "Converting ARC/INFO Coverages to Macintosh Mapmaker Format" C Holdesson £3.00.

For further information regarding the above, including the costs of postage and packing, please contact Dr J Shepherd, Department of Geography, Birkbeck College, 7–15 Gresse Street, London W1P 1PA; tel: 071 631 6483.

Associated site: London School of Economics

Working Papers

1 "An Evaluation of Desktop Mapping Packages for the Apple Macintosh" R R Hershey, D C Whitehead (April 1990) £10.00.
2 "A Review of Transport Infrastructure: Impact Studies in Britain" R Drewett, T Hoehn, B Linneker (Forthcoming).

For further information regarding the above, including the costs of postage and packing, please contact Rachel Hershey, Department of Geography, London School of Economics, Houghton Street, London WC2A 2AE; tel: 071 405 7686 ext 2591.

Wales & South West England Regional Research Laboratory

Publications

1 "Some Theoretical Considerations in the Design of Disaggregated Geographical Information Systems" C Webster (November 1987) £5.00.
2 "Distributed Software for Access to Census Data" S Wise, I Bracken (November 1987) £5.00.
3 "Map Manager: Intelligent Software for the Display of Spatial Information" I Bracken, S Holdstock, D Martin (December 1987) 10.00.
4 "Rapid Access to a Large Cartographic Database using CAFS: A Feasibility Study" S Wise, R Pellat (December 1987) £5.00.
5 "An Approach to Surface Generation from Centroid-Type Data" D Martin (May 1988) £5.00.
6 "Exploratory Computer Modelling of Urban Housing Markets" (May 1988) £5.00.
7 "A Unified Statistical Computing Environment for the Analysis of Shopping

Behaviour" C Guy (May 1988) £5.00.

8 "A Systems Architecture View of GIS Functionality" I Bracken, C Webster (January 1989) £5.00.

9 "Computer Packages for Survival Analysis" S A Matthews (June 1989) £5.00.

10 "Informative Planning: The Intelligent Use of Information Systems in the Policy-Making Process" M Batty (September 1988) 10.00.

11 "The Rural Wales Terrestrial Database: An Example of Inter-Agency Co-operation" G Higgs (October 1988) £5.00.

12 "An Introduction to the Use of Logic Programming in Spatial Analysis" C Webster (February 1989) £5.00.

13 "Gastric Cancer and the Environment. A Review of the Epidemiological Literature" S A Matthews (February 1989).

14 "Further Developments in Online Access to Large Cartographic Datasets" H Edmunds, S Wise (June 1989) £5.00.

15 "Cities as Information Networks. The Evolution and Planning of New Computer and Communications Infrastructures" M Batty (January 1990) £5.00.

16 "Information Systems for Planning in Developing Countries: A Review" M Batty (January 1990) 10.00.

17 "A Logic-Programming Formulation of the Point-in-Polygon Problem" C Webster (September 1989) £5.00.

18 "Text Animation or Knowledge Engineering?: Two Approaches to Expert System Design" C Webster, H C Siong, T Wislocki (December 1989) £5.00.

19 "Towards a General Attribute Data Linker for Geographic Information Systems—Concepts and Software" I Bracken (November 1989) £5.00.

20 "'Intelligent' Digitizing for Complex Map Bases" I Bracken (November 1989) £5.00.

For further information regarding the above, including the costs of postage and packing, please contact Christine Phillips, Department of Town Planning, University of Wales, College of Cardiff, PO Box 906, Colum Drive, Cardiff CF1 3YN; tel: 0222 874000 ext 5272.

Urban Research and Policy Evaluation Regional Research Laboratory

Working Paper Series

1 "Geodemographics: A Review of Recent Developments and Emerging Issues" (October 1989) £5.00.

2 "The Study of Poverty and Deprivation in Developed Countries: A Selective Review of Data Sources, Approaches and Analytical Techniques" A Hirshfield (October 1989) £5.00.

3 "A Bibliography of Address Referencing" R Barr, K Campbell (October 1989) £5.00.

4 "Giving Users What They Want: The Evolution of the Merseyside Address Referencing System (MARS)" J Willis (Forthcoming) £5.00.

5 "The Postal Address as a Geo-Reference" K Campbell, R Barr (Forthcoming) £5.00.

6 "Data Sources and GIS Methodology for Organisational Research in the UK" P E Lloyd (May 1990) £5.00.

7 "An Evaluation of the PAF on CD-ROM" R Barr, K Campbell (Forthcoming)

£5.00.

8 "Economic and Demographic Impact Multipliers for Merseyside and Greater Manchester" P W J Batey (Forthcoming) £5.00.

9 "Skills for GIS: A Survey for the Association of Geographic Information" J Willis (Forthcoming) £5.00.

10 "Drug Abuse Cases in Liverpool and South Sefton Health Districts: An Exploratory Analysis" C Fazey, P J B Brown, P W J Batey (Forthcoming) £5.00.

11 "Mortality and Morbidity in Blackpool, Wyre and Fylde: A Statistical and Spatial Analysis" S Bowman, P J B Brown, P W J Batey (March 1990) £5.00.

12 "Applications of Address Referencing" R Barr, K Campbell (Forthcoming) £5.00.

14 "The Spatial Epidemiology of Food Poisoning in Blackpool, Wyre and Fylde: Preliminary Results" A Hirschfield, P W J Batey, P J B Brown, J Marsden (March 1990) £5.00.

15 "An Inventory of Data Held in Greater Manchester and Merseyside" (Forthcoming) £5.00.

16 "Geodemographics and the Construction of Individual-Level Market Classifications: the Case of the Home-Shopping Industry" P W J Batey, P J B Brown (April 1990) £5.00.

17 "Issues in the application of GIS in Urban and Regional Policy Making" L Worrall (May 1990) £5.00.

18 "Poisson Chi Square Mapping, GIS and Geodemographic Analysis: the Spatial and Aspatial Analysis of Relatively Rare Conditions" P J B Brown, P W J Batey, A Hirschfield, J Marsden (September 1990) £5.00.

19 "The Evaluation of Local Economic Policies: A Review of Theoretical and Methodological Issues" R L Stephenson (Forthcoming) £5.00.

20 "An Inventory of Data Holdings on Merseyside: Introduction to the Survey and Notes for Accessing the Database" H Pattison (October 1990) £5.00.

21 "Household Structure and Extended Input-Output Models: A New Formulation" P W J Batey (Forthcoming) £5.00.

For further information regarding the above, including the costs of postage and packing, please contact Dr Alex Hirshfield, URPERRL, Department of Civic Design, University of Liverpool, PO Box 147, Liverpool L69 3BX; tel: 051 794 3110.

AGI Annual National Conference 1990: "The key to managing information"

General account of the event

Data availability is no longer the major barrier to success for the GIS community. The major problem is the reluctance of British managers to face up to the radical changes needed to implement GIS in their own organisations in both private and public sectors. This was the challenging message put to delegates by Mike Brand at the opening session of AGI '90 in Brighton. He went on to suggest that this reluctance on the part of management will mean that opportunities will be lost for improving people's lives through better health planning, emergency service provision and better local government. He called on Government to take a stronger lead.

In his opening address to AGI '90, Viscount Astor, spokesman on the Environment in the House of Lords, said that Government recognised the importance of the efficient handling of geographic information. Government was keen to see that use of the new technology was more widespread. The industry, he suggested, represented a major business opportunity estimated at worth 200m in western Europe this year alone.

The scale of the exhibition at AGI '90 confirmed Viscount Astor's remarks. Some 58 organisations were represented at the exhibition. They included many of the major names in computing and GIS together with a large number of companies dealing with fields such as data capture, data conversion, training and personnel, and consultancy services. In addition to the 600 delegates registered at the conference, more than the same number again attended the exhibition as day visitors, confirming the importance of the event in the calendar.

Conference Review*

The character of the GIS industry in the 1990's was the theme taken up in the keynote address to AGI '90 by John Beaumont, Professor of Business at the University of Bath. He suggested that the market will begin to show signs of maturity, although it will come to be dominated by a number of "niche players", its strength will be based on diversity. The future is, however, very challenging. Professor Beaumont warned that the effective management of IT is often illusive. There is no relation between the scale of investment and likely benefits. At present there is often insufficient up-front planning and developments are managed on a more *ad hoc* basis. Better coordinating is generally required between GIS initiatives and IT strategies.

The theme of "market maturity" came across very strongly during the exhibition. The Metropole Centre proved an excellent venue, with visitors circulating freely around the exhibits. The chance to see the exhibition during the "icebreaker reception" on the day before the conference opened was an excellent innovation, giving delegates a less pressured chance to see what was on offer. Compared with earlier events the impression of many of those involved was that it was mainly the people that had changed rather than the systems. One vendor commented that the visitors were asking more difficult and intelligent questions than before, questions not so much abut the technology, but about its management and business implications. They were less impressed by the technology and more aware of its challenges.

The exhibition did show, however, that the technology had also improved. Operational object-oriented systems, for example, are now available, and ease of customisation by the customer is another new feature. In addition to technological changes, however, the number and diversity of those offering support services also seems to have grown. There was no shortage of advice on data conversion, training and consultancy. One stand even offered advice on where to eat in Brighton, based on data captured during the conference!

"Do more data make for better decisions?" asked John Beaumont. Not always, since too much data can often lead to overload and confusion. Systems not only need to be more friendly so that data can be obtained by the user rapidly, but users also need to know what to do with that information. Success depends upon the information culture within an organisation. Beaumont suggested that not enough is known about what the key factors for success are.

* The Editors are grateful to Roy Haines-Young, of the University of Nottingham, for this personal review of AGI '90.

Some organisations are, however, clearly adapting to the challenges of the new technology. Many delegates were impressed by the approaches to handling geographic information by Safeway plc and the Nationwide Anglia Building Society. Both Chris Wilkinson, Store Planning Director at Safeway, and Richard Hornby, Market Analysis and Planning Manager at Nationwide, described how geographic information can be used to determine the success of a new site. The Safeway case study showed the value that can be added to existing data held within the organisation by better integration of information. He described not only a very successful pilot, but went on to outline the lessons which can be drawn from the experience. He emphasised the importance of tight project definition, an appreciation of the scale of investment involved in developing a GIS, and the management of "user expectation". Both Richard Hornby and Chris Wilkinson noted how access to the technology has begun to transform the initially fairly simplistic models about site evaluation into much more complex analytical tools.

Case studies such as those of Safeway and Nationwide Anglia illustrated Beaumont's earlier remarks about the capacity of GIS and IT more generally to "automate" and "informate" processes within an organisation. The power to "empowerate" and "transformate" are likely to be features of the 1990's, suggested Beaumont. By making information available more generally within an organisation more employees are given the capacity to take effective decisions. Organisations may therefore be transformed from rigid hierarchical structures to networks of decision-makers.

The need to understand the impact of GIS and other tools for handling information on organisations came across very strongly during the conference, emphasised particularly by the session dealing with GIS projects which had failed to live up to expectations. In this session, papers by John Antenucci, Nick Pearce and Stan Openshaw highlighted the need to move the focus away from the technology towards the management of its introduction and use. The importance of staff training, and not just in the narrow technical sense, was a constant theme. Understanding the one or two hundred commands needed to drive a proprietary GIS did not make an employee a GIS expert. Education and training must capture much broader perspectives if the technology is to be used effectively. Nick Pearce, like Rob Mahoney and Helen Mounsey in papers delivered in later sessions, emphasised that success depends upon a mix of human, business, financial and technological factors. The management of these factors at every stage is essential.

The conference organisers attempted to illustrate a range of application areas for GIS by inviting papers from a number of subject areas. Topics covered included the use of GIS by civil engineers, the utilities, the environment and local authorities. Two themes emerged; the first concerned the need to find better ways of sharing data, and the second concerned the need to develop better tools for spatial analysis.

Michael Blakemore, of the National Online Manpower Information System (NOMIS), presented a challenging perspective of data sharing from the perspective of a major supplier of data. The effective development of GIS in the UK depends not only on effective systems and skilled personnel, but also properly available and integrated information. The latter issue has been highlighted in recent years by criticisms of the availability and quality of Government statistics. Blakemore argued that issues of data demand and supply are not the only issues to be addressed. We must concern ourselves with trying to develop a much deeper understanding of the character of the data themselves. A major issue for the future is in training people to understand data, limitations, errors, and the way in which data are used to solve problems. Just as suppliers of information must have an appreciation of the needs of users, so users must have a better understanding of the data they are seeking to acquire. "The acquisition of data", Blakemore suggested "involves a two-way partnership that benefits both supplier and user...".

Richard Dunn, in his account of GIS in the analysis of countryside data, reported on the development of a demonstrator project which aimed to perform the educative role which still seems to be necessary. The project, undertaken for the Directorate of Rural Affairs in the Department of the Environment, was based on the use of GIS in planning context of the county of Devon. The demonstrator illustrated to the user not only data deficiencies, but also problems of bringing together data from different sources within a single system. Dunn concluded that data issues were still a major barrier for many policy applications.

Case studies such as those described for Devon also illustrated a second theme which emerged from the conference, namely that a major limitation of GIS currently available concerned the lack of adequate spatial modelling capabilities. Indeed, this was one of the major factors responsible for the failure of the system developed for Northumbria Police described by Stan Openshaw. Although the system had the appropriate data, it failed to deliver the appropriate level of analysis for the investment to be of value to the user. There were few signs in the exhibition that such modelling tools will become widely available in the near future. This topic must remain an important item on future research agendas. Such modelling tools will be essential if GIS are to be used effectively and with confidence in a business environment. The implications for education and training, however, are perhaps even more profound than those which arise out of the need to understand the nature of data. On this score, at least, remarks about the market maturing seem rather premature.

If anything can be concluded from AGI '90 then it is perhaps that we are approaching the point where users have a wide choice of systems that will meet a good proportion of their needs. The emphasis for the future must therefore be on ensuring that users have the wider range of skills necessary to unlock the potential which the technology offers for their particular business applications.

AGI-sponsored seminars

GIS and strategic planning

"Why provide this service?" is a question being asked of many in local government. John Burnet, Director of Economic Development at Cumbria County Council, addressed the issue and explored the role which GIS might have in producing some answers during the North West Regional Meeting of the AGI at the University of Salford in February 1990.

The meeting, which was hosted by the Salford University Business Unit and the Department of Geography at the University, attracted some 50 delegates. It was the second of its kind designed to take the AGI into the regions. In his talk John Burnet described the experience gained in Cumbria using GIS technology as a tool for policy analysis and development in the County's Corporate Information Unit.

The Unit was set up in September 1988. In many ways it reflects the cultural change taking place throughout local government. The environment is now one of greater devolution of management and the formation of independent business units.

The function of the Corporate Information Unit at Cumbria is to provide information, research and intelligence on Cumbria's population, economy and environment, not only to the County Council but also to district councils, and other clients in the public and private sector. The success of the Unit is measured largely by the ability to sell its services to this client group and in this task access to GIS technology has an important role to play.

In order to answer questions about service delivery and resource allocation one needs to understand where people locate their numbers and their needs. John Burnet explained that to do this one must integrate census information with other local authority and geographical data to map, analyse and model processes in order to provide the intelligence necessary for corporate decision-making.

To illustrate the use of GIS for policy development and evaluation John Burnet described several case studies, including one dealing with the local management of schools. An issue in this policy area is whether resources should be distributed in favour of disadvantage. In order to examine the implications of such a policy GIS can, for example, be used to rank schools in relation to patterns of social deprivation. When these data are combined with information concerning budgetary constraints maps of resource and pupil allocation can then be produced and examined.

Policy-makers and their advisers are clearly major agents for change in society. John Burnet's talk illustrated very clearly the important role which GIS can play in strategic planning. He showed the value of the technology in the kind of administrative environment now developing in many areas of local government, where decisions have to be taken in more consumer- and business-orientated ways.

Standards for Handling Geographic Information

The AGI held an extremely successful one-day workshop at Nottingham University during September, led by Sara Finch of Logica Communications. The meeting, originally intended as a small gathering, attracted more than 120 delegates representing the user, vendor, consultant and academic communities.

The morning session was chaired by Mike Brand, Chairman of the AGI, who explained the role of the association's Standards Committee and introduced its Chairman, John Rowley. John began his presentation by explaining the needs perceived by the Chorley Committee for standards in handling geographic information, which promoted the setting up of the AGI's own committee. The Standards Committee has set up a number of working groups to study specific issues in detail, for example the NTF (National Transfer Format) Steering Committee and SQL (Standard Query Language) Working Party are two such groups. Their work is not carried out in isolation and regular contact is maintained with organisations such as NJUG (National Joint Utilities Group) and the local authorities' GISG (Geographic Information Steering Group).

The AGI recently carried out a survey of members to determine priorities for the evolution of standards. The results of the survey mapped very closely onto the AGI's own plans for study. The Standards Committee seeks to be a prime mover in evolving standards in the UK and to be responsive to users' activities.

The following presentations were given by the chairmen of a number of the working groups set up by the AGI. They described the scope of their work and some of the technical and political issues which impact on the business of setting standards.

Mandy Lane is a member of the group which looks at the standardisation of spatial units. She described a basic spatial unit as the smallest area to which data are associated; such an area may, for example, be represented as a unit postcode or a 1 kilometre grid square. Attempts to standardise spatial units are made complex by the fact that each type of basic spatial unit is suited to a particular type of data. Mandy explained that unit postcodes are effective when used in property databases since they represent a number of properties rather than a complete geographic area, and clearly they would not be appropriate for most environmental data.

A further difficulty occurs when different sets of spatial data are aggregated to create a new set of polygons. Mandy described the way in which most GIS will assign the original data values to new aggregated polygons, regardless of the fact that the polygon has changed its size, shape and location. Mandy also warned that users must be aware of the ways in which a GIS handles data. They will also look at ways of identifying the spread of data across areas to allow the more accurate combination of datasets.

John Turner leads the working group with responsibility for investigating data quality. The group has put together a set of introductory notes, which were the basis for his presentation. He began by describing a range of ways in which data quality can be expressed. Such measures of quality can be applied at different levels such as the dataset, subset, feature or co-ordinate level. It is also possible to express quality by textual comment or by using a measurable quantifier such as standard deviation.

John's working group has still to study the impact of incorporating update information on data quality. Full or partial datasets can be used for update, the use of either will result in a need to reassess data quality.

In the next session Mike Bosworth spoke about the local authorities' work to define a national street gazetteer. He stated the objectives of the Street Gazetteer Working Party as being to determine a standard definition for a street gazetteer, to look at the feasibility of devising a standard specification which should ideally be linked to the national grid, and to consider the options for a single body distributing and maintaining the gazetteer.

Mike described the Working Party's plans to provide unique identification to each street in the gazetteer. Their definition of a street encompasses all roads from motorways to paths, tracks and ways. The Working Party has already produced a format for the gazetteer, and plans to market the data as it becomes available, using the National Transfer Format.

David Gradwell chairs the SQL Working Party (SQL is an industry standard product used by programmers and analysts to design databases). He explained that SQL had been developed for handling textual information and thus, in its current form, is not suited to geographic data. His group plans to provide input to influence the development of the next version of SQL which will become a standard in five years' time. They have recently produced a paper analysing the requirements for database software for constructing a GIS.

In the final presentation Sam Sowton described the development of the National Transfer Format as a set of standards for the transfer of digital map data. NTF was initially administered and maintained by Ordnance Survey, but since 1989 has been the responsibility of an AGI Working Party. The NTF Steering Committee comprises members with a range of interests, wider than just digital mapping. However, Sam explained that they were keen to widen the membership still further by encouraging representation from the field of CAD/CAM, the British Computer

Society and the British Standards Institute. One of the Working Groups which has been set up to support the NTF Steering Committee is tasked with making NTF compatible with ISO 8211. The Steering Committee is also keen to have NTF adopted as a British Standard and eventually as the European Transfer Standard.

This session concluded with a lively debate on the merits of the various standards which have evolved for the transfer of digital map data. John Rowley closed the meeting, which was an undoubted success given the large turnout and the very active participation of the audience.

PART XI

Established Geographic Information Systems

This section contains a compilation of established GIS systems as complete as it was possible to make given responses received at the time of going to press. The page number indicated in the final column (headed **Ref.**) refers to a more extensive description supplied.

Established Geographic Information Systems

Trading name, phone and contact	Address	Availability for visits etc.	Brief installation details	Ref.
Anglian Water Services Limited 0223 235235 Richard Sage	Chivers Way, Histon, Cambridge CB4 4ZY	Not open to visitors.	INFORMAP from Synercom.	
ASDA Stores Plc. 0532 418027 Jon Towson	ASDA House, South Bank, Great Wilson Street, Leeds LS11 5AD		InSite and ATLAS* GRAPHICS.	
Bartholomew 031 662 4072 Robin Orr	12 Duncan Street, Edinburgh EH9 1TA	Open to visitors by appointment.	ARC/INFO linked to ORACLE (RDBMS) on networked SUN workstations. Suppliers of software are ESRI and Oracle (UK) Limited.	282
Bass Inns & Taverns 021 558 1481 Bill Riley	PO Box 27, Cape Hill, Birmingham B16 0PQ		InSite and ATLAS* GRAPHICS.	
Berkshire County Council 0734 234780 Chris Gower	Department of Highways and Planning, Shire Hall, Shinfield Park, Reading, Berkshire RG2 9XG	By appointment only.	ARC/INFO software running on a Digital Equipment platform.	

Trading name, phone and contact	Address	Availability for visits etc.	Brief installation details	Ref.
Birmingham City Council 021 235 3349 Paul Bayliss	The Council House Extension, Victoria Square, Birmingham B1 1BB	Selective and by prior arrangement.	GFIS with GeoManager supplied by IBM United Kingdom Limited with PIMMS property management supplied by Hoskyns GIS running on an IBM 3090 500J.	
Braintree District Council 0376 552525 D Molyneux	Causeway House, Bocking End, Braintree, Essex CM7 6HB	Open to visitors by appointment.	SysScan/ Digital System with BULL mainframe computer.	
Bristol Waterworks Company 0272 665881 PW Jones	Bishopsworth Road, Bedminster Down, Bristol BS13 7JN	Selective and by prior arrangement.	INFORMAP from Synercom on Digital VAX equipment all supplied by Laser-Scan.	
British Geological Survey 06077 6111 Kevin Becken	Nicker Hill, Keyworth, Nottingham NG12 5GG	By appointment.	Intergraph Microstation 32, MGE, MGA, IRAS, IVEC. Microstation-PC digitising stations. Laser-Scan Horizon—VAXstation. I^2S image processor. Various data translation software.	
British Market Research Bureau 081 567 3060 Martin van Staveren	Saunders House, 53 The Mall, Ealing, London W5 3TE		InSite.	
Cheshire County Council 0244 603136 David Challen	Geographic Information Unit, Room 8, County Hall, Chester CH1 1SG	Limited access.	In-house software, GIMMS, AUTOCAD, Ingres, ARC/INFO.	

Trading name, phone and contact	Address	Availability for visits etc.	Brief installation details	Ref.
Cleveland County Council 0642 248155 Geoff Burke	Technology Development Unit, First Floor, Melrose House, 1 Melrose Street, Middlesborough, Cleveland TS1 2HG	On written request.	PIMMS from Hoskyns GIS on an IBM 3090. IBM SQL/DS, GPG (GFIS) and Infoter.	
Clydesdale Bank Plc 041 223 2103 Alasdair Law	30 St. Vincent Place, Glasgow G1 2HL		InSite and ATLAS* GRAPHICS.	
Colne Valley Water Company 0923 223333 Bob Hollingsworth	Blackwell House, Aldenham Road, Watford, Hertfordshire WD2 2EY	Open to Visitors by appointment.	Software: ALPER Records, ALPER Scan, ALPER Enquiry, ALPER Maps. Hardware: SUN workstations, Calcomp electrostatic plotter, Dahle guillotine, scanning camera.	282
Cornwall County Council 0872 74282 /2287 Mrs Jane Reid	Treasurer's Department, Room 1, Fowey Building, Treyew Road, Truro, Cornwall TR1 3BD	In order to avoid disruption to the office participating in the pilot, contact and access will be by approval of the Steering Group.	METROPOLIS supplied by Laser-Scan as pilot system running on Digital VAXStation 3100's. These will be changed to UNIX workstations during 1991. Output is by Calcomp thermal transfer plotters.	282
Cumbernauld Development Corporation 02367 21155 /257 Sharon McAlister	Planning Department, Cumbernauld House, Cumbernauld G67 3JH	On written request.	GPG from Hoskyns GIS running on IBM PS2 Model 80 and 55 PC's and AST 386 in a Token Ring network. The AST is used with a NEC MGE-AT graphics engine (1024 × 768 resolution).	
Data Consultancy 0734 588181 Bryan Wade	7 Southern Court, South Street, Reading, Berkshire RG1 4QS	By arrangement.	Mapping and GIS software including MapInfo, spatial analysis, geodemographic, drivetime, modelling and digitising software. Hardware includes PC's, plotters, digitizers and colour printers.	

Trading name, phone and contact	Address	Availability for visits etc.	Brief installation details	Ref.
Directorate General of Military Survey 081 890 3622 /4134 Brian Hatherell	Feltham Garrison, Elmwood Avenue, Feltham, Middlesex TW13 7AH	Authorised visitors by prior appointment.	Digital VAX, Laser-Scan and Intergraph.	283
Dudley Metropolitan Borough Council 0384 452964 DA Cook	Director of Finance, The Council House, Dudley, West Midlands DY1 1HF	By arrangement with ICL.	ICL 3900—PLANES. SUN workstation—AXIS Mapping from AXIS Software Systems Limited.	
Duggal Associates Limited 081 204 8351 Andrew Shaw-Hamilton	73 Portland Crescent, Stanmore, Middlesex HA7 1LY	Data conversion from paper to vectors—demonstrations at your own or Duggal offices of CADCore tracing product. By arrangement.	IBM Compatible 386 with maths co-processor or 486 running PC DOS or MS DOS versions 3.1 and above. Hitachi CADCore tracer.	
East Midlands Electricity Plc GH Jackson	398 Coppice Road, Arnold, Nottingham NG5 7HX	By prior arrangement with the contact.	IBM GFIS.	283
East Sussex County Council 0273 482244 Mr AK Gower	Highways and Transportation Department, Phoenix Causeway, Lewes, East Sussex BN7 1UE	Not open to visitors.	ARC/INFO running on SUN workstations with EASINIS road network software (written in Fortran) running on an ICL mainframe.	284
English Heritage 071 973 3095 Bronwen Knox	Fortress House, 23 Savile Row, London W1X 1AB	By appointment only.	SUN workstations; ADVENT Systems Limited software.	284

Trading name, phone and contact	Address	Availability for visits etc.	Brief installation details	Ref.
ESR Limited 09323 48981 Mr K Brook	Rosemount House, Rosemount Avenue, West Byfleet, Surrey KT14 6NP	Open to visitors by appointment only.	Digital MicroVAX II and MicroVAX 2000, 1300 mbytes hard disk, VMS/ SysScan software, HP1000/A900 in-house software. IBM compatibles running MapData, AutoCAD and MOSS software.	284
Genasys (UK) Limited 061 232 9444 Peter Bayley	Manchester Science Park, Lloyd Street North, Manchester M15 4EN	Seminars held regularly.	Most UNIX platforms.	
Glasgow District Council 041 227 5813 C.S. McAllister	Planning Department, 231 George Street, Glasgow G1 1RX	By agreement.		
Grampian Regional Council 0224 664917 Allan M Harrison	Department of Water Services, Woodhill House, Westburn Road, Aberdeen AB9 2LU	Can be arranged through the contact.	ALPER Records, Maps and Scan Edit using SUN 386i workstations and Calcomp plotters/ digitizers.	
Great Grimsby Borough Council Allan Thompson	Devonshire House, Bull Ring Lane, Great Grimsby DN31 1ES	None—during six month Pilot Study period.	Bull DPSB Mainframe—Tektronix 4211 Terminal—Océ G1830 Plotter—Summagraphics Microgrid II Digitising Tablet.	
Hampshire County Council 0962 846765 David Couchman	County Planning Department, The Castle, Winchester, Hampshire SO23 8UE	By invitation only.	IBM GFIS on IBM 3081 mainframe running under VM/HPO (GPG, INFOTER, SQL/DS), GIWS, SPANS and PC-ARC/INFO.	285

Trading name, phone and contact	Address	Availability for visits etc.	Brief installation details	Ref.
Healey & Baker 071 629 9292 Dr Yvonne Court	29 St George Street, Hanover Square, London W1A 3BG		InSite and ATLAS* GRAPHICS.	
Hertfordshire County Council 0992 588695 John Alleyne or Bill Martin	County Hall, Hertford, Hertfordshire SG13 8DE	By appointment.	IBM 3081 KX3. IBM SQL/DS (RDBMS). GeoManager. GPG (GFIS). PIMMS supplied by Hoskyns GIS.	
Hillingdon Health Authority 0895 79145 Nick Blankley	St Johns, Kingston Lane, Uxbridge, Middlesex UB8 3PL		InSite and ATLAS* GRAPHICS.	
HM Land Registry 071 405 3488 Godfrey Hix	Peterborough District Land Registry, City Road, Peterborough PE1 1XN	Not open to visitors except by special appointment and as operational requirements permit.	IGDS graphics and DMRS database customised to HMLR requirements and supplied by Intergraph (UK) Limited.	285
Horsham District Council 0403 64191 /387 Stuart James	New Park House, North Street, Horsham, West Sussex RH12 1RL	Considered on application.	Siemens C40-G mini computer configured with 16 mbytes of main memory and 1 gbyte online storage running Sicad software. This is supported by 3 9735-10 UNIX work-stations with digitizing tablets.	286
Hove Borough Council 0273 775400 John Packman	Town Hall, Norton Road, Hove, East Sussex BN3 4AH	1. Complete Planning Application System: open. 2. Cemetery system.	1. ICL M80/ M40s with Novell Network and SIA dataMAP. 2. ICL M.60 with AutoCAD.	

Trading name, phone and contact	Address	Availability for visits etc.	Brief installation details	Ref.
Jones Lang Wootton 071 638 6040 Mike Dutton	Kent House, Telegraph Street, Moorgate, London EC2R 7JL		InSite and ATLAS* GRAPHICS.	
Kent County Council 0622 696984 Roy Laming	Information Systems, Springfield, Maidstone, Kent ME14 2LQ	By appointment only.	ARC/INFO on PC's and SUN workstations. Interface to INGRES.	286
Lancashire County Council 0772 264539 Geoff Howarth	County Surveyor's Department, Guild House, Cross Street, Preston, Lancashire PR1 8RD	By arrangement with system vendors: Doric Computer Systems.	ARC/INFO supplied by DORIC, Oracle relational database: Digital VAX 4000 plus VAXStation 3100 workstations.	
Lee Valley Water Company 0707 268111 Richard Giles	PO Box 48, Bishops Rise, Hatfield, Hertfordshire AL10 9HL	Selective and by appointment.	SysScan software using both OS digital maps and SysScan digitised maps and records on Digital VAX equipment distributed over four sites.	
London Borough of Barking & Dagenham 081 592 4500 /2345 Barry Capper	Borough Treasurer and Computer Services, Civic Centre, Heathway, Dagenham, Essex RM10 7BN	Open to visitors by appointment.	GGP from Hoskyns GIS running on IBM PS2 Model 80. Model 55 PC's linked to an IBM ES9000 320. GGP is also running on a token ring network and one copy is on a standalone PC.	287
London Borough of Barnet (Tech Servs) 081 446 8511 /4378 John Stannard	Barnet House, 1255 High Road, Whetstone, London N20 0EJ	By appointment only.	ARC/INFO supplied by Doric Computer Systems on IBM PS2 Model 80 with 640 mbytes storage. Also related HIMS running on SUN Sparcstation 1+. Currently under review.	

Trading name, phone and contact	Address	Availability for visits etc.	Brief installation details	Ref.
London Borough of Brent 081 862 3124 Tony Dickson	5th floor, Lakeside House, Lakeside Way, Wembley, Middlesex HA9 0LA	Consideration will be given to all enquiries.	GGP from Hoskyns GIS running on Compaq and Viglen 386 PC's linked to a Comperex IBM 3090 equivalent. GGP is used with the corporate street gazetteer and with the LRC Accstats data.	
London Borough of Harrow 081 863 1459 Ian Hinton	Department of Architecture and Planning, PO Box 39, Harrow, Middlesex HA1 2UJ	By appointment.	GGP from Hoskyns GIS.	287
London Borough of Hounslow 081 862 5469 John Polwin	Civic Centre, Lampton Road, Hounslow, Middlesex TW3 4DN	Open to visitors by appointment.	ADDMAPS supplied by Cobham Digital Services: digital mapping system. Networked 386/486 PC computers with Océ Graphics 36″ colour E/S plotter.	
London Borough of Lambeth 081 671 5542 Malcolm Forbes	Courtenay House, 9-15 New Park Road, London SW2 4DU	Visitors welcome—by appointment only.	Autocad running on SUN Sparcstation 1. Compaq Deskpro 386/25 with Autocad /FMS/AC.	
London Borough of Richmond 081 891 1411 Ian Trowbridge	Department of Technical Services, The Civic Centre, 44 York Street, Twickenham TW1 3BZ	On written request.	GGP from Hoskyns GIS running on a standalone Wang PC. GGP is being used with DBase 3+ databases.	
London Borough of Waltham Forest 081 527 5544 /4575 Alan Saville	West Block Annexe, Town Hall, Forest Road, Walthamstow, Essex E17 4SZ		Housing Services applications based on Hitachi CADCore TM/Hybrid supplied by Duggal Associates running on AST 33MHz PC and linked to an EDP Mentor 6000 mini computer.	288

Trading name, phone and contact	Address	Availability for visits etc.	Brief installation details	Ref.
London Research Centre 071 735 4350 /2089 Chris Feltham	LAAU, Parliament House, 81 Black Prince Road, London SE1 7SZ	By appointment.	GGP from Hoskyns GIS running on Viglen 386 PC's and also using a Calcomp Digitizer.	
London Residuary Body 071 633 5000 Barry Williams	Property Services Department, Globe House, 4 Temple Place, London WC2 3HP	On written request.	GGP from Hoskyns GIS running on standalone IBM PC's.	
Mercury Personal Communications 071 971 8136 Ciaran Coburn	1 Harbour Exchange Square, Docklands, London E14 9GE		InSite and ATLAS* GRAPHICS.	
Mid-Southern Water Company Limited 0252 835031 Mrs JA Hall	Sturt Road, Frimley Green, Camberley, Surrey GU16 6HZ	Open to visitors by appointment.	ALPER Records, Maps and Enquiry on SUN 2 × 3/60 and 2 × SparcStation 1 workstations and 3/280 server supplied by Alper Systems Ltd. Calcomp Electrostatic plotter, thermal plotter and 2 × digitizers.	
Midlands Regional Research Laboratory 0533 523825 Dr Alan Strachan	Bennett Building, University of Leicester, Leicester LE1 7RH		ARC/INFO supplied by Doric Computer Systems Limited & ERDAS, Earth Resources Data Analysis System supplied by Computer General Limited. Intergraph supplied by Intergraph UK Ltd.	
Milton Keynes Borough Council 0908 682223 Karen Andres	Civic Offices, 1 Saxon Gate East, Central Milton Keynes MK9 3HQ	Systems demonstrations will be provided by appointment through Alper Systems.	ALPER GIS software. SUN SPARC workstations (7), 3/60 servers, vector plotters, electrostatic plotters, thermal printers.	

Trading name, phone and contact	**Address**	**Availability for visits etc.**	**Brief installation details**	**Ref.**
Milton Keynes Development Corporation 0908 692692 Ken Robinson	Saxon Court, 502 Avebury Boulevard, Central Milton Keynes MK9 3HS	Visits to the system can be arranged through Mr J Wain at Midsummer Computing Exchange 0908 668866.	Intergraph VAX and server running 8 workstations in a mixed UNIX/ VMS environment, holding land terrier and property data.	
Motherwell District Council 0698 66166 Alistair McNeill	Planning Department, PO Box 14, Civic Centre, Motherwell ML1 1TW	By appointment only and from Summer 1991.	ALPER Maps and Records (part) EDIT Gazetter, OS Process on SUN SparcStation IPC with A0 plotter and A4 Laser printer.	
Nationwide Anglia Building Society 071 242 8822 Valerie Collyer	Chesterfield House, Bloomsbury Way, London WC1V 6PW	By appointment only.	GeoPin (ARC/INFO application) supplied by Pinpoint Analysis Limited. Mosaic Systems. Using Compaq Deskpro 386/25 PC.	
NJUG Dudley Digital Records Trial 021 476 1247 Norman Harrison	c/o Midlands Electricity Board, Bull Street, Dudley, West Midlands DY1 2DE	Open by arrangement to Utilities, and Local Authorities.	Prime 9755 CPU; Tektronix 4114/ 4109/ 4207; Benson 1332 plotter and 6201 digitizers. Micom communications equipment connected to British Telecom lines.	288
Northumbrian Water 091 284 3151 AR Keddie or P Bainbridge	IT Consultancy, PO Box 4, Regent Centre, Gosforth, Newcastle upon Tyne NE3 3PX	By appointment only.	FRAMME system (using UNIX) supplied by Intergraph with Ingres RDBMS. Phase 1 implementation began January 1991.	
NORWEB—Electrical Appliance Retailing 0204 28731 Mrs Gill Swan	PO Box 148, Manchester Road, Bolton BL1 1AA		InSite and ATLAS* GRAPHICS.	

Trading name, phone and contact	Address	Availability for visits etc.	Brief installation details	Ref.
Ordnance Survey 0703 792420 Neil Smith	Romsey Road, Maybush, Southampton, Hampshire SO9 4DH	By appointment in working hours.	1: Digital map production using Laser-Scan Lites2. 2: ARC/INFO supplied ESRI for research purposes. 3: Internally developed software.	289
Ordnance Survey of Ireland 00 01 213 171 Richard Kirwan	Phoenix Park, Dublin 8, Republic of Ireland	Selective and by appointment.	Georec, Cadig and Ginis software supplied by SysScan and running on Digital VAX 6410 and VAXStations with Altek digitizers.	289
Ordnance Survey of Northern Ireland 0232 661244 Mr MJD Brand	Department of the Environment (NI), Colby House, Stranmillis Court, Belfast BT9 5BJ	Open to visitors by appointment.	COMTOD (Computer Mapping & Topographic Database) developed by OSNI. Systems etc from Digital Equipment Company, SysScan (UK), LaserScan, Precison Image, Calcomp, Versatec and Leica.	289
Oxford Institute of Retail Management 0865 735422 Dr J Reynolds or Mr C Talbot	Templeton College, Kennington, Oxford OX1 5NY	Open to visitors by appointment.	PC-ARC/INFO and SPANS software supported in Market Research and analysis for the retail and consumer service industries.	
Portsmouth Water Company 0705 486333 M.R. Hedges	PO Box 8, West Street, Havant, Hampshire PO9 1LG	By prior appointment only.	Alper Records software. ICL SUN 4/6SC-8 workstations; Calcomp digitizer 95480 HA, Calcomp printer 5912 VSE screen dump; Calcomp Drawing Master D.I. printer 52224.	
Robertson Group Plc 0492 581811 Chris Parker	Llandudno, Gwynedd LL30 1SA	Open to visitors by appointment.	ARC/INFO, ERDAS, ORACLE and a suite of seismic interpretation and mapping software. Running on a SUN server, workstations and PC's with digitizers, Calcomp, Versatec and HP Laser plotters.	

Trading name, phone and contact	Address	Availability for visits etc.	Brief installation details	Ref.
Royal Borough of Kensington & Chelsea 071 937 5464 George Hulme	The Town Hall, Hornton Street, London W8 7NX	Open to visitors by appointment.	GGP from Hoskyns GIS linked to Pinpoint referenced databases running under ADABAS/Natural on Compaq 386 PC workstations.	290
Sandwell Metropolitan Borough Council 021 569 2027 DJ Vining	Town Hall, High Street, West Bromwich, West Midlands B70 8DX	Please contact Mr Vining to arrange demonstration. Car parking is limited; visitors usually welcome.	Microvax 3400, McDonnell Douglas Series 19045S using PLUS software and GDS, AMS, SITES and PMS on the VAX: 2 × Microvax II's on other sites.	290
Scottish Homes 031 337 0044 Karen Robertson	Thistle House, 91 Haymarket Terrace, Edinburgh EH12 5HE	Until June 1991 our address will be: Rosebery House, 9 Haymarket Terrace, Edinburgh EH12 5YA.	Software: PC ARC/INFO Starter Kit, PC Overlay, PC ARCPLOT. Hardware: Compaq 386/33PC, Roland DXY 1300 A38 Plotter, TDS Numonics LC Series II A4 Digitising tablet.	291
Sears Plc. 021 733 2233 Malcolm Kirkup	Mens and Childrenswear Division, Marshall Lake Road, Shirley, Solihull, West Midlands B93 9BS		InSite and ATLAS* GRAPHICS.	
Shropshire County Council 0743 25100 Peter Bingham	Department of Planning and Property Services, The Shirehall, Abbey Foregate, Shrewsbury, Shropshire SY2 6ND		PIMMS from Hoskyns GIS, IBM GFIS workstations (x2), plotter, digitizer and various GFIS applications.	291
Silsoe College 0525 60428 /273 Dr Chris Bird	Rural Land Use Department, Silsoe College, Silsoe, Bedfordshire MK45 4DT	Visits to the Remote Sensing and GIS laboratories can be arranged by appointment.	PC Systems—SPANS and ERDAS software. SUN workstation—GRASS and ARC/INFO software. Prime minicomputer—GEMS image processing.	292

Trading name, phone and contact	Address	Availability for visits etc.	Brief installation details	Ref.
Southern Water Plc 0273 606766 Peter Elkins	Southern House, Lewes Road, Brighton, East Sussex BN1 9PY	Southern Water Services Limited and I.T. Southern are open to visitors by appointment.	Alper Systems software running on 50 SUN workstations networked to an ICL 3980. Operational map base for the South East of England.	292
South Wales Electricity Plc 0222 792111 R.V. Gwinnutt	St. Mellons, Newport Road, Cardiff, South Glamorgan CF3 9XW	Subject to prior arrangement and approval from Chief Engineer.	Alper records on SUN SparcStations.	293
Strathclyde Regional Council 041 355 5333 Robert Bone	Water Department, 419 Balnore Road, Glasgow G22 6NU	Subject to certain constraints would allow visits by one/ two persons to discuss mutual interests or view current activities.	ARC/INFO running on SUN SparcStation and PC's. Calcomp 9100 A0 Digitizer.	293
Sunderland & South Shields Water Company 091 510 1050 Mr. S.W. Trow	29 John Street, Sunderland, Tyne and Wear SR4 1JT	Visits can be arranged in advance through Mr. S.W. Trow or Mr. A.J. Robinson.	Alper Records software on 2 × SUN SparcStations and 1 × SUN 386i with digitizers; SPANS; Calcomp A0 electrostatic plotter.	
Taff-Ely Borough Council 0443 407 811 Brian Stark	Personnel and Management Services Department, Municipal Buildings, Pontypridd, Mid Glamorgan CF37 2DP	On written request.	GGP from Hoskyns GIS running on a Novell 386 network using Elonex PC's.	
Taywood Data Graphics 0642 750515 Ralph Yardley	Dukes Way, Teesside Industrial Estate, Thornaby, Stockton-on-Tees, Cleveland TS17 9LT	Open to visitors by appointment.	Laser-Scan Lites, Anatech Eagle Scanner, Intergraph Microstation and AutoCAD in volume production operation.	

Trading name, phone and contact	Address	Availability for visits etc.	Brief installation details	Ref.
Taywood Data Graphics 081 575 9240 Pat Jarvis	Greenford House, 309 Ruislip Road East, Greenford, Middlesex UB6 9BQ	Open to visitors by appointment.	Intergraph Microstation, IGDS/ DMRS and Anatech Eagle Scanner, LaserScan Lites, VTRAK, ARC/INFO, Microfilm scanner.	
Thames Water Utilities 0734 593333 Roger Morgan	Nugent House, Vastern Road, Reading, Berkshire RG1 8DB	Visits and discussion by prior arrangement only.	GFIS and GeoManager supplied by IBM United Kingdom Limited and PIMMS property management application supplied by Hoskyns GIS.	
Wakefield City Metropolitan District 0924 296612 R.E. Aspinall	Planning Department, Newton Bar, Wakefield, West Yorkshire WF1 2TX	Demonstrations by arrangement.	GDS supplied by McDonnell Douglas with PMS/RCONNECT under VMS 5.1. Digital VAX 3800 with 5 × DEC 200 Servers. Tektronix, Pericom and Acer PC workstations.	293
Wales & South West Regional Research Lab 0222 874000 Dr. Gary Higgs	PO Box 906, Colum Drive, Cardiff CF1 3YN	Open to visitors by appointment.	Genamap supplied by Genasys; PC-ARC/INFO; PC-GIMMS; PC-MAPICS. PC's: IBM PS2 Model 80; OPUS V with output to Calcomp 5912 plotter and Pericom MX 7200 colour terminal.	
Welsh Water 0352 700845 John Morley	Penrhosgarnedd, Bangor, Gwynedd LL57 2EQ		InSite.	
Welwyn Hatfield Council 0707 331212 Cecil Elliston Ball	Council Offices, The Campus, Welwyn Garden City, Hertfordshire AL8 6AE	By appointment only.	ALPER Records and Maps on SUN 3/60 and SUN SparcStation 1+ work-stations, Calcomp 1023 plotter, Calcomp 91360 digitizer, Panasonic KX-P1124 printer.	294

Trading name, phone and contact	Address	Availability for visits etc.	Brief installation details	Ref.
West Somerset District Council 0984 32291 Kelvin Rufus	20 Fore Street, Williton, Taunton, Somerset TA4 4QA	Visits can be arranged for Local Authorities considering similar installations.	IBM 9370 M90 with 5080 + PS/2 user terminals attached. Operating GFIS.	294
Woolworths Plc. 071 706 5579 Matthew Murphy	242/246 Marylebone Road, London NW1 6JL		InSite and ATLAS* GRAPHICS.	

Bartholomew

The overall system is comprised of three basic units. One unit for graphics data capture, the central GIS unit for management of graphics and non-graphic data and the last unit(s) for output. The data capture system comprises networked Apricot P.C.s running on MS-DOS and MAPDATA application software linked to digitising tablets for capture of feature coded vector graphic data. The main GIS is a central SUN 4 fileserver with networked SPARC workstations and VT320 terminals running the ARC/INFO and ORACLE RDBMS software for integrated management of the graphic and non-graphic data. Output can go through the Scitex Response 280 system with its raster laser plotter/scanner, or through a DTP system to a Linotronic. The Scitex software is proprietary, whereas ARC/INFO and ORACLE can be customised using AML.

The Colne Valley Water Company Plc

The Colne Valley Water Company Plc purchased a digital records system in 1989.

Management commitment was such that instructions were given to the newly formed GIS office to computerise all 2,500 Km of pipe network and associated data in *12 months*. It was also agreed to computerise the Company's asset database, a 14,000 entry Roads Gazetteer and to plot 11,000 burst locations and add associated burst data, during the same period.

This work was completed in-house in April 1990, five weeks ahead of schedule, and data is in daily use throughout the Company.

It has been recognised that maximum benefit would be gained by connecting all 760,000 customers to underground assets.

The Colne Valley have embarked upon a project to complete three tasks under the heading of 'Connectivity Project':-

- To download 320,000 mainframe customer property entities into the asset database and assign a spatial reference based upon the Ordnance Survey seed point.
- To assign to each property entity the same UPRN as that used for each property on the mainframe customer database.
- To connect every property entity to its true point of connection on the correct pipe without breaking the network.

The Company has paid for the development of necessary software and transfer files and as before will undertake all data capture works in-house, using Company and Agency staff.

The project will commence in December 1990 and be complete within 12 months.

Cornwall County Council

The County Land Charges project represents the first corporate venture into GIS for Cornwall. Data used in the production of Land Searches is supplied by the Planning, the Surveyor's and the Clerk's Departments.

It was decided to focus the pilot on the Carrick District of Cornwall which covers some 500 1:2500 and 100 1:1250 maps but does not have complete digital coverage. Thus the map base is a combination of OS digital, raster and vector data purchased from a local utility. The Council's data is a mix of proprietary data sets from originating bodies, data captured by a bureau and some in-house data capture. The data sets used represent a variety of scales and their use has required great care. The preparation, input and checking of data has been and will continue to be a major function.

Because of the nature of the background maps (a mixed raster/vector base) the graphical output for the pilot is produced via a thermal transfer plotter.

The pilot will run for a period of a year starting in the spring of 1991. During that time there will be a transfer from VMS to a UNIX platform. At the end of the pilot period a decision will be made on expansion. It is possible to extend throughout the County's Departments on the existing map base or to increase the Land Charges system into other districts or both depending upon available resources.

Directorate General of Military Survey (DG Mil Svy)

The Director General of Military Survey (DG Mil Svy) has responsibility for the provision of geographic data in all its forms to support the current and future needs of UK armed forces.

DG Mil Svy will satisfy these requirements with output from a Defence Geographic DataBase (DGDB) which will support multi product generation. The study to define the structure of this database began in 1984. The data will be held in a topological structured vector format, although the system will also have a raster data handling capability. Evaluation of the data structure and the proposed levels of information to be held in the DGDB is being performed against a Pilot Terrain Analysis (Pilot TERAS) system fielded in 1BR(Corps). Procurement of the DGDB system hardware is now in hand with the aim of commencing database population in 1992.

At the same time work has proceeded on the definition of the products which will be supported by the contents of the database. A range of digital products is being specified and agreed by international working parties taking note of civil activities. The aim is to achieve compatibility of geographic data used by military systems and to facilitate the exchange of geographic information in digital form between nations.

East Midlands Electricity Plc

Following completion of a pilot, GFIS is being used for 132kv circuit records on a production basis, covering both overhead lines and underground cables. During 1991, the system will be extended to circuit records at the 33kv, 11kv and Low Voltage levels. Other corporate applications are under consideration, including non-engineering applications.

East Sussex County Council

EASINIS is a suite of programs written in ICL mainframe FORTRAN to handle the plotting of a network of roads together with data relating to sections, points, a range of grid references or a range of chainages.

A pilot study of ARC/INFO on a SUN is about to start to move this system to a graphic workstation and give screen displays in addition to plots. We also hope to automatically cross reference and analyze data with different types of geographical reference eg. section based with point based.

English Heritage

English Heritage adopted a combined Raster/Vector system in 1987 to produce maps showing the location and extent of Scheduled Monuments. Full map coverage for England at 1:1000 scale was on line by January 1988 and the system has recently been enhanced to cater for large scale maps.

The current hardware configuration is as follows:

Network file servers—	2 Sun File servers (3/180 and 3/280) 2 (Wyse) terminals 6.1Gb storage (8 SMD disks (Sun 575Mb) and 2 SCSI disks)
Workstations—	6 (Sun 3 series)
Printers—	Versatec Electrostatic (AO) QMS Laser 2 (A4)

The principal software packages (Advent Systems Ltd) include:

- 'Rectify' for the accurate re-scale of scanned images
- 'Paint' for raster editing.
- 'Designer40' —a CAD package combining raster and vector data.
- 'MMS' —the 'Map Management System' (utilising the UNIFY DBMS).

A number of other packages are used extensively.

Initial input of 14000 sites was completed by April 1989 and a quality control check was completed in 1990. A total of 50000 sites should be added over the next few years.

Possible areas of development for the system are:

- Direct link with DEC RISC: ULTRIX/ORACLE system recently installed
- Additional hardware
- Addition of other constraint areas eg. Listed Buildings
- Integration of OS digitized maps
- Data exchange

ESR Limited

Our established and experienced digitising department operates a versatile range of GIS/CAD software to provide a variety of digitising options to meet the requirements of an expanding market.

With the growing demand for digitised map data as the basis for

Geographic Information System (GIS) within the utilities sector, we offer a complete service for digitised software of basemaps, overlays and client's specific database records. Using the latest in SysScan technology and a double shift working arrangement, we can service the needs of a wide range of users, supporting most international and industry standard formats such as OSTF, NTF, SIF, DXF etc.

Our renowned cartographic facility produces conventional and digital mapping commissioned by publishers, tourist authorities, government departments etc, requiring high quality products for international markets.

Recent projects include digital mapping for 3 major atlas publications and a national database for the British Tourist Authority. An unusual feature of one of the atlases is the incorporation of LANDSAT Thematic Mapper imagery, designed to give an effective and accurate means of relief depiction.

Hampshire County Council

The County Council has developed three GIS applications:

- a Routine Maintenance Management System (using GFIS) for Motorways and Trunk Roads;
- a GFIS system for aggregating 'small area' population, dwellings and unemployment information to produce base information and forecasts for larger user-defined areas;
- a system using PC ARC/INFO for use in association with tendering for grounds maintenance work. A digitised plan of each contract site has been produced using aerial photographs as source information.

SPANS is a recent acquisition for use in the Council's Planning and Social Services' departments.

Plans for the year 1991/92 include:

- the establishment of a corporate map unit.
- the development of county-wide highways-related applications (road traffic accidents, roads and bridges maintenance, traffic counts) based on the Ordnance Survey's road centre-line data, OSCAR; and
- feasibility studies in respect of other potential applications.

HM Land Registry

The digital map record system at Peterborough is operational for the local authority districts of Corby, Kettering and Wellingborough, and is a development from an experimental system originally developed between 1984–1986. Further development work on GIS sub-systems for land registration is being conducted by the Land Registry's Computer Services Division at Drakes Hill Court, Burrington Way, Plymouth PL5 3LP, Tel. 0757 779851. A programme of implementation of a computerised application processing system (CAPS) primarily related to the register (text) records is currently in hand, and is planned to be installed in all land registries by the end of 1993. An exercise is currently in hand to identify a suitable state-of-the-art GIS product capable of holding and processing all of the Land Registry's public index map and title plan records, and interacting with the CAPS system. It is the intention to customise a suitable system to the Land Registry's requirements, and install and use it in a working environment at the Plymouth District Land Registry. Provided the system meets cost and operational performance criteria, it would be the intention to extend the system to other district land registries.

Although there are no immediate plans to provide direct on-line access facilities to Land Registry data, this is a feature of the Registry's long term strategy. In the meantime, the introduction of improved access methods such as searching by telephone, and by the use of FAX is in hand.

Horsham District Council

Having undertaken a pilot study into the use of GIS within the authority based within the Planning Department we are now entering the final stages. The successful completion of this will include connectivity to and outline data transfer between the authorities AS400 B70, the development of several prototype applications to store planning constraints and search of planning applications lying within them as well as being able to search and display all properties within the property database on an Ordnance Survey backdrop.

The policy is to integrate the GIS throughout the authority linked by a corporate property database referenced by Unique Property Reference Numbers. The graphic data will be held on the Siemens C40 and the alphanumeric data on the AS400. This will involve development of applications for such areas as Environmental, Health, Land Charges, Local Plans and ultimately others such as Finance where graphical data and analysis is or will be used.

Kent County Council

The development of applications in Highways and Planning, the two departments in which the GIS pilot project was located, are now well advanced, particularly in the areas of highway inventory and maintenance in Highways, exploiting the ARC/INFO–INGRES link, and the capture of wildlife habitat information in Planning.

GIS has now taken on a more corporate focus with the establishing of a three-man Corporate GIS Support Team to coordinate and assist in the development of key applications in other departments. This team will also be responsible for maintaining the large volume of Ordnance Survey large scale and other digital map data which will be acquired. A 'Map Management System' is due to be installed in early 1991 to aid the management of this data and make it available to ARC/INFO and other graphical systems in use.

Links with District councils, utilities and other authorities and organisations are being established to ensure that benefits of using GIS and exchanging information are exploited wherever possible.

Lancashire Highway Information System

A highways information system using ARC/INFO GIS interfaced to the ORACLE relational database. An in house digitised highway centreline network operates as the central integrator to a variety of specialist modules. This network is fully structured and referenced to include several administrative 'labels' which serve as possible enquiry parameters.

Initial Applications

These are primarily maintenance oriented and include facilities to handle an extensive highway inventory (ie signs, gullies etc), off road drainage, road condition surveys and a Routine Maintenance Management System. Output is available either as a schedule, network map, strip map or as business graphics (ie histograms or pie charts) where appropriate.

Further Development

Future development includes modules for the management of accident data, traffic flows, bridge data, operational routing for the routing of vehicles and construction records. In addition there is considerable interest in extending the system capability to provide full cartographic details.

London Borough of Barking and Dagenham

The Borough has been active in GIS for approximately three years. We use Hoskyns GGP together with AutoCad on both standalone and networked PC's. Ordnance Survey base maps for the whole Borough are held on our IBM mainframe. Most systems are installed in the Development and Technical Services Department where the more efficient production of maps and plans alone has almost justified the expenditure. Additional links to databases have been developed to further capitalise on this investment, for example, Street Lighting, and Planning. Our Education Dept will be using GGP for administration and to support in-school projects. A pilot project for our Legal Department aims to prove that we own the land we claim to! In 1991 we will be using aerial photographs obtained from both the Ordnance Survey and the London Research Centre. All addresses in the Borough have been geocoded and a major project is under way to examine the implications of establishing a Corporate Property Database. GIS/Digital Mapping activities are enhanced by a variety of PC packages and have been of particular use in the preparation of tenders for CCT.

London Borough of Harrow

Harrow's first phase towards the introduction of a Geographic Information Systems involves three departments—Architecture/Planning, Engineering and Estates/Valuation. There are ten workstations (Compaq 386) running Hoskyns GGP to which are attached a wide range of printers, plotters and digitisers.

The Authority has full Ordnance Survey digital coverage and the 1:1250 scale maps are stored on the central mainframe system (IBM compatible: DOS/VSE and CICS) and downloaded to departmental workstations as required.

Immediate benefits have been obtained from the digital mapping facilities in areas such as disposal of council properties by way of sale or lease, large scale highways redevelopment, preparing competitive tendering contracts, site surveys and maps for planning committees. Some projects also include links to Autocad.

The next phase will be to coordinate (i) the map updating procedures and (ii) the borough's information on property which is currently held on a number of different systems across departments. The basis of linking property data will be through a digitised reference for each property.

London Borough of Waltham Forest

The Housing Services Dept of the London Borough of Waltham Forest has the complete Grounds Maintenance details of Housing Properties and Estates on CADcore/HYBRID supplied by Duggal Associates. By windowing, one can display a map of the property or estate concerned, together with text giving all the relevant information, with detailed costs, areas, etc.

In the future it is envisaged that localised cost centres can be itemised in a similar way with data links to the Repair and Maintenance system, RAM, on the Departmental mini computer running under PICK on an EDP Mentor series 6000.

The above is achieved by the use of HITACHI CADcore TM/HYBRID which has full-featured VECTOR design (CAD) and fully integrated RASTER scanning and editing. It also utilises "TRACER", a special feature of CADcore/HYBRID which speeds the conversion process.

The whole system was installed because of the efficiency of CADcore/HYBRID and it's relatively low cost of installation and operation.

National Joint Utilities Group
Dudley Street Works Register Trial
(formerly Dudley Digital Records Trial)

National Joint Utility Trial, funded by British Gas, British Telecom, Mercury Communications, The Electricity Council and the Water Research Centre.

Commenced 1982/3 and went live in 1985, with all participants (Midlands Electricity Board, South Staffs Water Co, British Gas West Midlands, British Telecoms North & West Midlands District and Dudley Metropolitan Borough Council, as agents for the Severn Trent Water Authority) mains in adopted streets exchanged on a common map base created from Ordnance Survey digital data and maintained up to date from Developer's plans by Dudley MBC. The Trial covers an area of 50 square kilometres and a population of 150,000 and operates on a shared central PRIME computer using Pafec software. Development is now concluded and an Evaluation is available. The system is retained by the local participants as an operational system.

A Street Works Register was introduced in 1987 for the whole of the Dudley MBC area (100 square kilometres and 300,000 population), covering all Public Utilities Street Works Act (PUSWA) activities. This "active" system allows all "works" to be directed to interested parties by computer boundary analysis. The system uses a Prime/ORACLE data base in conjunction with the Pafec mapping software, providing considerable management facilities, from the works inception to the settlement of the final invoice. This non-paper system is being developed further and will be used as a test bed for a future national computerised street works register system with financial support from the original parties and the Department of Transport. Initial studies have been carried out, funded by the Department of Transport, to determine the most suitable approach to the creation of spatial references (in National Grid Format) for use in locating street works within the future national register.

Ordnance Survey of Great Britain

As Britain's national mapping agency, Ordnance Survey is responsible for the production of maps of the entire area of Great Britain. The basic mapping scales are 1:1250 in urban areas, 1:2500 in rural areas, and 1:10 000 in mountain and moorland areas with smaller scale maps being derived from these basic maps. Over 100,000 of the basic scale maps have already been converted to digital form, and good progress is being made on the remainder. In addition, digital data is available for the smaller scales of 1:250 000 and 1:625 000.

To keep this mass of digital data up to date, and to supply both data and maps to customers, Ordnance Survey uses Laser-Scan software together with software written in-house. The latter is for specific purposes such as update of the digital data in the field offices.

Ordnance Survey sees its role in the GIS world as a principal data provider. To help to fulfil this role a VAX-based ARC/INFO system has been installed to investigate present and future customer requirements for data structures and specifications. ARC/INFO also provides Ordnance Survey with an internal GIS facility, and the ability to offer specialised customer support. In future, Ordnance Survey aims to expand this facility to offer support and advice in the use of digital map data in a GIS environment.

Ordnance Survey of Ireland

Data capture and compilation for all scales of mapping is fully digital. Major investment in technology for digital mapping and photogrammetry is set to continue in 1991.

Data capture methods include blind and interactive digitizing and all data capture is in a structured format. Scanning and pattern recognition software is increasingly being used in production. A new small scale database is being produced and a totally new 1:50,000 mapping series will be produced.

Ordnance Survey of Northern Ireland

COMTOD is a fully structured digital topographic database being produced and maintained from the official survey archive of Northern Ireland. It is forming the fundamental locational element of the distributed but fully integrated Northern Ireland Geographic Information System (NIGIS), the initial stages of which are presently being put in place.

NIGIS is the means whereby the major Government and public utility functions within Northern Ireland can be linked, the georeference of the particular feature, be it house, land parcel, etc, providing the unique link between the graphic and associated textual information.

NIGIS development is being progressed through Liaison Committees involving Ordnance Survey of Northern Ireland and its partner organisations. Some of the main bodies involved are the Roads Service, Water Service, Planning Service, Land Registry, Valuation and Lands Office, Department of Agriculture, Department of Health and Social Services, Northern Ireland Electricity, Northern Ireland Housing Executive and British Telecom (NI).

Major progress to date has resulted from an on-going series of joint pilot projects, hosted by OSNI, which have enabled the partners involved to assess the potential of GIS for their own applications. A dedicated NIGIS Unit is presently being created at OSNI to act as the focal point for future progress.

Royal Borough of Kensington and Chelsea

A number of departmental pilot GIS systems are in the early stages of operation, running concurrently with the formulation of a corporate GIS strategy.

It is intended that the GIS pilots will assist the Directorates of Finance, Engineering and Works, Services and Planning and Transportation in:

- Providing a spatial distribution of the status of Community Charge payers at a small area level.
- Displaying the location and status of street lamps for routine maintenance.
- Displaying the location and details of street trees for contractual works.
- Providing a spatial framework for holding and analyzing on-street parking data.

Three Compaq 386 PC workstations have been installed, running GGP software from Hoskyns. An IRMA mainframe to PC link provides access to Ordnance Survey digital map data and the Pinpoint referenced Corporate Property Index. Twin objectives of the pilots are to raise awareness of the added value of GIS and to assess the potential for departmental and corporate benefits, particularly from the interchange of spatially referenced data.

Sandwell Metropolitan Borough Council

At present this Authority has 3 separate sites using digital mapping systems from McDonnell Douglas. It is our intention to link these together as soon as possible. We also have an ICL mainframe on which we are running a number of PLANES topic sets.

The McDonnell Douglas systems are currently being used for Land Surveys, site modelling, highway design, grounds maintenance, land terrier and property management. A land charges application is being developed.

PLANES topic sets include compulsory purchase orders and council house sales.

We are currently looking to form links between ICL's PLANES and Mcdonnell Douglas's mapping system.

The land terrier, property management and land charges applications are using a McDonnell Douglas 19045S computer to handle the database, while a

Digital Microvax 3400 handles the mapping. Plus software is used on the 19045S and is supplied by M.I.C. Computer Services Limited.

A variety of terminals are in use including Tektronix workstations, Vaxstations and P.C.'s.

Training, tailoring and consultancy have been received from McDonnell Douglas and the systems are all "live" with the exception of land charges.

The huge task of data capture has been started and McLintock Limited have been appointed to capture some of the graphical data.

It is hoped to make progress on all existing applications and spread the use of this technology throughout the Authority. Its spread will depend on the resources available, data capture and forming links between the different systems and databases.

Scottish Homes

Scottish Homes is a national housing development agency providing a national strategic function, with a remit to improve the level of understanding of the Scottish housing system, and a housing management service.

When Scottish Homes was set up, a corporate IT strategy review was undertaken by Coopers and Lybrand Deloitte. During this review, it was suggested that GIS technology might address a number of functional requirements.

Subsequently, a 6 month pilot was undertaken with the following objectives:

- To evaluate: (i) the extent to which GIS can enable tasks to be performed more efficiently; (ii) the costs associated with introducing GIS on a corporate basis; (iii) the potential benefits obtained through the introduction of a corporate GIS;
- to increase awareness within Scottish Homes of the potential uses of GIS; and
- to broaden specific GIS experience within Scottish Homes.

The two pilot projects currently being evaluated are:-

- The construction of a corporate research database driven by the needs of the Market Analysis Department of Strategic Development in head office;
- the provision of a housing research and management tool at district/regional level in the Ayrshire, Dumfries and Galloway Office.

A Technical Information Paper outlining the pilots will be available from Scottish Homes in the new financial year.

Shropshire County Council

GIS is being introduced within the Authority on an incremental or phased basis. First priority has been given to the introduction of a Property Information System using the PIMMS software. For this purpose, two workstations, digitising tablet and printer were installed at the end of 1989. Data capture and input is now proceeding for the urban area of Shrewsbury for which Ordnance Survey digital map coverage has been purchased. The need for a third workstation, InfoTER software and digital map coverage for Telford is being considered. The use of micro-based workstations is also under consideration for the development of other applications within the Planning and Property Services Department. However, it is intended that GIS will be a corporate facility with the coordinating or lead role being taken by Planning in close collaboration with the Computer Division which is currently developing a demonstration facility.

Silsoe College

Silsoe College is establishing a comprehensive suite of software packages within the Applied Remote Sensing Laboratory for training, research and consultancy. SPANS software is used for the main GIS activities with support from the ERDAS and GEMS image processing facilities. A SUN workstation supports GRASS and ARC/INFO has been acquired under the CHEST agreement. The IDRISI training package is available on the PC network of the Information Technology Laboratory.

A major database has been developed for the Countryside Commission which holds landscape maps for the National Parks. The maps were drawn up from aerial photograph interpretations at 1:10,000 scale. These were then digitised, mosaiced and analyzed using the SPANS software. The database is equivalent to nearly 900 1:10,000 maps and provides a comprehensive assessment of each Park.

Research projects involving GIS have included applications in forestry, geology, agriculture and soil conservation. Visitors to the laboratory are welcome by appointment and lecture/demonstration visits can be arranged for groups.

Southern Water Plc

Southern Water has installed one of the largest operational GIS networks in the UK, with over fifty SUN Workstations, distributed and interconnected throughout the south east home counties.

The many thousands of digital maps, in both raster and vector format, held centrally on an ICL-Series 39-Level 80 and on locally networked SUN systems, continue to grow and provide the geographical basis for development of other information systems needed by Southern Water Services.

Southern Water, having completed the digitising of all 11500 km of the water mains by early 1989, commissioned IT Southern to install, over the summer of 1989, Alper GISs in eight water supply operational centres across the Region. The Newport (Isle of Wight), Andover, Southampton, Hardham, Brighton, Hastings, Chatham, and Broadstairs operational centres have, since then, been fully supported by GIS.

These GIS's are networked together and linked to IT Southern's ICL Mainframe at Falmer and further application links are now under development. Full GIS support is available from IT Southern's development and technical systems teams via the communication network.

An in-house, thirty man data conversion team, is now many months into the task of computerising the 24600 km of Southern Water's sewers.

South Wales Electricity Plc

The system is installed at five locations within South Wales and is being used to capture the position of all the company's electric lines and cables. The records database will form a crucial part of the company network information system and provide the circuit information for design studies, operational diagrams and an asset register of the mains.

The information will be used in the future to schedule our maintenance programmes and provide the data for trouble ordering (fault and incident tracing) in a totally integrated database.

Strathclyde Water

The Water Department of Strathclyde Regional Council serves a population of around 2.2 million customers. The Department has recently completed a broad study to identify options, issues and constraints associated with the deployment and implementation of GIS within the organisation.

Four application areas have been selected: Distribution System Records, Distribution System Failures (Bursts), Water Quality Regulations, and Customer Complaints. These cover both the Utility and Strategic uses of GIS. A small pilot exercise is under-way for some of these applications, and there is a proposal (still subject to approval) that a larger scale trial be conducted.

Other Departments within the Council are also actively involved in GIS and there is recognition that a Corporate approach should be adopted on key datasets such as background mapping and demographic information.

Wakefield Metropolitan District Council

The GIS was purchased early in 1990 with the aim of providing a corporate resource to integrate, access and retrieve property related information held on existing databases on the Council's 5 McDonnell Douglas Series 129 mini computers which use the REALITY operating system.

The host VAX machine is located in the New Technology Unit whose role is to maintain the machine and its resident data and to assist in system development. Three departments are involved in developing applications using a variety of Tektronix, Pericom and Acer PC workstations, digitisers and plotters.

The Planning Department will manage the corporate district-wide digital map base and property gazetteer. It has a high quality electrostatic plotter. The initial development involves linking GIS to the computerised land availability database to demonstrate the system's capabilities.

The initial development in Transportation and Engineering is to provide an interface to the text based highways management system (HERMIS) which runs on INGRES.

Housing Department's main requirement is to link existing text files based on property addresses via the property gazetteer.

The first corporate application will involve computerising the Land Charges search process in association with McDonnell Douglas. This will include capturing all property boundaries, planning applications and other relevant data.

Welwyn Hatfield Council

This is an operational GIS for recording planning applications and reporting on the policy zonings and physical features which affect each one. Work is currently in progress on capturing existing planning application records dating back to the beginning of 1975. New applications will be plotted directly into the GIS as they are received from around July 1991.

Information entered into the GIS is recorded over an Ordnance Survey vector map background. The system is in use to provide an Ordnance Survey map extract service for the Authority as a whole.

West Somerset District Council

The Council has developed its own GIS using the IBM-GFIS environment.

The principle of development employed was that all information maintained by the Council related to land/property, thus the primary feature of the GIS is that all land/property parcels are identified to the system.

Departmental applications have been developed overlaying the primary land/property parcels. Data from the primary land/property feature are passed to all succeeding applications and, equally, common data between departmental applications is passed to each relevant application.

End users are provided with flexible access to all geographical and textual data whereby any enquiry can be made by users setting their own criteria for selection through menu driven panels.

The Council's development plan is over a five year timescale. Within the two years of development now accomplished applications have been developed relating to:-

Primary Land Parcels
Land Charges
Planning Applications
Sewer and Associated Records
Housing
Land Terrier
Private Water Supplies
Public Footpaths and Highways
Public Utilities

Further development of applications will continue to provide a total Geographic Information System for the Council within the five year programme.

PART XII

Trade directories

Introduction

One of the primary aims of the *Yearbook* is to inform the user as fully as possible on many and varied aspects of GIS/LIS. With the help of Geobase Consultants a series of trade directories has been constructed broken down into the following seven categories:

Within each category the entrants are listed alphabetically. In addition to brief tabular entries, certain entrants have elected to supply a more full description of their services and products. These have been written by the companies concerned; they do not constitute impartial reviews by the Editors or the AGI. The extended entries are indicated by page number in the final column of each table (headed *Ref.*), and follow the tabular entries in each category.

A complete alphabetical index of all entrants is given on pp.418–420, in order that search may be made by name of entrant as an alternative to activity or product type.

Companies or organisations wishing to be included in these directories in future editions of the *Yearbook* are welcome to contact:

Geobase Consultants Ltd.
28 Church Road,
Epsom,
Surrey KT17 4DX

GIS systems and supporting software

Trading name, phone and contact	Address	GIS system function	Hardware supported	Ref.
Action Information (Management) Limited 0225 777288 John Page	Ashton Road, Hilperton, Wiltshire BA14 7SZ	MAPS IN ACTION. The map driven operational information system for decision makers. Easy manipulation and display of geo-referenced data including AVLS reports in real time.	Standalone on IBM PC compatible or as an intelligent terminal to a Network or Mainframe.	
Alper Systems Limited 0223 420464 Anne Conder	Cambridge Science Park, Milton Road, Cambridge CB4 4WG	Alper GIS offers the full range of GIS functionality, including a unique blend of raster & vector maps: RECORDS—data management, ENQUIRY—decision support, AD2000 line following, conversion.	ALPER MAPS runs on networked SUN workstations linked to mainframe databases for high performance GIS. Services include: data conversion, systems integration, consultancy, installation/training.	
D'Arcy Race Limited 071 631 4841 Jon J Gibbons	96 Portland Place, London W1N 3HD	DRUID: digital mapping, attribute data, SQL interface, comprehensive editing and updating facilities, ESTATE MANAGEMENT MODULES.	RM Nimbus VX386, IBM compatibles, Cambridge Computer Graphics Screen, Summagraphics digitising tablet.	308
Auto-trol Technology 021 456 2022 Barry Jones	53-54 Calthorpe Road, Edgbaston, Birmingham B15 1TH	Graphics SQL, RDBMS	SUN Microsystems, Hewlett-Packard, Digital Equipment.	
Axis Software Systems Limited 0604 720477 Eric Malcolmson	Park House, 88-102 Kingsley Park Terrace, Northampton, NN2 7HJ	Full function single/ corporate GIS system also capable of integration with external D/BS and user customisable.	SUN, APOLLO, HP, X-Windows PC-MSDOS/ VGA workstations. Most plotters and digitisers. Most Local Authority applications plus demographic and 3D analysis.	

Trading name, phone and contact	Address	GIS system function	Hardware supported	Ref.
Bradly Associates Limited 0344 779381 Kevin Bradly	Manhattan House, 140 High Street, Crowthorne, Berkshire RG11 7AT	Supplier of GINOSURF (subroutine library) and VIEWSURF (interactive program) for visualising 3D data as contour maps, perspective views and cross-sections.	A wide range of hardware is supported from PC's and workstations to mainframes and Crays. Output includes HPGL, Postscript, DXF, CGM and many others.	
CACI Limited 071 528 7980 John Rae	Regent House, 89 Kingsway, London WC2B 6RH	ATLAS*GRAPHICS—Mapping program for designing and generating maps showing geographic based information.	IBM and 100% compatible PC's. Wide variety of Hewlett Packard and other plotters and graphics printers.	
CACI Limited 071 528 7980 John Rae	Regent House, 89 Kingsway, London WC2B 6RH	INSITE—Storage, analysis, reporting and mapping of geographic information. Flexible database, area searching, polygon/ circle creation, report generator.	IBM and 100% compatible PC's. Wide variety of Hewlett Packard and other plotters and graphics printers.	
CES Limited 071 388 7187 Andrew Broadbent	5 Tavistock Place, London WC1H 9SS	AEGIS is an Economic and Geographic Information System for strategic planning, forecasting, impact and location studies (drivetimes, socio-economic indicators, consumer spending etc.)	IBM compatible PC's, Hewlett Packard plotters and graphics printers etc.	308
Cimage International Limited 0344 860055 Ewan Botterill	Centennial Court, Easthampstead Road, Bracknell, Berkshire RG12 1JZ	ImageMaster combines DIP and RDBMS for applications where seamless maps and visualisation of geophysical phenomena are important, i.e. Resource planning, radio wave propagation.	ImageMaster supports industry standard platforms, protocols and operating systems including DOS based PC's and UNIX workstations. No proprietary hardware is required.	309
Cobham Digital Services Limited 0932 68133 Bob Coote	Vine House, 41 Portsmouth Road, Cobham, Surrey KT11 1JQ	Digital Mapping on PC's linking to information databases to provide full GIS for Local Government/ Utilities/ Private Sector. Application and Design Consultants.	IBM and 100% compatible PC'c. SUNSparc computers. High resolution digitisers and graphics. Océ Graphics and Calcomp peripherals.	

Trading name, phone and contact	Address	GIS system function	Hardware supported	Ref.
Computer General GeoData Services 0225 482744 Dr Pauline A Waters	The New Church, Henry Street, Bath BA1 1JR	PCI EASI/PACE: User-adaptable, easy to use, integrated suite for image processing, analysis and mapping of geographic information. GRASS—Public Domain raster GIS.	SUN, Silicon Graphics and IBM RS/6000 workstations; Digital, Stardent, IBM and AT compatible PCs. Supports multi-vendor networks with full networking transparency.	309
Data Consultancy 0734 588181 Bryan Wade	7 Southern Court, South Street, Reading, Berkshire RG1 4QS	MapInfo GIS, drivetime isochrone system, geo-demographic analysis system, store location modelling system. Map and data files for all systems.	IBM PC's and compatibles. Apple Macintosh computers.	310
Digital Equipment Company Limited 0256 843333 Andrea White	Hampshire House, Wade Road, Basingstoke, Hampshire RG24 0PL	Integrated GIS for all markets—Local Government, Utilities, Private Sector. UNIX and VMS supported.	Digital VAX Computers, VAXStations, DECStations, UNIX, VMS, DECNET, Ethernet.	311
Doric Computer Systems 0923 210450 Peter Paisley	Doric House, 23 Woodford Road, Watford, Hertfordshire WD1 1PB	ARC/INFO is a database oriented GIS integrating map and attribute data and providing sophisticated analysis and modelling capabilities. Output is as high quality maps or tabular reports.	ARC/INFO runs on a wide range of computer hardware platforms—micro's, UNIX workstations, mini's and main-frames, and supports a large number of peripherals.	312
Duggal Associates Limited 081 204 8351 Malcolm Shelbourn	73 Portland Crescent, Stanmore, Middlesex HA7 1LY	CADCORE, developed & maintained by Hitachi, captures and integrates map & attribute data in a spatial data reference system for Utilities, Local Authorities and Environment Agencies.	CADCORE is MSDOS based & real time linked to an Oracle database—on most PC's, mini's & mainframes. Open archi-tecture & comprehensive software dev-elopment tools for total customisation.	
Earth Observation Sciences 0252 811181 Matthew Stuttard	Branksome Chambers, Branksomewood Road, Fleet, Hampshire GU13 8JS	ILWIS: raster GIS and satellite image processing on a PC. Includes vector digitising and conversion, table calcula-tion/ graph plotting. Developed by earth scientists at ITC.	AT compatible PC or 386 under DOS with maths co-processor. Supports a range of VGA graphics boards, plotters, printers and digitisers. Used for applica-tions and education worldwide.	

Trading name, phone and contact	Address	GIS system function	Hardware supported	Ref.
Eurecart 0621 851114 Andy Wolfe	Ulting Lane, Hatfield Road, Langford, Essex CM9 6QA	APIC—Full object-oriented GIS solutions with outstanding ability to build systems to meet user needs. 50+ sites across Europe in the Utility, Local Authority, Telecomms & Private Sector markets.	APIC supports a wide range of workstations, servers & X-terminals operating in standalone or networked environments, including SUN and Digital (VMS/ ULTRIX).	313
Genasys (UK) Limited 061 232 9444 Peter Bayley	Manchester Science Park, Lloyd Street North, Manchester M15 4EN	Open systems suite of GIS products inc. GENAMAP/GENACELL, vector & raster GIS, GENACIVIL, civil eng., GENASCAN/ GENARAVE, scanning & vectorising systems all with GENIUS OSF/MOTIF 'GUI'.	HP, IBM, Silicon Graphics, MIPS, SUN and PC's under UNIX.	313
Geographic Management Systems Limited 0454 618618 Bill Knowles	Woodlands Grange, Woodlands Lane, Almondsbury, Bristol BS12 4JY	GS-MAP & PC-MAP are multi-application GIS developed specifically for the PC market, running on IBM AT and PS/2 micros with two way links to mainframes and associated databases.	M.A.P. software provides 90% of large system functionality at a fraction of the cost. Networking and direct linking capabilities will support all system integration requirements.	314
GeoMatrix Limited 0742 724272 Edwin Guiton/ Geoff Beacon	The Cooper Buildings, Sheffield Science Park, Arundel Street, Sheffield S1 2NS	* ProSpex Market Analysis System * 1:1250 mapping database * Ground Modelling * Architectural Modelling/ Facilities Management	* PC AT compatibles * PC ATs and workstations * PC ATs and workstations * Unix workstations	
Gimms Limited 031 668 3046 Marlene A. Ferenth	30 Keir Street, Edinburgh EH3 9EU	GIMMS mapping and graphics software for thematic/ statistical mapping.	Available for most mainframes and minicomputers. Now also available for IBM PC/AT (& 100% compatibles) and APPLE Macintosh. Supports a wide range of output drivers.	
GIS Applications Unit 031 557 5824 Gavin Park	Department of Geography, University of Edinburgh, Drummond Street, Edinburgh EH8 9XP	On-line map and air photograph cataloguing and retrieval systems (CARTONET and PHOTONET) using an interactive graphical interface to a relational database.	VAX/ VMS and IBM/ PCDOS.	

Trading name, phone and contact	Address	GIS system function	Hardware supported	Ref.
Graphic Information Systems Limited. 0506 39251 Mrs. A Burggraaff	Pentland House, Almondvale, Livingston, West Lothian EH54 6NG	MapInfo is a low-cost, user friendly desktop mapping package. Its analysis functions include: locate, query, filter, search within radius or polygon and a range of thematic calculations.	IBM PC XT, AT, PS/2 or compatible with 640kb, MS DOS 2.0 or higher or Windows 3.0+2MB. Apple Macintosh+ 1MByte, system 4.1 or later. SUN SPARC Station 1 or 2 with Open Windows.	
Ground Modelling Systems Limited 0908 667799 Steve Gould	Rockingham Drive, Linford Wood, Milton Keynes MK14 6PD	PanTerra land survey, GIS, ground modelling and civil engineering design.	AT compatible PCs, graphics screens, plotters, printers and digitisers.	
Grove Consultants Limited 081 846 2468 A.H. Palmer	28 Hammersmith Grove, London W6 7EN	Digital mapping; land surveying including network adjustment and feature coding; ground modelling; road design; earthworks; estate layouts; general purpose draughting.	Data General Eclipse AOS/VS. IBM PC-AT & compatibles; digitisers; graphics displays; pen plotters and printers.	
Hoskyns GIS 0962 844188 Julie Hoskyn	Technology House, Victoria Road, Winchester, Hampshire SO23 7DU	A Range of GIS (PIMMS, GEONET and GGP) for Local Government, Utilities and private sectors.	IBM 9370, 43XX, 3090, RS6000, 286/386 PC's. Full support: Helpline, Consultancy and Training.	315
Howard Humphreys & Partners 0372 376190 IT Applications Group	Thorncroft Manor, Dorking Road, Leatherhead, Surrey KT22 8JB	Roads/ Structures Mngmnt. Systems. Highway network referencing. RMMS inventory + inspections. Cyclical maintenance; accidents. Traffic flow; lighting. Structures	Digital VAX + Ingres; SUN + Ingres; IBM PC + Ingres; PRIME + Oracle; SUN + Oracle; PC + Oracle.	
IBM United Kingdom Limited 0926 332525 Preet Dhillon	GIS Market Development, PO Box 31, Birmingham Road, Warwick CV34 5JL	IBM, with its Business Partners, offers an enterprise approach to GIS—addressing functional needs (mapping & facilities management) & also the integration of GIS with corporate information systems.	IBM's GIS is implemented on: * IBM RS/6000 tech. workstations * Mainframe+, enterprise systems * IBM PS/2 workstations * Low cost enquiry terminals	316

Trading name, phone and contact	Address	GIS system function	Hardware supported	Ref.
ICL (UK) Limited 0532 441111 Tony Grimbley	ICL House, 5 South Parade, Leeds LS1 5QZ	PLANES centralised GIS providing data integration and application development. DINIS Network analysis.	ICL Series 39 mainframe range from standalone departmental systems to large DP centre systems. ICL SUN workstation range.	317
Intergraph (UK) Limited 0793 619999 NJ Price	Delta Business Park, Great Western Way, Swindon, Wiltshire SN5 7XP	Intergraph supply a comprehensive range of interactive GIS systems based on graphics linked to SQL standard databases and full topological structures.	Intergraph design both hardware and software to optimise performance, whilst complying with standards such as UNIX, Ethernet, XNS, TCP/IP, OSI etc.	318
IT Southern Limited 0273 606766 Peter Elkins	Southern House, Lewes Road, Brighton, East Sussex BN1 9PY	Planning, integration of all GIS components to establish extensive operational systems, inc. the new MANTELL sewer data conversion software package. Consultancy, data conversions, implementation.	* SUN Workstations * ICL * CALCOMP	318
Kingswood Limited 081 994 5404 M. Short	19 Kingswood Road, London W4 5EU	Map-based information systems, custom built or using commercially available software.	IBM—DOS, OS/2, Windows IBM—RS/6000, Apple Macintosh, UNIX	
Landplan West Country Limited 0803 842065 Mrs Karen Pearson	The Studio, Alston Lane, Churston, Brixham, Devon TQ5 0HT	VIRIDIAN GIS for planning, environmental and land use applications. High quality raster mapping and scanned images, vector overlays, graphics, full relational database. Turnkey GIS.	All SUN workstations.	
Laser-Scan 0223 420414 Doug Cross	Cambridge Science Park, Milton Road, Cambridge CB4 4FY	Laser-Scan GIS enable Local Authorities, Utilities, Environment Agencies, Defence customers and Commercial users to record, analyse and display a variety of spatial and related data.	Solutions are based on industry standard hardware workstations and servers, with the option to attach other workstation types or PC's as well as many other peripherals.	319

Trading name, phone and contact	Address	GIS system function	Hardware supported	Ref.
Market Profiles Limited 0423 566755 Andrew Kelly	14/15 Regent Parade, Harrogate, North Yorkshire HG1 5AW	Agents for market analysis systems: PROSPEX & MOSAIC Systems. Owners of GEOPLAN digitised postcode sector boundary file.	Systems run on IBM PC and compatibles. GEOPLAN is available in ASCII & IBM EBCDIC: other formats on request.	319
Mason Land Surveys 041 222 7733 David Weatherill	14 Newton Terrace, Glasgow G3 7PJ	SLIMPAC: property map information system for terriers, facilities management and control of building complexes (factories, estates, hospitals, colleges, shopping centres).	AUTOCAD linked to database. 386 based microcomputer. IBM compatible.	320
MOSS Systems Limited 0403 59511 Ken Livingstone	MOSS House, North Heath Lane, Horsham, West Sussex RH12 4QE	MOSS is a surface modelling system for civil and mining engineering, land survey and cartography; used for recording and designing surface details and associated data.	Engineering workstations: Digital, SUN, HP/Apollo, IBM, Intergraph. Also Prime, IBM 6150.	320
Nextbase Limited 0784 460077 Peter Stansbury	Unit 18, Central Trading Estate, Staines, Middlesex TW18 4XE	AUTOROUTE PLUS is an easily used, low cost, powerful GIS for route planning, database analysis and business planning on maps of Great Britain, France and Europe.	AUTOROUTE PLUS requires an IBM PC or compatible with a hard disk and a graphics screen. It works with most standard printers.	
NPA Group 0732 865023 Nigel Press	Edenbridge, Kent TN8 6HS	MIPS: an integrated raster and vector GIS processing software package. DIAD: software for satellite image processing.	IBM PC and related workstations plus wide range of peripheral scanners, plotters, printers.	
Pinpoint Analysis 071 836 1511 Martin Higgins	Tower House, Southampton Street, London WC2E 7HN	GEOPIN is based on ARC/INFO, a modular GIS. Functions include interactive query, map and report display; network analysis and modelling; incorporating data transfer.	GEOPIN will run on a wide range of hardware platforms, including PC, workstation, mini and mainframe versions. A wide range of peripherals are also supported.	

Trading name, phone and contact	Address	GIS system function	Hardware supported	Ref.
Prime Computer (UK) Limited 0276 682821 Mike Cleghorn	Beech House, 373-399 London Road, Camberley, Surrey GU15 3HR	SYSTEM 9 is an advanced, comprehensive GIS ideally suited to capturing, analysing large volumes of land-based data as it relates to economic, environmental or demographic factors.	SYSTEM 9 runs on the SUN family of workstations from Prime, operating stand-alone or linked together via an Ethernet Local Area Network.	321
Ramtek (UK) Limited 0256 469541 Jonathan Shears	The Old Coach House, The Avenue, Farleigh Wallop, Basingstoke, Hampshire RG25 2HS	Sophisticated Remote Sensing package seamlessly integrated with ARC/INFO providing simultaneous vector and raster processing. Supports virtual imaging and 64 bit data types.	UNIX platforms include SUN 4, IBM 6000, Digital and HP/ Apollo, accelerated version available on Ramtek M-series. All with 8 or 24 bit X-Windows display.	
Remote Sensing Services Limited 0672 20226 Duncan Milnes	Lychgate House, 24 High Street, Ramsbury, Wiltshire SN8 2PH	ERDAS—raster based remote sensing, ARC/INFO, vector based GIS, and Live Link, to combine both images on one screen.	PC's and PS/2, workstations including SUN, Solbourne, UNIX systems including Digital, HP etc.	322
SIA Limited 071 730 4544 Janice Mabert	Ebury Gate, 23 Lower Belgrave Street, London SW1W 0NW	DataMAP: GIS solutions and PC based software for Local and Central Government; Utilities; Highway Authorities; Emergency Services and the private sector.	GIS applications developed for a wide range of areas including land registration, planning, crime analysis, sewer records, street furniture, property/ land terrier management, retail & distribution.	
SIAS Limited 031 557 0200 Sarah McRae	4 Heriot Row, Edinburgh EH3 6HU	BROWSER (Crown Copyright) provides a user-friendly graphical interface to spatial data which may be downloaded from other computers and databases. Map, graph & report formats.	Digital VAX & VAXstation (VMS) & DECstation (ULTRIX), IBM PC-AT & compatibles (DOS), IBM PS2 (DOS), Logicraft 386ware (DOS).	
Smallworld Systems Limited 0223 460199 Richard Green	Burleigh House, 13-15 Newmarket Road, Cambridge CB5 8EG	Smallworld GIS—Modern system based on strong fundamentals with applications for utilities and local authorities. Highly customisable multi-user GIS.	DIGITAL VAXStation/ VMS, DECStation/ Ultrix, SUN Sparcstation/ SUNOS. X-Windows, Openlook and Motif. PC/ X-Server access. Heterogenous network support.	323

Trading name, phone and contact	Address	GIS system function	Hardware supported	Ref.
Strategic Mapping Inc. 081 994 2780 Clive Humby	C/o Dunn Humby Associates, 9 Chiswick High Road, London W4 2ND	ATLAS*GIS: Full featured GIS for storage, analysis, mapping and reporting. Direct links to DBase data files. Up to 250 geographies.	IBM PC and Compatibles. Network or stand-alone. Wide range of plotters, digitisers and output file formats.	324
Structural Technologies Limited 0527 854819 DF Schindler	Woodside, The Slough, Studley, Warwickshire B80 7EN	STRUMAP: Interactive GIS, Topology and Attributes. Line following, assisted digitising. Raster/ vector backgrounds. QA features. Import/ Export for NTF0-4, IBM GFIS, AutoCAD etc.	MS-DOS, Xenix, Unix running on 286, 386, Apollo, RS6000 etc. Windows, X-Windows, support for HPGL, Tektronix.	
SysScan (UK) Limited 0344 424321 Stuart Maxwell	SysScan House, Easthampstead Road, Bracknell, Berkshire RG12 1NS	Full GIS with vectorising, editing, management, plot output and database facilities. Also PC-based TERRASOFT GIS and GEOREC data conversion software.	Based on full range of standard Digital processors running under VMS. TERRASOFT runs on standalone or networked 80386 based PCs.	324
Systems Options Limited 0252 334383 Mr Jim Pedroza	45 Victoria Road, Aldershot, Hampshire GU11 1SJ	WINGS is a PC based GIS using MS-Windows/ Presentation Manager. It supports scanned and digital maps. Interfaces to external databases.	PC AT compatibles and networks.	
Tydac Technologies Limited 0491 411100 Adla Worobec	Chiltern House, 45 Station Road, Henley-on-Thames, Oxfordshire RG9 1AT	SPANS, modular GIS with advanced data modelling and analysts functions; integrates point, vector, raster and quad-tree data; menu driven and may be customised.	SPANS runs on DOS, OS/2 and AIX platforms including 386 PC's, IBM PS/2's and RISC 6000 workstations. A wide range of digitizers, plotters and printers are supported.	
Unisys Limited 0532 428444 Mike Kowalski	Central Park, New Lane, Leeds LS11 5EB	ARGIS 4GE is a UNIX based GIS providing a continuous integrated map and attribute database. ORACLE RDBMS and powerful 4GE facilities enable rapid development of end-user applications.	UNISYS U5000, U6000, ranges of UNIX processors. SUN workstations. A wide range of peripheral devices including digitisers, plotters and printers.	325

Trading name, phone and contact	Address	GIS system function	Hardware supported	Ref.
URPI Group Limited 0734 588181 Bryan Wade	7 Southern Court, South Street, Reading, Berkshire RG1 4QS	MapInfo GIS, drivetime isochrone system, geo-demographic analysis system, store location modelling system. Map and data files, including Census of Population, for all systems.	IBM PC's and compatibles. Apple Macintosh computers.	
Wang (UK) Limited 081 568 9200 David Robson	1000 Great West Road, Brentford, Middlesex TW8 9HL	GDMS, a true GIS featuring object orientated Geographical model, Virtual database, 4GL Application Language, 65535 layers, non-dedicated CPU plus much, much more.	GDMS runs on the Wang VS range of servers. Graphic workstations are any PC-AT. A wide range of Digitisers, Printers and Plotters are supported. Supports most standard exchange formats.	326

D'Arcy Race Limited

Estate management is the process of continual optimisation and adjustment of an organisation's land, buildings, utilities and equipment resource in relation to its present activities and long term goals. The way in which this resource is managed in conjunction with the corporate resources can have a major impact on the organisation's economy and effectiveness.

D'Arcy Race Limited offers assistance to organisations in implementing cost effective estate management. We provide a co-ordinated range of services, techniques and computer systems based on DRUID our own PC-based geographical information system. DRUID's powerful graphics and database system compare favourably with mini and main-frame systems costing more than twice the price. DRUID is also a true multi-user system in a networked environment. DRUID can be customised to individual requirements by using the inbuilt application programming language.

As well as computer systems we can offer advice and services on any estate-related problem. If you think we may be able to help you then please contact us. We will be pleased to discuss your problems and requirements.

CES Limited

AEGIS—an 'Economic and Geographic Information System' is designed for strategic planning, forecasting, impact studies, and report production, rather than for handling the massive and often obscuring detail of a 'corporate GIS' used for day-to-day operations to track every metre of sewage pipe or dwelling over a huge area. But AEGIS can be used at any geographic level from region to street.

With its tailored menus, and bundled datasets, including road networks, demographic and economic data, it can be used quickly and easily without huge investment in trained staff, and data capture. A novice user can produce drive-time boundaries, travel routes for anywhere in the country, demographic tables, consumer spending estimates etc.

AEGIS has a command language which can be used to build applications, and its analytical capability is state of the art. Modules include:

I Isochrones—drive time areas
II Area summaries and demographic profiles
III Database and model builder
IV Road network editor
V Consumer spending
VI Distribution and competition
VII Maps
VIII Graphs

Implemented applications include a FORECASTING SYSTEM, which produces socio-economic profiles for regions, counties, towns etc. and a SITE IMPACT SYSTEM, showing catchment areas and traffic loadings on road links leading to a site. It has been used in applications with property Researchers, Commercial Estate Agents, Data Consultancies, Transport and Planning Consultants and in Academic Research.

Cimage International Limited

Cimage Corporation provides Technical Document Image Management Systems that combine raster/vector graphics with relational database technology. Applications where 'seamless' maps or visualisation of geophysical phenomena are important are covered. These include:- emergency planning, resource planning, radio wave propagation imaging, underground service alerts, utility network planning.

Data take-on rates are dramatically improved by scanner input of maps and plans etc. and by the automatic population of databases. ImageMaster supports industry standard platforms, protocols and operating systems including IBM and IBM compatible PCs and PS/2s using DOS, IBM RS/6000 (AIX), Sun SPARC-stations (SunOS), DECstations (ULTRIX). The system requires no proprietary hardware. It provides multi-system, multi-product transparency for the User.

Cimage offers a full complement of support services to assist in implementation. This is backed by a team of computing and engineering professionals dedicated to working closely with customers on all aspects of application design and development, integration to major line of business applications, training and document image processing technology.

Computer General GeoData Services

Computer General GeoData Services (GDS) has been established to meet the highly specialised image processing needs of organisations involved in Geographic Information System (GIS) implementation and management. GDS aims to maximise the operational flexibility of GIS users in terms of facilities and operating schedules.

Combining nearly 15 years of remote sensing experience, with 10 years computer systems expertise, GDS offer a total solution. Services include Systems Consultancy (fully configured turnkey solutions or bundled solutions comprising hardware, software and pre-processed data sets), Research & Development, Training and data processing bureau facilities. All services are tailored individually to complement the availability of in-house hardware, expertise and budget.

The GDS total solutions approach enables organisations to save months of system development time and cost, or even achieve results without purchasing hardware, ensuring cost effective system implementation and maintenance. Provision of bureau services extend the available processing facilities of organisations to meet short deadlines.

The leading image processing and GIS software suite, PCI EASI/PACE, supports a wide range of industry standard workstations as well as IBM and compatible personal computers. EASI/PACE integrates both raster and vector data and operates with full networking transparency over mixed vendor networks. DIRIGO from PCI is available for the Apple Macintosh and is data compatible with EASI/PACE.

The Data Consultancy

The Data Consultancy is an established provider of GIS related systems including drive time systems and geo-demographic analysis systems and is an authorised value added reseller of MapInfo the leading desktop GIS and mapping system. The Data Consultancy also provide a range of other software products designed for market analysis and site assessment studies. Although available as separate systems each system interfaces to the others, including MapInfo, to provide a full range of GIS, market analysis and mapping capabilities.

- **MAPINFO:** the leading desktop GIS and mapping system which sets the standard for cost effectiveness. It is particularly suitable for marketing, site assessment, planning and business applications. Includes display, editing, overlay, analysis, query, searching, geocoding and digitising in one easy to use package. Fast display, analysis and querying of user data files in dBASE format. Ideal for mapping and analyzing Census data at all spatial levels, including enumeration districts.
- **DRIVETIME:** an extremely powerful and flexible interactive system to generate drive time isochrones around any location in Great Britain. Outputs maps and plots on dot matrix printers and HP plotters and generates boundary files suitable for input to mapping and other programs.
- **ILLUMINE:** a full geo-demographic analysis and reporting system giving direct access to the Census of Population small area statistics for all 130,000 Census enumeration districts and/or postcode sectors in GB. Analyses can be undertaken for any area however defined. The modular design allows users to start with a simple entry level system with a few variables and add additional variables, population updates, neighbourhood profiling, expenditure estimates, retail business turnover potential, drive time isochrones etc. as and when required.
- **MARKETS:** our widely acclaimed and widely used store location model which simulates and predicts shopping patterns and expenditure flows. Model results include store or scheme turnover levels, turnover to floorspace ratios, trade draw rates and market penetration by drive time bands or zones.

The software can be purchased on its own or fully configured with UK boundaries and data. A wide range of databases, boundary files, maps and analyses can be supplied.

Digital Equipment Company Limited

Digital Equipment Company Ltd. is the world's leading manufacturer of integrated, networked computer systems.

The Company's products are used worldwide, in all major industries and together with its third party business partners, provides more business applications than any other supplier in both UNIX and VMS environments.

Digital supplies Integrated GIS systems that are based on open and flexible system architectures that can be easily tailored to suit all customer requirements.

This successful approach has been used in the development of some of the world's largest GIS projects, where Digital has prime contracted a total solution, starting from requirements analysis, aligning GIS technology to business needs, through to education and training of staff.

Digital's approach allows Geographic Information Systems to be integrated with existing information technology employed by an organisation, so that map-based systems can incorporate data already held within diverse and different databases. This clearly protects investments already made in data and allows full use to be made of it in a GIS analysis and planning environment.

Digital is committed to providing successful GIS solutions which can easily grow to meet changing business needs.

Doric Computer Systems

ARC/INFO is the complete Geographic Information System solution for the management and analysis of spatially referenced data. A market leader in GIS with many hundreds of installations worldwide. ARC/INFO applications include local government planning, property management, highway information systems, facility management, emergency planning, vehicle routing, marketing and many more.

ARC/INFO offers:

- Efficient data entry with input from digitizers, scanners, Ordnance Survey digital data and output from other CAD & mapping systems.
- Generate new maps of high quality with attribute information linked to map features such as properties, roads, post codes etc.
- Query the system to extract information on map features.
- Exchange digital map and attribute date with utilities and other digital map users.
- Carry out 'what if' modelling on development scenarios to assist in planning of facility siting, road improvement, welfare considerations etc.
- Overlay data from various sources to identify the geographic relationships and combine the attribute information.
- Carry out network analysis to plan routes or allocate demand to centres such as schools, fire stations and hospitals.
- Create 3-Dimensional models for water pollution, soil erosion etc.
- Raster display and editing of scanned maps and documents with vector data overlaid on the raster background.

Hardware platforms for ARC/INFO include IBM AT and PS/2 micros; Sun, Digital, IBM, Apollo, Data General and Hewlett Packard workstations; Digital VAX, Prime and Data General mini computers; and IBM mainframe.

Doric provide application development services, consultancy, database design, and full training and support for ARC/INFO users.

Eurecart

Eurecart is part of the Lyonnaise des Eaux-Dumez group marketing the GIS software APIC in the UK. Eurecart is directly related to Polilog, the developers of the software; and Eurecart in France who have been established in Europe for the last 5 years providing fully operational GIS solutions to the Utility, Local Authority, Telecommunications and Private Sector markets. Eurecart manage complete projects from consultation to defined user requirements through to customisation and integration with existing systems and procedures.

APIC is an object-oriented GIS which provides the user with ease of customisation and a powerful query language. APIC systems are built around standard modules providing the following features:

- Data capture and import
- A fully continuous database
- Collation and analysis of statistics
- Thematic Mapping/ Polygon Processing
- Network Management
- Comprehensive reporting and map presentation
- Integration with SQL databases
- Standalone or networked environments
- Distributed Database

We offer a solution providing real business benefit. Our approach is customer oriented, APIC is performance oriented.

Genasys (UK) Limited

The Genamap Family

Genasys can supply GIS products to address a wide range of applications. The core product, **Genamap** is an Open Systems based, topological 3D GIS with extensive data management and spatial analysis capabilities. **Genacell** extends Genamap's facilities to include cell and raster management and analysis. **Genacom** provides tightly-coupled access to multiple external databases. **Genacivil** is a full Civil Engineering design system, integrated with Genamap. **Genascan** supports the scanning, import and editing of documents and maps. **Genarave** is a semi-automatic vectoriser that enhances the capture of graphical features from the scanned maps. **Genindex** is a comprehensive document management system for coordination of vector, raster, image and database information. **Genaris** is an image processing system based on the DISIMP package from Australia's Commonwealth Scientific and Industrial Research Organisation (C.S.I.R.O.) Finally, **Genius** is an OSF/Motif-based Graphical User Interface that allows users to develop customised windows-based applications on any platform simply by pointing with a mouse. Genius is used by the entire product family, ensuring an integrated, consistent interface compatible with other OSF/Motif applications the user may have.

Geographic Management Systems Limited

GS-MAP (Geographic Systems Multi-Application Product) is a system developed specifically for the MS-DOS PC market, running on IBM AT and PS/2 microcomputers with two way links to mainframes and associated databases.

PC-MAP is a smaller textual and integrated graphical management development tool.

Graphical data can be used as a powerful enquiry medium linked to unlimited text.

M.A.P. software provides:-

- Data accepted from two or more external databases, as well as its own in-built files;
- Unlimited manipulation screen display and high quality output of thematic maps and attribute hardcopy;
- Easy data entry via digitizers, scanners and mouse; direct input from Ordnance Survey digital data, CAD and mapping systems plus satellite;
- Two-way search facility between graphical and textual information databases;
- Unrestricted zoom capability, 9^9;
- Polygon overlay and comparison from up to 255 layers with automatic text carry-over when elements move from layer to layer;
- Exceptional data compression features storing 12 times more information than larger systems;
- Full data handling including editing tools for comprehensive information management;
- Controlled access with password entry;
- Modular construction permitting system expansion when required;
- Vector overlay and editing of raster scanned images;
- Menu driven system with minimal training requirements, all resulting in:-
- **90% of large system functionality at a fraction of the cost!**

Hoskyns GIS

Hoskyns Group plc is one of the leading computer services companies in the UK. Founded in 1964, the Group celebrated 26 years of success this year and is now a substantial organisation with offices throughout the UK and, increasingly, across the world. There are over 3,800 people within the Group which has current worldwide revenues of £189 million.

Hoskyns GIS division is one of the leaders in the GIS field, offering a range of products and services to provide a cost-effective implementation of GIS.

Hoskyns GIS have three GIS products available ranging in size from a PC based system to a mainframe system.

PIMMS is a full function, high level, generic GIS that runs in an IBM VM or VMS mainframe environment. It is designed as a series of modules centred on the Core Module. Further PIMMS modules allow you to integrate textual data from sources outside of the GIS environment and to produce plots and reports in various forms. PIMMS also allows you to analyze your assets and display the results in a clear, concise format using presentation graphics and spreadsheet. With its wide range of application uses PIMMS can be taken up by all departments in an organisation who need access to spatially related data.

GeoNet is a workstation-based GIS that can be implemented quickly and at an affordable price. It has fully integrated graphic and textual databases and maintains full network connectivity. GeoNet allows centralised corporate control over database definition, validation tests, symbols, command structure and report formats ensuring corporate consistency is maintained, even in a decentralised organisation.

GGP is a low cost, entry level, digital mapping system that runs on standalone PCs or on a Local Area Network. In addition to the usual digital mapping facilities GGP has a standard data interface giving access to data in existing databases. These features enable you to take advantage of existing investment in current hardware and software, resulting in substantial cost savings.

Hoskyns GIS software products are complemented by a comprehensive range of support services. Our consultants have a wealth of experience from high level strategic planning to technical consultancy and application development covering all aspects of GIS implementation.

IBM United Kingdom Limited

IBM's Enterprise approach to GIS is designed to meet three key objectives:

- Provide the sophisticated GIS functionality required for geographic analysis, graphic display, network modelling, mapping and facilities management applications within individual departments.
- Integrate geographical information with the corporate information systems within an organisation.
- Provide access to both spatial and attribute information to anyone who needs it through low-cost, general purpose terminals.

This enterprise approach is based upon the IBM Geographic Facilities Information System (GFIS) architecture. GFIS allows conventional maps and descriptive data for objects on the maps to be stored in a relational database and subsequently retrieved, viewed, updated, interrogated and analyzed by personnel throughout an organisation. A variety of terminals can be used—full function GIS workstations, general purpose IBM PS/2's or enquiry only, business graphics screens.

Three primary components of the GFIS architecture are:

- The geoManager product, which handles database management and the integration with non-GIS applications.
- GPG which is used to create a 'digital model' of an organisation's operating environment.
- The Interface Format File (IFF), which provides an object-oriented data definition for the transfer of all geographic-facility data.

IBM also has alliances with a number of GIS Business Partners who provide complementary applications which conform to the IFF specifications. This enables IBM to provide a complete GIS solution—covering consultancy, implementation services and tailored applications designed to streamline departmental operations.

IBM GIS is used throughout the world, in various areas including:

- Utilities
- Local and Central Government
- Communications
- Transportation
- Asset/Property Management
- Mining and Oil Exploration
- Health and Environmental monitoring
- Retail

There are well-established UK, European and Worldwide User Groups which provide a forum to exchange ideas and share knowledge.

IBM and geoManager are registered trademarks of International Business Machines Corporation.

ICL (UK) Limited

PLANES—The heart of ICL's strategic GIS development consists of three closely linked information systems. These cover property and land parcel related information, network related information together with the spatial representation of this and other data. This modular approach allows the processing of these different information types to be optimised, especially important when considering the very large geographically-based information sets that will be commonplace in the future, whilst allowing complete integration where necessary.

PLANES has been purchased by over 110 customers worldwide.

The ICL philosophy is one of integration and co-ordination and although PLANES provides full database facilities, great emphasis is placed on its ability to link to external databases and applications. This enables PLANES to act as the hub for corporate information, providing common access to all data in textual or graphical form and gaining the maximum benefit from the geographic content that is an integral part of up to 80% of an organisation's information.

PLANES **Property** is essentially concerned with address based information. It allows for a central directory of this information to be linked to the internal PLANES database, (Information Manager), external databases and external applications and to the Network of Spatial modules.

PLANES **Network** is concerned with network information such as highways or utility networks. It provides a central directory of all network elements and connectivity with flow and time stamping information and is linked to Information Manager, external databases, external applications and to the property and Spatial modules.

Both these modules allow the incremental development of functional applications which are automatically integrated with GIS. Applications may share common data and can be developed by users using 4GL facilities without the need for scarce computer staff expertise.

PLANES **SPATIAL** is concerned with user-specific geographic data and general topographic data. It provides a topologically-structured, object orientated database that allows the geographical objects to be linked to either Property, Network, Information Manager or external systems. Spatially-based enquiry to all PLANES modules and linked systems is available through PLANES Spatial Desktop Access an MS-DOS/Windows3 based application.

The ICL approach to GIS also includes the provision of related workstation based applications, working either standalone or linked to PLANES. These include data take-on and management systems for digital mapping via a range of third party solutions, network analysis through DINIS Electricity and Gas and structured data editing and spatial analysis systems.

ICL is committed to providing the hardware, software and services needed to make the best use of geographically related information and through this provide integrated, corporate-wide geographically-based, business information systems.

Intergraph (UK) Limited

Intergraph is independently acknowledged as one of the world's leading suppliers of GIS/ digital mapping systems. In the UK more than 500 workstations are in daily use working with spatially referenced data.

Intergraph design, manufacture, market, deliver, install and support interactive GIS systems, based on standard platforms, with a choice of SQL relational databases.

Intergraph GIS customers in the UK include British Telecom, British Gas, CEGB, Electricity Boards, Water Companies, Local Authorities, Universities and conversion vendors.

Intergraph installations range from single user to large multi-discipline, multi-user systems and offer full facilities for data capture via scanning and digitising, data editing in both raster and vector formats and output to many standard peripherals.

Workstation features which are unique to Intergraph include dual-screen graphics and a 27″ 2 megapixel workstation with a performance of 10 million instructions per second—these provide unrivalled flexibility and performance for the GIS user.

Intergraph are sponsors of the AGI and are committed to the support of GIS through active participation in all the activities of its Council and the various committees.

IT Southern Limited

IT Southern is a leading GIS implementor having successfully implemented an extensive Alper based GIS for Southern Water. The system comprises some 50 networked workstations, interlinking 8 operational sites across the four counties of South East England and Isle of Wight. IT Southern provides proven skills and in-depth experience to a developing external customer base.

IT Southern blends and packages skills and facilities in consultancy, planning and design, data conversion, implementation, and data communications, to a client's individual needs. Service packages can range from data conversion to complete GIS Facilities Management, whereby the whole GIS can be planned, implemented, and if appropriate managed for a client.

IT Southern's GIS development and technical support teams produce tailored solutions to match a wide range of needs, spanning from stand alone workstations to large integrated network implementations.

For those clients with sewer network data to convert, the new MANTELL product is now available.

Laser-Scan Limited

Laser-Scan Limited are recognised world leaders in Geographic Information Systems and Digital Mapping technology. The company has developed a range of products, innovations and services for customers such as national mapping, environmental and defence agencies, local government, emergency services and the service industries. Combining a strong commercial awareness with leading edge technology, Laser-Scan has consistently provided cost-effective, practical solutions for an ever increasing range of industry needs including those of clients such as Ordnance Survey, British Gas and Cornwall County Council.

Laser-Scan's proven track record in GIS development, in both product and research work, has resulted in systems now being used in more applications than ever before. The company provide a total service, from initial consultancy and project management, through installation, training and after sales support and are both POSIX compliant and committed to open systems enabling the most stringent requirements to be met.

In data capture, Laser-Scan have a world leading capability providing unparalleled productivity while maintaining quality and integrity, cutting project costs dramatically. GIS such as **Metropolis** are being used in local government, service industries, emergency services and in market research. **Horizon**, the latest GIS from the company, helps solve environmentally specific problems, avoiding conflict arising from proposed developments, ensuring more effective planning.

Laser-Scan prides itself on the quality of its products, its history of technological innovation and its ability to both identify needs and provide the best solutions. Organisations worldwide are now benefitting from Laser-Scan's expertise, finding a return on their investment which will ensure that they remain leaders in their field.

Market Profiles Limited

Market Profiles Limited are the owners of GEOPLAN, the most widely used digitised postal boundary file in Britain. GEOPLAN is available as postcode sectors, districts or areas and in either polygon or line segment format as required. GEOPLAN is digitised at 1:50,000 scale, is fully compatible with OS data and is used on a wide variety of platforms and systems.

MPS are also a leading independent market analysis company. We are able to offer the widest range of geodemographic systems in the country, including

- Super Profiles
- Mosaic
- Pinpoint (Geopin/Finpin)

Our analysts utilise both syndicated market research data and/or clients own data where available. The analysis may be bought in report or map format. Alternatively it may be integrated into a project by our in-house consultants to aid:-

- Outlet selection and evaluation
- Distribution planning
- Merchandising (Matching product profile to outlet profile)
- Media Selection
- Direct Mail Targeting
- Promotional cost benefit analysis

Mason Land Surveys Limited

SLIMPAC has been developed by the University of Newcastle and is a management information system built on the most successful CADD package available today, Autocad, combined with a powerful relational database.

The package retains all the standard Autocad drawing and plotting functions but extends the facilities, referencing data by features such as land parcels, a section of road, a building or any map or drawing feature. The textual records associated with the map features are held within a relational database which can be structured to the Clients requirements. Complex enquiries of both map data and textual data can be made. For example information could be requested on buildings of certain design before a specified date and with a pre-determined value.

The package is an ideal IBM compatible computer application system for the management of map and associated textual records. It has application as a desktop information source for surveyors, engineers, architects, and managers involved in the administration of maintenance of physical assets.

Examples of existing applications and users include:-

- land terriers for Development Corporations
- management of a leisure complex
- facilities management in central government
- management of underground services for TML
- grounds maintenance and engineering management

The facility exists to take in data from other databases.

Moss Systems Limited

MOSS is a surface modelling system for civil and mining engineering, land survey and cartography used for the recording and designing surface details and associated data.

Used worldwide by more than 650 organisations, its applications range through highway and railway engineering, site development, landscape design, mining and reclamation.

Using the system, detailed land information can be automatically recorded by digitising or ground survey. Further facilities allow data capture from aerial survey, borehole information and hydrographic survey.

This information can be uniquely identified and recorded in layers as required. New projects can be created in similar format for integration with existing data. Full interactive graphics are used to edit, interrogate and analyze the stored information.

MOSS's unique features for interactive alignment and 3D geometrical design can be used to develop the geometry of the most complex civil engineering projects, as full surface models. Analysis of models can be undertaken to produce volumes and setting out data. Extensive graphics facilities produce plan, profile, section and perspective drawings, photomontage and fully rendered colour views.

The system is available on HP/Apollo, SUN, VAXstation, Intergraph INTERPRO, IBM 6150 and RS6000 and Prime hardware. In addition, MOSS software can be easily linked to complementary software systems.

Prime Computer (UK) Limited

SYSTEM 9 is an advanced geographic information system (GIS) ideally suited for capturing, managing, analyzing and displaying large volumes of land-based data as it relates to economic, environmental or demographic factors. Sophisticated GIS technology combines the power and speed of SUN-based workstations with the data management capabilities of relational database software.

The relational database provides flexible data access for ad hoc queries; stores the shared boundaries of features only once, resulting in more accurate, easily maintained data; uses SQL (Structured Query Language) support for database queries with extensions for spatial analysis; enables related attribute data and geographic data to be held in one single database. In addition geographical data is stored and displayed as "features"—real world objects such as roads, streams, land parcels and buildings.

Inherent, too, is the ability to model the relationship between objects. This object or feature-oriented approach enhances the ease-of-use, database integrity and efficiency of spatial analysis.

The SQL database comes with fourth generation tools to handle, for example, report writing and screen input forms. Customised user interfaces can be created using the Application Control Language (ACL).

SYSTEM 9 offers continuous mapping which escapes the limitations of sheet based technology.

Geographic or spatial analysis is the key to maximising the benefits of GIS and SYSTEM 9 has the most advanced analytical facilities available today in its Analytical Toolbox (ATB). Complete applications can be written in ATB using spatial functions that can interactively identify inter-relationships between features, for example, find if there are SSSI's or Nature Reserves in the path of a proposed new road. ATB also enables the integration of data from other databases such as ORACLE or INGRES.

The Network Trace module allows the analysis of networks for transportation or utility applications.

SYSTEM 9 has been applied to Public Sector applications such as Property management, Highways planning, maintenance and analysis, Planning, Socio-Economic studies and Emergency Service planning. In Utilities, including Local Government acting as an agent, SYSTEM 9 can be used for Asset management, Planning and design (networks), monitoring and maintenance.

SYSTEM 9 runs on the SUN Sparc family of workstations, supplied and maintained by Prime, operating on individual workstations or via an ETHERNET network to promote an information sharing environment.

SYSTEM 9 is a state-of-the-art GIS which offers:

- SQL relational database
- Continuous mapping, not sheets
- Topological data storage
- Feature-oriented data modelling
- Spatial Analysis Tools (ATB)
- Network analysis
- SUN-based workstation technology

Remote Sensing Services Limited

ERDAS™ is a complete GIS system for satellite and aerial survey image data processing, enhancement, analysis and presentation. It is flexible and modular, allowing the user to tailor the power of the system to their needs. With an installed base of over 2,000 users world wide ERDAS is one of the most extensively used GIS systems. Versions available range from fully featured production systems to laboratory and teaching systems.

ARC/INFO is a vector based GIS system with full data base facilities providing users with all the tools necessary for complicated spatial analysis, network analysis and cartographic quality map plotting.

Through **Live Link**, both ERDAS and ARC/INFO can coexist in the same host machine, and information from both systems can be displayed on a single image.

ERDAS runs on PC and compatible machines, PS/2 and compatible, IBM RISC machine, the full Sun family, Solbourne workstations and UNIX workstations including DECstation, Hewlett Packard and Silicon Graphics machines. A full range of input and output peripherals and display systems are supported.

Smallworld Systems Limited

Smallworld GIS is a modern system which comprehensively addresses all of the major fundamentals essential for the successful implementation of a corporate GIS.

These issues include:

- Multiple-user access to a large seamless mapbase, with good performance for all users. The Smallworld Datastore, which supports versioning, is one of the few systems that is truly multi-user.
- Support for raster and vector data in the same environment. Vector data is topologically structured and is treated as part of an organisation's object data model.
- Integration with corporate systems using a defined interface between concurrent processes such as database management systems from ORACLE and INGRES.
- Support for standards allows access to one database over a heterogeneous network of workstations, including PC/X-servers.

An interactive object-oriented language, Smallworld Magik, yields large savings in customisation costs.

The basic technology supports higher level systems for map generation, report writing, polygon and network analysis, object editors and constructive drafting which then support applications such as data conversion, network management, planning, facilities management and vehicle tracking.

Smallworld was founded in 1988 in Cambridge, UK, and has subsidiary companies in Germany, Holland and Sweden. Customers in the UK and the rest of Europe include large industrial companies, utilities, local and central government.

Strategic Mapping Inc.

ATLAS*GIS extends SMI's widely acclaimed range of PC software. It combines all the features of ATLAS*GRAPHICS and ATLAS*DRAW with a wide variety of sophisticated spatial analysis capabilities including a built-in database manager. The software has superb presentation quality and ease of use that has become the ATLAS trademark. The result is an advanced, fully featured GIS that brings personal computer users into a new era of data analysis.

ATLAS*GIS puts the capabilities of a mainframe computer into the hands of personal computer users. Instead of months of instruction required for operation of mainframe systems, users of ATLAS GIS can have the program up and running in a matter of days.

Government analysts, corporate planners and market researchers have a tool that saves time, allows for sharing of maps and data files and ensures fast turnround—all at a break-through price of £1,750.

ATLAS*GIS was chosen by the Association of American Geographers as Corporate Software Product of the Year (1990) and was rated Best Desktop Geographic Information System by PC Week.

Over 15,000 public agencies and private organisations are already using the ATLAS line of software in planning, analysis and teaching applications.

SysScan UK Limited

Still the only mapping specialist with systems in all three Ordnance Survey organisations (GB, Northern Ireland, Ireland), SysScan develops and markets Geographic Information Systems (GIS) and digital systems for spatial and map information. SysScan is a pioneer in the technology of raster/ vector products and automated data capture.

SysScan's digital mapping software combines maps of any scale with records and attribute data, and includes interactive editing and map management facilities.

For utilities, our **DNMS** application software keeps all the network data in Digital's widely-used relational database, Rdb.

For the PC user, **TERRASOFT** provides a low cost entry into all GIS applications. TERRASOFT is believed to be the only program of its kind able to uniformly support vector and raster polygons. This gives the unique advantage of providing quick and easy access to analysis previews. Watch out for UNIX announcements in 1991!

GEOREC automated data conversion software delivers maps or map related information, to these or other vendors' GIS, dramatically faster than manual digitising methods. GEOREC takes scanned and vectorised input (from a variety of scanners) and delivers structured data via a process of pattern recognition algorithms. Cost effective conversion flowlines exist in a number of cartographic institutions and service bureaux throughout Europe.

Unisys Limited

ARGIS 4GE is an Advanced Relational Geographic Information System from Unisys.

ARGIS 4GE incorporates many recent advances in GIS and database technology, and combines these with state-of-the-art hardware, to provide a system that is able to respond to your demand for geographic information management, whatever the size or nature of your organisation.

ARGIS 4GE is based on open systems architecture and industry standard components including UNIX, ORACLE RDBMS and X-Windows. The system is available on a range of UNISYS processors and workstations, and can be supplied with many different data input and output devices. The hardware components can be flexibly configured and integrated using industry standard LAN and WAN protocols such as Ethernet and X.25 to provide a distributed processing environment.

ARGIS 4GE operates within a fourth-generation development and processing environment. It offers both a standard SQL and an advanced Geographic Query Language (GQL) interface for database query, and an application development toolkit (ADF) for product customisation and integration with existing GIS application software.

ARGIS 4GE provides a comprehensive range of functions for GIS activities including data capture, editing, report generation and analysis. Advanced features of ARGIS 4GE include:

- a fully integrated relational database management system that can respond to the varied and growing GIS demands of an organisation
- a continuous geographic database with no artificial partitioning according to map sheets
- comprehensive query and analysis capabilities available using graphics-capable PCs
- quadtree spatial indexing for efficient retrieval of geographic data irrespective of database size
- support for three-dimensional coordinate data, and coordinate referencing systems such as TM and UTM
- support for the storage, edit and display of raster imagery derived from remote sensing platforms, or by scanning paper maps, plans or photographs
- the ability to associate and register raster images to database geographic features
- a comprehensive set of database security and management facilities
- complete Digital Terrain Modelling facility
- retrieval and query of geographic features for analysis, edit and display, based on a combination of spatial and attribute criteria
- a data model that supports the storage and analysis of topological data structures
- polygon generation and analysis software including parcelisation, automatic centroid allocation, polygon overlay and sliver area removal
- graphic symbolisation facilities that allow geographic data to be flexibly displayed dependent upon application or associated attribute information
- support for standard interchange formats including UK NTF (levels 2 and 3), OSTF and UTF

Wang UK Limited

Wang UK Limited, worldwide suppliers of office automation and imaging, market the Geographic Data Management System, GDMS. This purpose written GIS features an object orientated geographic model, scaleless and edgeless maps, a true virtual database (you can integrate with nearly any database on any system), a purpose written 4GL application builder and over 65535 layers of information.

GDMS is not a CAD package with some added GIS type functions; it is a true GIS allowing every type of spatial search to be carried out, including common areas and adjacency. In use worldwide, GDMS has applications in local and central government, utilities and emergency services. The 4GL enables these to be quickly adapted to other users or new applications to be generated. This work may be done by Wang or the users.

GDMS will fit into your existing environment and communicate with your existing data processing equipment and databases. This avoids much of the data conversion and cross training needed with other GIS's. GDMS will support from 1 to 1000 users and does not require a dedicated CPU, so you can concurrently use WP, electronic mail, image processing etc. on the same system and incorporate these into your GIS.

GDMS supports a wide range of plotters and digitisers and most standard formats ie DXF, SIF, OSTF, HPGL, POSTSCRIPT etc. Normal PC-AT's are used as the graphics workstations allowing concurrent, mouse, digitiser and COGO editing.

Independent consultants

Trading name, phone and contact	Address	Market sectors	Applications, services	Ref.
Apple Mapping Centre 071 631 6577 Jonathan Raper	Department of Geography, Birkbeck College, 7-15 Gresse Street, London W1P 1PA	Apple Macintosh solutions to mapping and GIS. Experts in system design, communications and interfaces to host-based GIS.	Information centre for mapping software on the Macintosh, system choice, data acquisition and GIS projects. Custom Hypercard application development.	
WS Atkins Planning Consultants 0372 726140 Bernard Denore	Woodcote Grove, Ashley Road, Epsom, Surrey KT18 5BW	Operating for Industry, Commerce and Public Sector clients including Government in the UK and Overseas.	GIS advice and development for infrastructure planning and design, land use, agriculture and transportation projects.	
BDO Consulting 071 583 3303 Michael Mainelli	20 Old Bailey, London EC4M 7BH	Independent Consultants to Utilities, Central and Local Government, Retail, Distribution and other private sector companies.	All GIS/LIS and AM/FM applications, strategic planning, requirements analysis, specification, implementation and quality management.	
Business Information Management 0273 515018 Rob Mahoney	14 Kings Avenue, Denton, Newhaven, Sussex BN9 0NA	Utilities and local government, private sector in UK, Europe and overseas; vendors; digital mapping.	General GIS applications; awareness/executive seminars; requirement study/strategy; market research; project management and support.	337
Cadam Consult Limited 0730 67781 Edward Stephens	Antrobus House, 18 College Street, Petersfield, Hampshire GU31 4AD	Independent consultancy services for GIS usage to Utilities, Local Government, Highways, Railways and Airports. UK and Europe.	* Education and training * System specification and evaluation * Implementation support * CAD, Surface modelling * Project Management	

Trading name, phone and contact	Address	Market sectors	Applications, services	Ref.
CDD Limited 0494 713769 DJ Pavely	Marlborough, Spurlands End Road, Great Kingshill, High Wycombe, Buckinghamshire HP15 6HY	Utilities, Resource, Environment, Mapping and Energy Agencies. Central and Local Government. Local Authorities, Commercial companies UK, Europe and Worldwide.	Identification of GIS applications and benefits. Advice on design and layout of data bases and on hardware for specific applications. Project management and support.	
CMG Computer Management Group Limited 071 233 0288 Peter Todd	Telford House, Tothill Street, St James's Park, London SW1H 9NB	Utilities, Local and Central Government, Airlines, Transport and Distribution.	Independent GIS Consultancy, Awareness Seminars, Feasibility Studies, Strategy Studies, System Selection, Implementation Support and Project Management.	337
Computer General GeoData Services 0225 482744 Dr Pauline A Waters	The New Church, Henry Street, Bath BA1 1JR	Oil, gas, mineral, local government, central government, defence, health, water, emergency services, environmental, civil engineering and architecture.	Hydrocarbon exploration; hydrogeological, environmental vegetation, hazard potential and health/disease cluster studies; temporal change modelling etc.	
Comsult 0234 342401 DR Foster	67 Goldington Road, Bedford MK40 3NB	Defence, Police, Emergency services, Utilities, Local Authorities.	Computer/ Display Systems. Database and AM/FM software packages; bespoke software.	
Coopers & Lybrand Deloitte 071 583 5000 Nick Pearce	Plumtree Court, London EC4A 4HT	Utilities, Local and Central Government, Private Sector and vendors in UK, Europe and Worldwide.	Independent consultancy in GIS, IT and other business areas. Executive awareness seminars, strategy, system selection, implementation, project management and support.	338
CSL Group Limited 071 982 0071 Gary R Levett	Consilium House, City Forum, City Road, London EC1V 2NY	Public Sector—CSL is particularly experienced in the application of GIS to Local Government and the Health Service.	• GIS strategy & info. planning • Investment appraisal • Product evaluation • Implementation assistance • Data capture services	

Trading name, phone and contact	Address	Market sectors	Applications, services	Ref.
DMR Group Inc. +31 1820 3 33 55 Mr B de Groot	Harderwiijkweg 7, 2803 PW Gouda, The Netherlands	Independent management and information systems consulting services for Public Sector, Financial Services, Manufacturing, Distribution & Retail Services, High Technology Services.	Management consulting & strategic planning, application design, systems evaluation, development & implementation, technical support & technology development & value-added services (seminars/education)	338
DWH Associates Limited 0789 268102 Derek Hilder	20 Rother Street, Stratford upon Avon, Warwickshire CV37 6NE	* Government and Agencies * Local Government * Utilities, Health Services * Environmental Bodies * Police and Emergency Services	* Feasibility Studies * Requirements Analysis * Functional Design * Implementation Support * Post implementation Review	339
Earth Observation Sciences 0252 811181 Matthew Stuttard	Branksome Chambers, Branksomewood Road, Fleet, Hampshire GU13 8JS	Environmental, Utilities, Remote Sensing, Space Agencies. UK and abroad.	Bespoke interfaces for world-wide access to spatial data bases. Applications developed for local systems using SPANS, ILWIS, ARC/INFO & hypermedia methods. Data gathering & spatial data know-how.	
Fairbairn Services Limited 061 976 3536 Martin Dibnah	Fairbairn House, Ashton Lane, Sale, Manchester M33 1WP	Public and Private Utilities, Local Authorities, Property Companies and General Practice.	GIS, Facilities and Asset Management systems. Independent hardware, software and Management Consultants.	
Gardline Surveys 0493 850723 David Pettit	Admiralty Road, Great Yarmouth, Norfolk NR30 3NG	Digital Mapping & GIS consultancy for: * Local and Central Government * Utilities * Vendors * Oil Industry	GIS consultancy based on practical experience of a number of systems. Requirements analysis, system evaluation, selection and implementation. Training/Project Management. Data capture solutions.	
Geobase Consultants Limited 0372 745261 John Rowley	28 Church Road, Epsom, Surrey KT17 4DX	Central Government and agencies, Local Government, Utilities, Transport, Academia, GIS and digital mapping supply or service companies, other private sector: UK, Europe and Overseas.	All LIS, GIS or digital mapping technologies, CAD, surface modelling: awareness / strategy/ audit/ implementation/ trials/ project management. Information service/ Market Research.	340

Trading name, phone and contact	Address	Market sectors	Applications, services	Ref.
Geonex UK Limited 081 640 1971 Robin Waters	92-94 Church Road, Mitcham, Surrey CR4 3TD	All GIS market sectors, particularly Utilities, Local and Central Government.	GIS consultancy, data conversion service, colour aerial photography, GIS feasibility, selection and implementation.	
Gimms Limited 031 668 3046 Marlene A. Ferenth	30 Keir Street, Edinburgh EH3 9EU	Customers have included local/central government agencies, utilities and marketing agencies.	Our consultants offer unbiased, professional advice on a variety of GIS applications including project management, software design/ implementation and benchmarking.	
Glen Computing Limited 0689 875577 Graham Bugler	309 High Street, Orpington, Kent BR6 0NN	* Central Government Agencies * Local Authorities * Public Utilities * Mapping Agencies * Construction Industry * Planners	GIS/ MIS: an unbiased service, independent of software houses, to analyse user requirements, with the experience to offer cost effective data capture, system and software solutions.	
GMAP, Geographical Modelling & Planning 0532 333450 Dr Martin Clarke	University of Leeds, 3F Springfield House, Hyde Terrace, Leeds LS2 9LN	* Consumer Goods Retailing—High Street and out of town * Financial or Leisure Services * Communications-media, Telecomms * Public Services * Motor Industry	Geographical modelling systems for geographical analysis and strategic planning, integrating mathematical modelling skills; intelligent information; mapping technology.	
Graphics Management Limited 0698 350843 Alan Giddins	9 Braefoot Court, Law, Strathclyde ML8 5HY	General mapping and cartography. Utility, central government and local authorities. Hydrographic charting UK, USA and Middle East.	Ten years CAD/ Mapping experience. Feasibility, system design, programming, implementation, support and training.	
Haiste Limited 0532 620000 James Gossage	Newton House, Newton Road, Leeds LS7 4DN	* Utilities * Local Government * Urban Development * Telemetry and Process Control	* Bespoke software for Intergraph * Production of overlay maps * Creation of Asset records * Upgrading of Asset records * Asset management * Housing stock database	

Trading name, phone and contact	Address	Market sectors	Applications, services	Ref.
Hoskyns GIS 0962 844188 Julie Hoskyn	Technology House, Victoria Road, Winchester, Hampshire SO23 7DU	Local & central government, public utilities, minerals extraction, retail and distribution, cable and telecommunications, other private sector.	All GIS applications; environment/ database design.	
Howard Humphreys & Partners 0372 376190 IT Applications Group	Thorncroft Manor, Dorking Road, Leatherhead, Surrey KT22 8JB	Engineering/ IT consultancy: * Highway maintenance * Water: clean and sewerage * Structures; geotechnical * Asset management	* GIS project management * SSADM analysis, planning * Specification preparation * Graphics; Ingres/Oracle RDBMS	
David Hughes 0206 572611 David Hughes	21 Baden Powell Drive, Colchester, Essex CO3 4SL	Local and Central Government, Private Sector including Vendors and Service Organisations. UK and Overseas.	IT & GIS consultancy including: Strategy and Needs studies, systems implementation and review, project management, marketing strategy and appraisals.	
Hunting Technical Services Limited 0442 231800 Mike Barr or Graham Deane	Thamesfield House, Boundary Way, Hemel Hempstead, Hertfordshire HP2 7SR	UK and Overseas. Remote sensing and GIS consultancy for all resource survey applications. Digital mapping for all market sectors.	Integration and analysis of spatial data for environmental monitoring/ land use planning. Remote Sensing/ GIS training including 3 month Resource Surveys course.	
IT Southern Limited 0273 606766 Peter Elkins	Southern House, Lewes Road, Brighton, East Sussex BN1 9PY	IT Southern provides GIS Technical and Business Consultancy to Utilities, Local Government and general areas of commerce. Experience at national and local levels.	General business planning for GIS. Technical appraisals. System specification/ evaluations. Strategic studies. Applications evaluation.	
Know Edge Limited 031 443 1872 Robin A. McLaren	33 Lockharton Avenue, Edinburgh EH14 1AY	Local and central government, utilities, environmental agencies, property sector: UK, Europe and overseas.	GIS/LIS: Awareness training, feasibility study, user requirement analysis, functional design, system evaluation & benchmark, project audit, market analysis.	

Trading name, phone and contact	Address	Market sectors	Applications, services	Ref.
KPMG Management Consulting 071 832 8888 Malcolm Leith	8 Salisbury Square, London EC4Y 8BB	Central and local government, non-departmental government bodies, utilities, private sector including retail, property, distribution & financial services; in the UK, Europe & Worldwide.	Independent consultancy: business IT and GIS. Awareness seminars, strategy studies, requirement definition, systems evaluation and selection, customisation & systems development, project management.	341
John D. Leatherdale FRICS 081 449 0123 John Leatherdale	10 Bartrams Lane, Hadley Wood, Barnet, Hertfordshire EN4 0EH	Independent consultant in GIS and digital mapping to users worldwide: * Government and Agencies * Local Authorities * Private Sector	* Feasibility Studies * Defining user needs * Implementation strategies * System and data procurement * Land information management	
Logica Plc 071 637 9111 /1270 Valerie Exton	68 Newman Street, London W1A 4SE	Water, Electricity Utilities, Oil Industry, Local/Central Government, Defence, Mapping Agencies, Space, Remote Sensing, Retail and Private Sector in UK and overseas.	Feasibility Studies, Requirements Analysis, Product Evaluation, Project Management, System Design/ Development/Integration, Data Take-on Management, Communications, Applications Development.	342
Manpower 071 224 6688 Angela Williams	International House, 66 Chiltern Street, London W1M 1PR	Consultants in GIS support staffing, providing permanent staff and temporary help for ad hoc work and project management, to all sectors in the UK and overseas.	High quality and flexible surveyors, data capture personnel, digitisers & data entry operators. Staff are tested in key abilities, skills, personal characteristics, ensuring immediate productivity.	
Mason Land Surveys 041 222 7733 David Weatherill	14 Newton Terrace, Glasgow G3 7PJ	Mason Land Surveys provides consultancy services to local authorities, district councils and development corporations.	The services provided have included identification and installation of appropriate software and hardware, staff training and database assembly.	
Midsummer Computing Exchange Ltd. 0908 668866 Ivan Bush/ Steve Gill	Midsummer House, 429 Midsummer Boulevard, Central Milton Keynes MK9 2HE	* Local Authorities * Development Corporations * Government & Utilities * Private Sector	* Feasibility Studies * Benchmark Generation * System Selection * System Analysis & Implementation * System Management & Operation	343

Trading name, phone and contact	Address	Market sectors	Applications, services	Ref.
Modulo 4 0532 771344 Martin Horner	Suite 27A, Concourse House II, 432 Dewsbury Road, Leeds LS11 7DF	* Utilities and all Government * Private Sector: UK and worldwide * Organisations needing corporate integrated systems * Land administration system users in developing countries	* Vendor independent services * Ad Hoc Consultancy * Project Management * Educational events * Recruitment * Systems Development & ICL PLANES	344
MVA Systematica 071 836 9381 Miles Logie	115 Shaftesbury Avenue, London WC2H 8AD	* Government & Local Authorities * Development Corporations * Utilities * Transport Sectors * Private Sector	* Feasibility Studies * Implementation plans * Design: structured methods, data modelling, interfaces. Application development, integration with GIS. * Integration of diverse data.	345
North East Regional Research Laboratory 091 222 8014 Mike Coombes	Centre for Urban & Regional Development Studies, The University, Newcastle upon Tyne NE1 7RU	Wide ranging experience of GIS for public/ private sector organisations: specialisms in crime pattern analysis, health, demographics and marketing research.	Internationally recognised expertise in developing new spatial analysis tools to enhance GIS's for both academic and 'real world' problem solving. Doctoral programme also provided.	
NPA Group 0732 865023 Nigel Press	Edenbridge, Kent TN8 6HS	Satellite Imagery and Remote Sensing.	Map revision and updating, geology and engineering.	
Harry Orenstein Limited 081 203 7475 Harry Orenstein	PO Box 1807, London NW4 2PF	Town and Country Planning Acts: development plans, conservation areas and listed buildings; urban regeneration: townscaping and infrastructure schemes; Environmental Protection Act plans.	Consultancy, data capture, project management, database resource management.	
P-E International Plc 0784 434411 Guy Pullen	Information Technology Group, Park House, Wick Road, Egham, Surrey TW20 0HW	Utilities, local and central government, commercial sector.	As a leading management and computer consultancy, P-E's services include strategy studies, system evaluation and selection, implementation and project management.	

Trading name, phone and contact	Address	Market sectors	Applications, services	Ref.
PA Consulting Group 0763 261222 Phil Jeanes	Cambridge Laboratory, Melbourn, Royston, Hertfordshire SG8 6DP	All market sectors.	PA is the leading management and technology consultancy. The GIS team offers complete services in strategy, systems, technology and implementation.	345
Performance Management Services (MK) Ltd 0908 211921 John Stokes	Lovat Bank, Silver Street, Newport Pagnell, Buckinghamshire MK16 0EQ	Local Authorities and other Public Sector organisations.	Provision of consultancy services in respect of systems specification, systems design and implementation of GIS in Local Authorities.	
Polytechnic of East London 081 590 7722 /2540 Roger Fisher	Department of Land Surveying, Longbridge Road, Dagenham, Essex RM8 2AS	A strong research and consultancy record in Land Information Management and Digital Mapping. Special courses provided on request.	Facilities include Intergraph (Microstation), ARC/INFO, GIMMS, Autocad and LS10 image analyser.	346
Remote Sensing Services Limited 0672 20226 Duncan Milnes	Lychgate House, 24 High Street, Ramsbury, Wiltshire SN8 2PH	Consultancy in Cartography, Photogrammetry and Geology, software development and implmentation.	Full consultancy from Project identification, through evaluation, pilot, system implementation, data analysis and interpretation.	
Robertson Group Plc 0492 581811 Chris Parker	Llandudno, Gwynedd LL30 1SA	Data capture, database design, data processing, specialist interpretation and analysis, map production and data presentation for natural resource and environmental applications.	ARC/INFO, PC-ARC/INFO, ERDAS, SUN, Apple Macintosh.	
Scott Wilson Kirkpatrick 0256 461161 James Cutler/ Trevor Beaumont	Scott House, Basing View, Basingstoke, Hampshire RG21 2JG	Worldwide in natural resources, engineering, regional planning and environmental appraisal. Remote sensing and GIS applications for project management.	GIST and SPANS in-house plus GIS selection on project specific basis; system development, installation and training for institutional strengthening.	346

Trading name, phone and contact	Address	Market sectors	Applications, services	Ref.
SCS (MVM Strategic Consultancy Services) 0483 426465 Bill Murray	Mill Pool House, Mill Lane, Godalming, Surrey GU7 1EY	All services. Experience with most leading GIS products, hardware configurations, communications protocols and relational databases.	Independent GIS consultancy to the public and private sectors in the UK and overseas. Particular expertise in the property, utility and retail sectors.	347
SIAS Limited 031 557 0200 Sarah McRae	4 Heriot Row, Edinburgh EH3 6HU	Central & Local Government organisations and private sector. Concentrating on highways, transportation & utilities.	Application design, implementation & consultancy. INGRES, X & MS Windows specialists, SSADM experience. Custom software in C & Fortran. Graphical interfaces to spatial data.	
Silsoe College 0525 60428 /273 Dr Chris Bird	Rural Land Use Department, Silsoe College, Silsoe, Bedfordshire MK45 4DT	* Rural Land Management * Agricultural Monitoring * Environmental Monitoring	* Satellite image analysis * Aerial photograph interpretation * Production of GIS databases * Thematic mapping	
South East Regional Research Laboratory 071 631 6483 Dr John Shepherd	Birkbeck College, 7-15 Gresse Street, London W1P 1PA	Central & Local Government— Economic Development, Planning & Property Sectors—Market Research—Integrated Land Use, Infrastructure & Demographic databases—Management Structures.	Education and training in GIS applications.	
Spatial Solutions Limited 0934 732421 Terry Robinson	Glenmoor, Cheddar Road, Axbridge, Somerset BS26 2DL	All Sectors	* Training Workshops * Project Management * Policy & Strategic Studies * Market Studies * Directory of Data	347
Structural Technologies Limited 0527 854819 DF Schindler	Woodside, The Slough, Studley, Warwickshire B80 7EN	Specialist transfer definition consultancy provided to Gas, Electric and Water Utilities or organisations concerned with the capture or transfer of structured foreground maps.	Detailed definition of digitisation or transfer specifications to ensure full integrity of topology and attribute data. Mechanisms include NTF 3/4 and STL's STRUMAP	

Trading name, phone and contact	Address	Market sectors	Applications, services	Ref.
Summerside Associates Limited 0480 407240 Mike Quarmby	Atlas House, 17a Huntingdon Street, St Neots, Cambridgeshire PE19 2BU	Local and Central Government, Government Agencies, National Mapping Agencies, Utilities, Private Sector and Digital Mapping supply or service companies in the UK, Europe and worldwide.	Awareness Training, Data Capture Strategy, Implementation Advice, Project and Quality Management for all LIS/GIS and Digital Mapping applications.	
Survey & Development Services Ltd 0506 825121 John E. McCreadie	3 Hope Street, Bo'Ness, West Lothian EH51 0AA	Provision of spatially referenced map data either from direct survey or by vector digitising from existing maps.	Information can be produced in any format and on any media to suit a Client's requirements. Original data capture using ECLIPSE software suite.	
Survey International 0525 718900 John G Callaghan	Dickens House, Off Maulden Road, Flitwick, Bedfordshire MK45 5BY	Local and Central Government, Utilities, Industrial and Commercial, Environmental Agencies: UK and Overseas.	Consultancy for analysis of user requirements. Development of application software for most GIS.	
Taywood Data Graphics 081 575 9240 Pat Jarvis	Greenford House, 309 Ruislip Road East, Greenford, Middlesex UB6 9BQ	Utilities, local, county and central government, defence, architectural, engineering, environmental bodies in UK and Europe.	Extensive hands-on experience in a wide range of data capture and GIS systems provide the base for pilot projects and consultancy in systems selection.	
Terraquest by Sutcliffe and Associates 021 500 5818 John Gannon	Kenrick House, Union Street, West Bromwich, West Midlands B69 6DB	* Local & Central Government * Utilities & Transport * Retail * Development Corporations * Commercial * Private Sector	LIS/GIS: requirements studies, system selection. Implementation management. Data capture and Conversion. Systems design, development and integration.	

Business Information Management

Business Information Management provides a wide range of services to meet the ever growing market demand for independent GIS consultancy. The company has provided services to a variety of organisations in both the public and private sectors and has a proven track record in delivering genuinely impartial advice to clients. The company's philosophy is always to work closely with its clients ensuring that they receive the best advice and assistance in defining their business needs and the most appropriate solution within an agreed time frame.

A modular approach has been developed to all GIS services that allows the client to select any combination of our services appropriate to their needs. Joint project modules are an important aspect of our services that allow our clients staff to work alongside Business Information Management to develop their own in house skills.

A selection of our services are as follows:

- Executive overviews
- Awareness seminars
- Feasibility studies
- Requirement studies
- Market research
- Project support
- Digital map use and management

Business Information Management takes a pride in its independence and the services it provides.

CMG Computer Management Group Limited

CMG is Europe's largest provider of independent GIS consultancy and is a major force in the IT and business world. CMG can provide GIS solutions to clients in the utilities, local government and private sectors. Our services range from awareness seminars, through strategy studies to project managing corporate GIS implementations.

CMG's experience in the field of GIS enables us to provide effective value for money services and reduces the risk to clients by tailoring solutions to their particular requirements. We have a corporate strength of more than 60 GIS consultants in the UK and Europe who can give practical advice based on experience and a structured approach that addresses clients' needs, manages projects and produces relevant quality solutions.

CMG's strengths are in its ability to identify the most appropriate methods of integrating data in a GIS and to cost justify that implementation. CMG are rapidly developing a reputation for delivering pragmatic solutions that fulfil users' requirements whilst realising appropriate financial returns.

Coopers & Lybrand Deloitte

A leading GIS consultancy in the UK, Coopers Deloitte has an existing client list covering all utilities, the oil and gas industry, central and local government, transport, communications, retail and marketing and financial sectors. Our GIS clients include, amongst others, Welsh Water Plc, Severn Trent Water Ltd, Southern Electric, British Rail, The National Trust, British Coal Opencast, Geographic Information Steering Group, North Bedfordshire Borough Council, Mid Glamorgan County Council and Hertsmere Borough Council.

With unambiguous independence and a proven reputation, we can bring our wealth of experience to ensure clients' success in their GIS projects.

The services offered include, but are not limited to:

- Executive awareness seminars
- Feasibility studies
- Business justification of GIS
- User Requirements Specification
- ITT preparation and tender support
- Implementation management
- Systems Integration
- Client project review and audit

Coupled with Coopers Deloitte's extensive expertise in areas such as IT, Financial Management, Economics and Business Strategy, Organisation Management and Human Resources, we can focus a team with the relevant skills needed for the largest, smallest or integrated project. As GIS projects become more complex and corporate in nature, clients will need more than just technical GIS support.

DMR Group Inc.

DMR Group has grown quickly and steadily since its founding in 1973. Today, DMR is an international group of consultants with offices across Canada, the USA, Australia and Europe. DMR has executed projects in information management and information technology in numerous countries throughout the world.

DMR Group has the capabilities and experience to undertake a wide range of assignments: strategic to operational; business-oriented or technological; simple or complex; large or small; involving mainframes, minicomputers, or microcomputers; and using conventional or leading-edge technology.

Our services span the breadth of the information management field:

- Management Consulting & Strategic Planning
- Systems Evaluation, Development, Integration & Implementation
- Technical Support & Technology Development
- Integrated Management
- Seminars, Education & Training

The acquisition, implementation, operation and maintenance of a Geographic Information System represent a major investment for your organisation. DMR can assist you in managing the risk factors associated with achieving each milestone in the GIS life-cycle in order to reduce the size of the investment and gain benefits sooner.

DWH Associates Limited

DWH Associates was formed in 1984 to meet the growing need for Consultancy in Information Systems and Technology. This includes the identification, development and management of information systems as well as the changes of approach to the use of information technology. We offer specialist consultancy services for broad ranging and specific issues alike; in all areas seeking to help our clients exploit the new tools that are evolving for the collection and analysis of data. Foremost among these are Geographical Information Systems (GIS).

Our consultancy success is based upon the employment of well experienced personnel with an aptitude for the work and in the disciplined use of proven methodologies. We have taken Information Engineering concepts and principles to develop an integrated approach to the execution and management of the significant number of tasks required to establish a GIS within an organisation. This structured methodology covers the complete range of activity from strategic planning through to practical implementation. To date it has been effectively applied in both the private and public sectors.

The Company prides itself on being able to offer truly impartial services through its deliberate policy of total independence from computer hardware and software suppliers, distributors, dealers or agents.

Geobase Consultants Limited

Geobase Consultants provides a broad range of services to users and intending users of Geographic Information Systems (GIS). We also service the needs of the industry's system suppliers and data services contractors. To all these groups we offer the benefit of our impartial advice and considerable wide-ranging experience in a spirit of independence from vested interests.

Our services are tailored to individual client needs and our policy includes for 'technology transfer' through the formation, where possible, of a joint project team with client's staff. Our services are broadly as follows:

- Education: awareness seminars at executive and more technical levels
- Preliminary and Feasibility studies
- Requirements and detailed Specification
- Technical issues such as mapping technology
- Availability of Standards
- All tender documents: conditions of contract, specification etc.
- Acquisition: pre-qualification, formal tenders, evaluation
- Implementation: project management, data take-on management
- Project audits
- Market Research
- Product (application or 'toolbox') planning and design

Local Authorities: we have recognised that whilst GIS have been shown to be feasible for all classes of Local Authority, this does little to set individual authorities off in the right direction for **them**. Our **Preliminary Low-cost Study** will identify the individual need or otherwise for GIS, the further resources and plan required to develop the system.

Geobase Consultants was formed in early 1988: our extensive capabilities are based on a background in design, project and operational management of Land Information Systems, Utility Records Systems, various digital mapping and DTM projects.

KPMG Management Consulting

KPMG Management Consulting is one of the UK's largest providers of professional services to both the public and private sectors. KPMG's GIS group offers a range of services designed to assist clients with development and implementation of GIS.

Our approach to strategic planning provides a systematic evaluation of the opportunities presented by GIS within the framework of an organisation's corporate objectives and overall business development plan and examines the possible implications of GIS on organisation structure, operations and data management. The approach involves the use of established techniques to resolve key issues including:

- Cost justification of GIS
- Phased implementation management
- Merging of GIS and global IT strategies
- Data ownership and data management
- Human resources and management of change

We have a strong, multidisciplinary team of consultants who are able to assist at every stage of evaluation, planning and implementation of GIS and our range of services include:

Awareness	•executive briefings •management seminars	
Feasibility/Strategy	•business strategy •data audit •EEE reviews •IT existing/required	•IT/GIS strategy •cost benefit analysis •organisation review •systems integration
Implementations	•specifications •business process design •benchmarks •customisation •training	•ITT •selection •project management •change management
Continuing Strategic Advice	•monitoring •horizon technology •review	

Logica

Information systems—systems for the storage, processing and interpretation of data—are now widely used and universally accepted. Increasingly, users are becoming aware that, in many applications, the spatial aspects of data are of significance, and the ability to store, process and display this data appropriately is of value. Such applications, generally known as Geographic Information Systems (or GIS) are rarely an end in themselves. Greatest benefits can be derived when these spatial aspects are combined with all others in a single, unified system.

Logica is an established computer systems, software and consultancy company, now employing 3600 staff in offices in twelve countries. Its reputation is based largely on its success as a systems integrator, that is, in taking a system approach, looking at business requirement and incorporating whatever elements (hardware, software, communications etc) are required into a unified whole.

As well as systems integration, Logica staff are also available to provide application specific technical support, and for strategic and technical consultancy.

Logica's involvement in a project which includes GIS would then take the form of:

- commercial and technical evaluation of available GIS products
- advice on the integration of GIS into an existing or future IT strategy
- technical or functional audit or existing facilities
- analysis of requirements to produce functional and technical specifications
- implementation of GIS functions within an existing system
- supply of specialist GIS staff to assist in systems development
- project management of GIS development
- design, development, implementation and support of systems which include GIS elements

Logica's client list for GIS related applications covers a range of sectors:

- utilities
- defence
- government
- industry

The nature of GIS work that Logica has carried out includes:

- location and management of plant and property by means of digital mapping
- interpretation of satellite remote sensing data
- definition of base map and asset data capture methodologies
- oceanography

Logica is currently assisting two major water companies with their GIS applications. Following completion of a digital mapping strategy Logica is assisting one of these utilities with the management of the implementation of a large digital mapping system. Such a system will provide the organisation with more effective statutory records maintenance and asset management.

GIS can provide significant benefit when integrated with other corporate computer systems by providing additional decision support and operational capabilities. Thus, the future of GIS depends upon systems integration companies such as Logica, with technical skills allied to specialist understanding of the business sector, and the capability to integrate them into a successful solution.

Midsummer Computing Exchange Limited

Midsummer is a leading provider of Information Technology Services to both the public and private sectors. Midsummer was formed in 1987 and currently employs over 60 staff supporting a wide range of services including GIS, LIS, Local Government Systems, IT Strategies and General Consultancy.

Each member of the GIS Consultancy team has over 10 years experience and can assist, not only in planning and selection, but can provide practical help during the implementation and post-implementation phases of a project. This experience combined with an indepth knowledge of the Information Technology Industry allows Midsummer to provide a complete range of services which include:

- Feasibility studies, system specifications and Invitations to Tender
- Benchmark design and supervision
- System selection and installation
- Network design and installation
- Database creation and digitizing services
- System management and staffing services
- Training

Midsummer is completely independent of all GIS hardware and software suppliers and can therefore offer impartial advice to its clients.

Modulo 4

Modulo 4 was formed in the United Kingdom in 1988 following similar activities in Australia. It aims to address the emerging requirements for large computerised systems that integrate geographically based Asset, Plant, Land, Property and Network information (GIS) in the form of Text, Maps and Drawings.

The Company has first hand experience with land administration in developing countries and designs large computerised land information systems (LIS) to improve the effectiveness of the administrative process. A number of the Company's principal consultants were initially drawn from personnel involved in local and national government land and property systems in Australasia.

The Company's activities encompass the following:-

Consultancy—preparation of feasibility studies either directly for a client organisations, or as part of a larger project structure. Production of Functional Requirements, System Design and Detailed Specifications for LIS and GIS.

Implementation—corporate systems for national government, local government, water and electricity distribution authorities.

Systems Development—Administrative systems supporting maintenance of data within LIS. Interfacing software enabling high volume data transfer between digital mapping systems and central database. Productivity software to facilitate rapid system building.

Training and Education—Specific tailored courses covering implemented land and property systems. Broad IT Awareness courses and IT strategic planning for senior management. General technical courses for IT personnel and end users. Bespoke courses developed for an organisation needing to improve IT knowledge and skills.

Typical Projects:-

Botswana—Feasibility study determining suitability of software for a development environment for a Land Allocation System. Implementation of the Land Allocation System as a turnkey project for the Ministry of Local Government and Lands.

Papua New Guinea—Design and implementation of national land information system to administer all state leasehold and freehold land in the country.

Australia—information technology training courses for public and private enterprise. Implementation of land and property systems for local government.

Hong Kong—Preliminary study determining feasibility of establishing GIS pilot projects relating to network asset management, architectural and civil engineering systems.

Great Britain—asset database implementation for Southern Water GIS Project. Implementation of Phoenix Project (Property Management System) for Property Holdings. Implementation of mainframe database and bulk loading of digital mapping data as part of NORWEB GIS pilot. Preparation of Requirements Analysis Report for GIS systems integration at Welsh Water. Implementation of GIS pilot at YEB. Development of software suite to enable bulk transfer of geographic data between applications using the National Transfer Format.

MVA Systematica

MVA Systematica draws together wide experience in *geographic information, information systems* and *database technologies, applications software development* (often technical in nature), as well as *project management* and *consultancy* in information technology. This combination of capabilities provides effective answers to questions of the most efficient way to obtain data, to exploit information as a resource, to make GIS serve applications, and to organize projects and designs so that future changes are anticipated and accommodated.

Major projects for MVA Systematica have included integrating diverse sources of data for use with GIS and related information systems, design and implementation of a large highway information system and software development for processing 1991 Census data. MVA Systematica has been active in the field of standards for European data interchange and for SQL3. The Network Information Workbench (NiW) design includes many technical advances and has demonstrated flexibility for applications areas and computing environments. Many of these applications are in the transport sector, including for traffic data, accident analysis, and bus route optimisation.

PA Consulting Group

PA Consulting Group is one of the UK's leading management and technology consultancies. Located in 20 countries our 1600 consulting staff are of thc highest calibre, each is at the forefront of his or her own field. We combine a unique breadth of specialist expertise with the experience of working with our clients in truly interdisciplinary teams to address key business issues.

The Geographic Information Systems consultancy team is based at PA's Cambridge Laboratory and offers a complete consultancy service for GIS, including:

- definition of corporate data policy
- requirements definition and justification
- system and supplier selection and implementation
- organisation and manpower planning
- project management support throughout the life cycle
- integration with existing databases and systems
- implication and application of technology developments

Successfully implemented Geographic Information Systems are rarely off-the-shelf; our team's ability to use extensive expertise in strategy, systems, technology and implementation is thought to be unique among major consultancies.

The Polytechnic of East London, Department of Land Surveying

The Department offers consultancy in education and training in the fields of Land Information Systems and Digital Mapping. It also has a well developed programme of research and consultancy in Cadastre and Land Information Management, particularly as applied to the needs of developing countries. There is significant experience in training students for work overseas and the Department can boast ex-students in most English speaking countries. It has a history of short courses to meet the needs of the surveying and mapping profession. There is the expertise and hardware to satisfy the majority of requests for such courses, whether they be at an introductory or advanced level. The Department offers a one year post-graduate diploma and a sixteen month MSc in Cadastre and Land Information Management. Research into digital mapping and spatial information systems has included feature classification, property data bases and problems associated with data sharing.

Scott Wilson Kirkpatrick

Scott Wilson Kirkpatrick are leading international Consulting Engineers, Transportation and Environmental Planners providing a wide range of independent consultancy based at its UK Headquarters and offices in over 30 countries worldwide. SWK's Remote Sensing and GIS Consultancy represents an integration of remote sensing techniques into the wide range of engineering activities. Specialist professional staff provide a comprehensive service covering all aspects both of the analysis and interpretation of remotely sensed data and of the applications of GIS for civil engineering, transport and regional planning, hydrology, resource inventory, rural development and environmental appraisal.

SWK have their own GIS, the Geographical Information System Toolkit (GIST), to satisfy the firm's demands for spatial data handling. GIST integrates raster and vector data with attribute data from an ORACLE RDBMS. This is particularly appropriate for natural resources and environmental applications where information handling and analysis are fundamental to effective decision making. Together with our in-house SPANS facility and system selection on a project and applications specific basis, SWK offer an essential tool for project management, the merit of which has been demonstrated on projects worldwide.

Particular emphasis is placed on the provision of specially tailored training courses to meet client requirements in the UK, Africa and Asia.

SCS (MVM Strategic Consultancy Services Limited)

Strategic Consultancy Services (SCS) has established a reputation for innovative and knowledgeable consultancy, based on a great deal of genuine practical experience of systems design, development and implementation.

SCS is a member of the MVM Consulting Group. The Company has an extensive and wide-ranging UK customer base, particularly within Local Government and related property-based GIS users.

Our experience covers a wide range of GIS application areas, and includes use of most leading GIS products, relational databases, communications protocols, and all major mainframe, mini, workstation and PC hardware/ software environments. This background equips us ideally to advise on all aspects of system selection and procurement, dealing with GIS in the wider context of a corporate Information Systems Strategy. SCS is also prepared to be fully involved in the process of putting GIS to work, from initial pilot project management to operational systems support.

Spatial Solutions Limited

Spatial Solutions Limited has been formed to provide a service to those organisations who are using geo-referenced data and are considering the implementation of Geographic Information Systems (GIS). Our practical experience gained from many years of using a number of GIS on a wide range of applications can assist your organisation in choosing and implementing an efficient and practical solution to your particular problems.

Particular areas in which we can assist include:

- Organising and presenting Workshops to promote awareness and give opportunities for "Hands-on" experience. Customised courses can be provided to suit particular application or organisational needs.
- Providing Project Management expertise to assist in the planning and progress monitoring of projects to ensure that adequate controls are implemented to achieve the objectives within stated time and cost constraints.
- Carrying out Policy and Strategic Studies to identify the implications and benefits of using GIS technology.
- Carrying out Market Studies, developing and implementing Market Strategies.
- Sourcing and advising on data availability from both commercial and government sources.

Data capture and conversion services

Trading name, phone and contact	Address	Services provided	End-user systems supported	Ref.
D'Arcy Race Limited 071 631 4841 Jon J Gibbons	96 Portland Place, London W1N 3HD	Surveys to measure & record space utilisation, building condition & furniture/ equipment assets. Data-base loading & graphic displays. Planning & design services to rationalise use of space & assets.	DRUID system for Estate Management.	
Atkins AMC 0672 63922 David Stevens	Fordbrook Business Centre, Marlborough Road, Pewsey, Wiltshire SN9 5NU	Preparation of fore- & background data for GIS. Capture of drawings for CAD use/archiving. Assembly of data for computer modelling systems. Aerial, land, industrial and architectural surveying.	AutoCAD, Intergraph, GDS and MOSS (internal systems) formats supported including DXF, IGDS, SIF, MOSS, GDS, OSTF together with most raster formats.	
Bartholomew 031 667 9341 Kevin Dickson	12 Duncan Street, Edinburgh EH9 1TA	Data sales: small-scale data sets of World & GB for market research & spatially related applications. Contract mapping. Bureau work: scanning/ plotting customer data.	Most standard transfer formats possible such as DIME, DXF, MOSS, SIF, Gerber and variety of ASCII formats.	360
BKS Surveys Limited 0265 52311 R.H. Blackwood	47 Ballycairn Road, Coleraine, County Londonderry, Northern Ireland BT51 3HZ	Fully digital Aerial, Terrestrial, Ground Surveys. Data conversion from existing maps and utility records. GIS design and structure. Data translations.	Informap, KORK, MOSS, LSS, ARC/INFO, Intergraph, GDS, Eclipse, OSTF/NJUG.	360
CA Design Services Limited 0493 440444 P Beales	CAD Centre, Riverside Road, Gorleston, Great Yarmouth, Norfolk NR31 6PX	Data Capture, digitising, digital maps, foreground records, utilities, database creation, training, customisation, consultancy. Interactive linkage between Microstation & DBase IV.	* Intergraph—VAX IGDS & DMRS * Intergraph—Microstation 32 * Intergraph—Microstation PC * AUTOCAD * Ashton Tate—DBase III/IV	

Trading name, phone and contact	Address	Services provided	End-user systems supported	Ref.
Cad-Capture Limited 0254 583534 Simon Watts	Whitebirk Estate, Blackburn, Lancashire BB1 5UD	Scanning services for raster & vector GIS or CAD systems. Specialising in small scale & contour mapping. Clients include Ordnance Survey, County Councils & Public Utilities.	Any GIS or CAD system: OSTF, NTF, TIFF, DXF, IGES, SIF, IRF, GENIO via 0.5″ magnetic tape, floppy disk or optical disk.	
Cartographical Services (So'ton) Limited 0794 390321 John Waterman	Landford Manor, Landford, Salisbury, Wiltshire SP5 2EW	Aerial/ ground surveys, digital mapping, photogrammetry, digitising services.	MOSS, AUTOCAD, INTERGRAPH, ECLIPSE, OSTF, LITES.	
CDD Limited 0494 713769 DJ Pavely	Marlborough, Spurlands End Road, Great Kingshill, High Wycombe, Buckinghamshire HP15 6HY	Data capture. Database design, linking to specific map graphics by customised programs and menus. Integration into GIS of aerial photography, photogrammetry and GPS.	Supports ARC/INFO GIS and Autocad.	
Cobham Digital Services Limited 0932 68133 Bob Coote	Vine House, 41 Portsmouth Road, Cobham, Surrey KT11 1JQ	OS digital data translation to DXF format. Manual digitising output in Autocad and DXF format. Database/ Graphics interface.	Ordnance Survey formats; Autocad; systems accepting DXF format; database ASCII I/O formats.	
Computer General GeoData Services 0225 482744 Dr Pauline A Waters	The New Church, Henry Street, Bath BA1 1JR	Image Processing: sourcing, processing and interpretation of data sets. Development of classification and manipulation algorithms.	PCI EASI/PACE on industry standard workstations and PCs. PCI DIRIGO on Apple Macintosh II. GRASS Public Domain raster GIS on industry standard workstations and PCs.	
Comsult 0234 342401 DR Foster	67 Goldington Road, Bedford MK40 3NB	Digitising large scale OS maps to OS88/ NJUG13. Digitising other maps, plans and overlay's to customer's specification.	OSTF, NTF. Other formats by arrangement.	

Trading name, phone and contact	Address	Services provided	End-user systems supported	Ref.
CSL Group Limited 071 982 0071 Gary R Levett	Consilium House, City Forum, City Road, London EC1V 2NY	CSL provide teams of GIS data capture staff with experienced professionals to manage the process. Data capture strategies are also undertaken.	Services are normally provided on site, therefore any system is supported. Remote data capture considered. Particular expertise in GFIS, Alper, Axis and Planes.	
Data Consultancy 0734 588181 Bryan Wade	7 Southern Court, South Street, Reading, Berkshire RG1 4QS	Specialists in the provision of spatially referenced datasets, digital maps, boundary files, geo-demographics, marketing and other data. Mapping, analysis and Survey Research.	Wide variety of systems.	361
Data Validation Surveys Limited 0709 581465 Peter G Davies	Kilnhurst Business Centre, Victoria Street, Kilnhurst, Rotherham S62 5SQ	Software for text and graphical data capture, validation and conversion. Multimedia applications.	Various hand-held and desktop computers. Conversion to most PC/ mini/ mainframe formats. Multimedia capture and delivery cards.	
Datatrak Limited 0793 722549 Robert Peel	Hargreaves Road, Groundwell Industrial Estate, Swindon, Wiltshire SN2 5AZ	DATATRAK are providers of real time data, in particular vehicle location. Supply PC 'Street Atlas' mapping facility where required.	Real time vehicle location and status, of particular value to maintenance service vehicle management in utilities and local government.	
Doric Computer Systems 0923 210450 Peter Paisley	Doric House, 23 Woodford Road, Watford, Hertfordshire WD1 1PB	Doric GIS services include data capture, database design and creation, application development and consultancy.	Supports ARC/INFO GIS.	
Duggal Associates Limited 081 204 8351 Malcolm Shelbourn	73 Portland Crescent, Stanmore, Middlesex HA7 1LY	CADCORE developed & maintained by Hitachi for interactive, selective data capture, raster to vector, by semi-automated techniques. Result is high productivity over manual & auto. conversion methods.	Compatibility with most existing AM/FM GIS and CAD standard formats. Hitachi is actively working on further direct interfaces.	

Trading name, phone and contact	Address	Services provided	End-user systems supported	Ref.
Engineering Surveys Group 0932 348981 Bill Stedman	Incorporating Clyde Surveys, Rosemount House, Rosemount Avenue, West Byfleet, Surrey KT14 6NP	Land & aerial analytical photogrammetry, digital data capture, raster scanning/ conversion, format conversion, digital cartography.	Any.	
ESRC Data Archive 0206 872323 Kathy Sayer	University of Essex, Colchester, Essex CO4 3SQ	ESRC Data Archive: the largest repository of computer-readable, economic & social data from commercial, government & academic sources in the UK, offering access to similar datasets overseas.	The Archive has a large number of spatially referenced datasets and offers a (dis)aggregation service across geographical area for local interest users. There is a full catalogue of such datasets.	
Fairbairn Services Limited 061 976 3536 Martin Dibnah	Fairbairn House, Ashton Lane, Sale, Manchester M33 1WP	Digitizing and data capture services. Royal Mail Postcode Analysis. General structured data services.	Most CAD and GIS systems. DXF, SIF and general structured data.	
Gardline Surveys 0493 850723 David Pettit	Admiralty Road, Great Yarmouth, Norfolk NR30 3NG	Digital Mapping and GIS services inc: * Automated Map Data Capture * Records Conversion * Consultancy and Training High Quality Standards (OS88, NJUG)	Interfaces to most systems supported including OSTF, NTF, Intergraph IGDS and SIF, Autocad DXF, ARC/INFO, SysScan, Laser-Scan, Synercom.	361
Geografix Limited 0603 788940 Miles Turner	Hurricane Way, Norwich NR6 6EW	Acquisition, analysis and presentation of geographically referenced data including video surveys and street inventories, utility asset conversion & raster or vector map production.	Data is supplied in OSTF, Intergraph, Autocad and other formats.	362
Geonex Air Survey Limited 0602 257990 RW Vango	92-94 Church Road, Mitcham, Surrey CR4 3TD	Digital colour aerial photographic imagery for use in GIS. Software for storing and viewing imagery on PC's.	Image library with graphical search capability, image display manipulation, data extraction, export of data to GIS.	

Trading name, phone and contact	Address	Services provided	End-user systems supported	Ref.
Geonex Air Survey Limited 0602 257990 RW Vango	92-94 Church Road, Mitcham, Surrey CR4 3TD	Geo-referenced aerial photographs, photogrammetric mapping and digitising services.	Most data exchange formats.	
Geonex UK Limited 081 640 1971 Robin Waters	92-94 Church Road, Mitcham, Surrey CR4 3TD	Digital map and records conversion including scanning and database building. New or updated GIS information from aerial photography.	All GIS systems through standard formats including DXF, SIF, OSTF, NTF, MOSS, ARC/INFO, Apple Macintosh etc.	
Glen Computing Limited 0689 875577 Graham Bugler	309 High Street, Orpington, Kent BR6 0NN	Digitising, GIS/MIS, Land Surveying, Photogrammetrical mapping, Terrain Modelling, CAD, Data Processing, Data Conversion, Data Transfer.	Output formats supported: Autocad, Acropolis, Cadam, BIPS, Calcomp IGS, GDS, Intergraph, MOSS (GENIO), OSTF, Rucaps, Eclipse, Panterra, DXB, DXF, ASCII, IGES.	362
Graphical Data Capture Limited 081 346 4959 Peter Klein	262 Regents Park Road, London N3 3HN	Contract digitising specialists for OS, ARC/INFO, Utilities, Terriers, Property Registers, Networks and Zones of all kinds. High Quality Assurance an integral part of the production process.	Data supplied in OSTF, NTF, 3D Models, ARC/INFO, MapInfo, MOSS and others. NJUG13 supported.	
Graphics Management Limited 0698 350843 Alan Giddins	9 Braefoot Court, Law, Strathclyde ML8 5HY	High production data conversion for graphic and/or database applications. Digitising of maps and engineering drawings.	NTF Format. Intergraph (IGDS, DMRS and FRAMME). OSTF and NJUG, AUTOCAD. IBM (GFIS), Digital.	
Ground Modelling Systems Limited 0908 667799 Steve Gould	Rockingham Drive, Linford Wood, Milton Keynes MK14 6PD	Processing survey data, digitising maps and plans: output to tape or disc.	Panterra formats, Ordnance Survey formats, other 2D/3D formats: information available on request.	

Trading name, phone and contact	Address	Services provided	End-user systems supported	Ref.
Grove Consultants Limited 081 846 2468 A.H. Palmer	28 Hammersmith Grove, London W6 7EN	Bureau service for processing survey data, infrastructure design and estate layouts. Digitising of maps, plans and drawings. I/O to tape or disc.	Grove proprietary system supports OSTF, Autocad and GDS.	
GTX Europe Limited 0256 843555 Robert Brown	GTX House, Cedarwood, Chineham Business Park, Basingstoke, Hampshire RG24 0WD	GTX Drawing Capture and Management systems utilise high speed parallel processing, various scanners, PC platforms, networking, raster/vector applications.	GTX Vectorisation for OS Map Conversion 12-20 mins. Output in DXF SIF IGES 2/3 major CAD formats. Raster Oracle database networked PCs for markup viewing.	
Hoskyns GIS 0962 844188 Julie Hoskyn	Technology House, Victoria Road, Winchester, Hampshire SO23 7DU	Data scanning & digitising; all OS transfer formats; text and graphics transferred via interface format file (IFF): existing digital data into GFIS.	Digitised data can be provided in any user specified format: eg OSTF, GIMMS, DXF, GFIS.	
Howard Humphreys & Partners 0372 376190 IT Applications Group	Thorncroft Manor, Dorking Road, Leatherhead, Surrey KT22 8JB	* Inventory/ condition surveys * Survey specification * Quality control/ manage surveys * Data capture devices, software	As required. Experience of Intergraph, GDS, Autocad, DOGS, ARC/INFO and TerraSoft.	
IT Southern Limited 0273 606766 Peter Elkins	Southern House, Lewes Road, Brighton, East Sussex BN1 9PY	IT Southern, in conjunction with Southern Water, has extensive data conversion facilities. Experienced teams in four counties across South East England.	Formats by arrangement.	
Kingswood Limited 081 994 5404 M. Short	19 Kingswood Road, London W4 5EU	Agents for AA Data GB—National coverage Central London—to street level Europe—Atlas coverage & postal areas Data consistent with AA Atlases	Data may be provided in format of Client's own choice for use in industry standard package or custom built systems.	

Trading name, phone and contact	Address	Services provided	End-user systems supported	Ref.
Landplan West Country Limited 0803 842065 Mrs Karen Pearson	The Studio, Alston Lane, Churston, Brixham, Devon TQ5 0HT	Land use and planning data services for collection and capture in GIS vector format. Raster mapping, vector mosaic base layers and datasets for VIRIDIAN GIS users. (See Vendors of GIS.)		
Laser-Scan 0223 420414 Doug Cross	Cambridge Science Park, Milton Road, Cambridge CB4 4FY	Laser-Scan's Mapping Services department digitises any type of map or map related data for GIS systems. The service works to the highest standards and is a contractor to the Ordnance Survey.	Laser-Scan's Mapping Services department has been in operation for over a decade. We can handle all map types including Topographic and Thematic.	363
Longdin & Browning (Surveys) Limited 0792 202244 Dr NJ Eales	Summer Fields, 50 Sketty Road, Swansea SA2 0LH	Precise plans capture as for OS / Transforming of distorted plans / Field survey data capture / Textual attributes & structuring / ACAD Reg. Applications Developer / Bespoke Software	* All GIS and CAD Systems * Conversion between most data formats including: OSTF, NTF, Autocad DXF, IGES, MOSS, Intergraph, McDonnell Douglas GDS and Eclipse.	
Loy Surveys Limited 041 333 0884 Jim Loy or Kenn Allan	3 Woodside Place, Glasgow G3 7QF	* Chartered Land Surveyors * Digital Data Specialists * Nationwide Service * Terrain Modelling	* Eclipse/ AutoCAD User * All formats supported * BS 5750 Quality Assured Firm * Survey Association Member	
Manpower 071 224 6688 Angela Williams	International House, 66 Chiltern Street, London W1M 1PR	Provision of immediately productive temporary and permanent staff for surveying, data capture, digitsing and data entry, on the basis of ad hoc work and project management.	All GIS systems.	
Mason Land Surveys 0383 727261 Kevan Mason	Dickson Street, Dunfermline KY12 7SL	Mason Land Surveys convert maps and plans into 2/3D digital data and provide an extensive field data capture service. Graphical and textual data may be integrated in predefined systems.	MLS uses Intergraph MOSS Autocad GRADE etc in house. Database software includes DBase3+ Superfile DMRS etc. These tools provide flexibility for all data & transfers.	363

Trading name, phone and contact	Address	Services provided	End-user systems supported	Ref.
McLintock Limited 0491 35351 John McLintock	PO Box 17, Crowmarsh, Wallingford, Oxon. OX10 8NP	* Graphics Capture * Text Capture * Data Conversion * Source Data Integration * Data Collection Consultancy	Hardware supported: Prime, Digital, ROCC, ICL, SUN, Tektronix Software supported: GDS, ARC/INFO, PLANES.	
Midsummer Computing Exchange Ltd. 0908 668866 Ivan Bush/ Steve Gill	Midsummer House, 429 Midsummer Boulevard, Central Milton Keynes MK9 2HE	Conversion of maps and plans with or without linked database records. Linework and polygon processing service for any GIS. Analysis and implementation of GIS.	Translated and converted data output in formats suitable for major GIS. Experience of OSTF, DXF, MOSS, Intergraph and ASCII.	
National Remote Sensing Centre Limited 0252 541464 Peter Fletcher	North Gate Road, Farnborough, Hampshire GU14 6TW	Specialising in the supply of satellite and airborne remote sensing data, image processing, consultancy and project management. Film writing from data held on digital storage media.	Data and products formatted to customer specification. Applications supported include: Mapping, Planning, Change Detection, Oceanography and Geology.	364
Norcom Technology Limited 0603 765252 Stuart Duncan	7A Friars Quay, Norwich, Norfolk NR3 1ES	Hydrographic and Environmental survey data capture, post-processing and display. Turnkey systems supply. Specialised I/O expertise.	All systems based on IBM PC and compatibles. Data translation services to GIS/ CAD formats.	
NPA Group 0732 865023 Nigel Press	Edenbridge, Kent TN8 6HS	Supply of satellite imagery, digitizing, scanning, film writing, colour electrostatic plotting.	Most data sources on magnetic tape, optical cartridge or floppy disc.	
Ordnance Survey 0703 792381 RM Bowditch	Romsey Road, Maybush, Southampton, Hampshire SO9 4DH	Britain's National Mapping Agency. Supply of OS digital maps including update service. Supply of road network data (OSCAR), digital contour data and DTM's.	Digital data supplied in NTF. Large scale map data also available in OSTF. All data available on 9 track magnetic tape, 3.5″ and 5″ floppy disk and UNIX cartridge.	

Trading name, phone and contact	Address	Services provided	End-user systems supported	Ref.
Oxford Cartographers 0865 882884 J Briggs/ D Roberts	Unit 13, Oasis Park, Eynsham, Oxford OX8 2BP	High resolution digital mapping in colour.	Image Mapper software producing Intergraph databases and screened film up to A0 size.	
Pinpoint Analysis 071 836 1511 Paul Winters	Tower House, Southampton Street, London WC2E 7HN	Pinpoint Address Code (PAC) database contains a 12 digit grid reference for every UK address. Pinpoint Road Network (PRN) contains digitised centre lines of all UK roads.	PAC and PRN are hardware/ software independent. Most systems and character formats are supported.	
Plowman, Craven & Associates 0582 765566 Mark Phillips	141 Lower Luton Road, Batford, Harpenden, Hertfordshire AL5 5EQ	* Digital Mapping * Terrain Modelling * Survey data processing * Data format translation * Attribute data capture	Full input and output from and to: * McDonnell Douglas GDS * Intergraph IGDS/ SIF * DXF * IGES and many others	364
QC Data 071 491 3813 Debbie Grieco	100 Park Lane, London W1Y 4AR	QC Data provides digitizing, scanning, vectorizing, symbol recognition, attribute encoding, hybrid conversion, on-site staff and consulting services.	ARC/INFO, Intergraph, AutoCAD, IGDS, DGN, ISIF, IGES, DXF, NTF, MOSS, Ingres, Oracle, R:Base, Informix, Paradox and Dbase are supported.	
Remote Sensing Services Limited 0672 20226 Duncan Milnes	Lychgate House, 24 High Street, Ramsbury, Wiltshire SN8 2PH	Digitising, scanning, raster to vector conversion, vector to raster conversion, specialist output to film recorders and writers etc.	Formats supported and converted from and to include AGN, SVF, DXF, DLG, SIF, DFAD.	
Reprocart BV Cartography Service Centre +31 10 4618883 Mr H. Cohen	Ceintuurbaan 213, 3051 KC Rotterdam, The Netherlands	Service bureau based on SCITEX colour scanner and ELP scanner plotter service. Database management.	Scan and plot service for cartographic and geophysical institutions.	

Trading name, phone and contact	Address	Services provided	End-user systems supported	Ref.
Robertson Group Plc 0492 581811 Chris Parker	Llandudno, Gwynedd LL30 1SA	GIS applications for all sectors in the environmental & natural resource fields in UK & overseas. Remote sensing, image processing & specialist interpretations. Digital mapping for all sectors.	Resources survey, evaluation & exploration. Environmental assessment & monitoring. Regional mapping and resource inventory. Vocational training in GIS and remote sensing.	365
Sales Performance Analysis Limited 0926 451199 Alan Pestridge	1 Warwick Street, Leamington Spa, Warwickshire CV32 5LW	* Support & analysis systems for sales management. * Postcode boundary & data files. * Bureau digitising service.	IBM PC.	
Satellite Positioning Company Limited 0865 883474 John Vivian	Davies House, Oasis Park, Stanton Harcourt Road, Eynsham, Oxon. OX8 1TP	Geodetic Control Densification Projects, Satellite based Digital Orthophoto Maps, Construction, Population and Maintenance of Geographically encoded Databases.	Support for virtually every GIS including: ARC/INFO, Autocad, Image Mapper, Intergraph, MapInfo, Synercom. Data provided in most relational database and data interchange formats.	366
Scanmedia Limited 071 250 3377 John Stammers	47-58 Bastwick Street, London EC1V 3PS	Bureau services for raster scanning, raster to vector conversion, supply of software for image storage and display, overtracing, document/ drawing distribution.	Input and output for most user formats and to meet user requirements, output to magnetic or optical media or to microfilm.	
Simmons Survey Partnership 0934 732122 GS Jones	5 West Street, Axbridge, Somerset BS26 1AA	Digital Mapping, for GIS/LIS or CAD of OS maps, existing plans & drawings, Land, Building & Engineering surveys, including 3 dimensional modelling and Visualisation. BS 5750 Registered.	NTF, OSTF & NJUG Mapping formats, Eclipse, Panterra, Moss, and LSS modelling formats. Autocad, DXF, IGES, Intergraph, GDS etc. CAD formats.	
Spectrum Energy & Information Technology 0483 730201 Sue Baker	Spectrum House, 56 Goldsworth Road, Woking, Surrey GU21 1LQ	High speed 200/ 400 dpi document scanning and editing. Output to tape, paper, film, disk or optical disk.	Raster image translated to required format and linked to databases.	

Trading name, phone and contact	Address	Services provided	End-user systems supported	Ref.
Structural Technologies Limited 0527 854819 DF Schindler	Woodside, The Slough, Studley, Warwickshire B80 7EN	Bi-directional translation between OS formats, IBM GFIS, AutoCAD, STL STRUMAP etc. Efficient User-selectable compression options. Validation of NTF level 3 structures.	OSTF, NTF levels 0, 1, 2, 3 & 4, AutoCAD DXF, GFIS IFF, GeoManager, Laser-Scan NTF interface, Viewmap NTF interface, IGES, various raster metafiles, most format easily supported.	
Summerside Associates Limited 0480 407240 Mike Quarmby	Atlas House, 17a Huntingdon Street, St Neots, Cambridgeshire PE19 2BU	Bureau data capture service: all forms of base and overlay mapping, structured data, attributes and DTMs using Laser-Scan VTRAK (raster to vector) or tablet based systems for clients worldwide.	Quality standards match OS 88, NJUG 13 or equivalent. Data conversion to/from most standards including OSTF, NTF (inc. level 3), SIF, DXF, Informap, MOSS, GFIS, ARC/INFO and ASCII.	
Summerside Associates Limited 0480 407240 Mike Quarmby	Atlas House, 17a Huntingdon Street, St Neots, Cambridgeshire PE19 2BU	Contract staff and services: a bespoke service including provision of mapping technicians, project/ facilities management to Clients Worldwide.	Comprehensively trained computer mapping technicians & professional staff experienced in Viewmap, Informap, Laser-Scan including VTRAK, Scitex, GFIS, GDS, MOSS, Intergraph, Medusa & MapData.	367
Survey & Development Services Ltd 0506 825121 John E. McCreadie	3 Hope Street, Bo'Ness, West Lothian EH51 0AA	Provision of vector map data either from original survey and mapping or by digitising from existing maps. All sectors of the market are catered for with quality and reliability assured.	ECLIPSE software suite used to produce all digital map data with in-house AUTOCAD facilities for CAD work. Output can be on any media and in most common transfer formats.	
Survey International 0525 718900 John Callaghan	Dickens House, Off Maulden Road, Flitwick, Bedfordshire MK45 5BY	Data capture by digitising to OS/NJUG spec. Aerial & terrestrial photogrammetry, land & building surveys. Data conversion, terrain modelling, database creation for GIS. BS5750 accredited business.	MOSS, AutoCAD and Medusa user. Data output to DXF, MOSS, ISIF, Medusa, OSTF, IGES, GDS, ARC/INFO.	
Taywood Data Graphics 081 575 9240 Pat Jarvis	Greenford House, 309 Ruislip Road East, Greenford, Middlesex UB6 9BQ	Digital mapping, raster scanning in colour and binary, also microfilm aperture cards, attribute & graphic records conversion, A0 colour plotting, systems selection and site survey.	Raster, vector and attribute data to any system: eg ARC/INFO, GDS, GFIS, DXF, Intergraph, OSTF, NTF (level 3), Scitex, Mapdata, SIA, SIF, Synercom.	368

Trading name, phone and contact	Address	Services provided	End-user systems supported	Ref.
Terraquest by Sutcliffe and Associates 021 500 5818 John Gannon	Kenrick House, Union Street, West Bromwich, West Midlands B69 6DB	Consultancy, data capture, data base design & creation, applications development, systems integration, facility management, bureau services & user training for GIS and Land Administration.	Dual Digital VAX 3900 and MDISL Series 19 processors in-house. GDS, PLUS, ARC/INFO, PC-ARC/INFO. Data formats: OSTF, DXF, SIF, IGES and ASCII.	368
URPI Group Limited 0734 588181 Bryan Wade	7 Southern Court, South Street, Reading, Berkshire RG1 4QS	Analysis, mapping and databases for market analysis including survey research, geo-demographics, large stores and shopping schemes, travel times and Census of Population.	Wide variety of systems.	
Walker Ladd and Company 0272 738414 Jonathan Davy	9 Park Place, Clifton, Bristol, Avon BS8 1JP	* Land Surveys * Building Surveys * Data Capture/ editing * Digitising * Underground location of services	* IBM—compatible PCs * AUTOCAD & DXF * MOSS * ECLIPSE * SURVPAK (Land Survey Software) * Digital Ground Models	

Bartholomew

Datasets available vary between the 1:250K topologically structured dataset of Great Britain, including basic topography, communications, admin boundaries, cultural information and postcodes, through to a 1:1M dataset of western and eastern Europe, a 1:5M dataset of Africa and 3M and 14M datasets of the whole world with varying levels of topographic and cultural layers. As the datasets are constantly being expanded, an up-to-date list of features and coverages available should be requested.

Small-scale datasets are currently used for market research, sales analysis and plotting of spatially related customer information in a wide variety of areas, such as utilities, local authorities, health services, emergency services and various commercial applications. The datasets are available through Combined Higher Educational Software Team (CHEST) for higher education establishments.

BKS Surveys Limited

BKS Surveys operate on a world-wide basis, specialising in the provision of all types of mapping, utility and other spatial information.

The Company has been providing surveying and mapping services for over 30 years and, during the last 15 years, has been supplying digital products ranging from simple terrain ground models to data for GIS systems.

BKS use INFORMAP, MOSS, KORK, AutoCAD and LSS to produce Digital Databases from Aerial and Terrestrial Photography, existing documents and field derived data.

BKS's impressive resources include 7 Photogrammetric Workstations, 7 Interactive Graphics Workstations using VAXstations, SUN SPARC and COMPAQ 386. An ETHERNET network combines the stations to produce a cost-effective and powerful database creation facility.

BKS are approved contractors for Ordnance Survey Digitising.

INFORMAP, AutoCAD, KORK, MOSS, and LSS are used in-house to prepare digital data. Three large, high accuracy scribing plotters, supported by Pen and Electrostatic capability are used for hard copy representations.

Other systems supplied include INTERGRAPH, ARC/INFO, GDS, Eclipse and OSTF.

The company supports its operation by its own specialised and experienced Software and Design Team and can provide translation services and consultancy.

The Data Consultancy

The Data Consultancy specialises in the provision and analysis of spatially referenced data sets for market analysis and site assessment. Analysis and mapping of data is undertaken together with a full and comprehensive service of help and advice on the sources, collection and use of data. Survey research and cross analysis of data is also undertaken. Most data sets are spatially referenced to the national grid, administrative and postal geographies. Major data sets include:

- **Map and boundary files:** roads, postcode areas, administrative boundaries, drivetime isochrones and other boundaries.
- **Demographic data:** 1981 Census small area statistics at enumeration district and postcode sector levels for the whole of Great Britain. Population updates to later years and Super Profile neighbourhood classification.
- **Consumer Retail Expenditure:** consumer retail expenditure and retail business turnover potential available at any spatial level down to enumeration district.
- **Stores:** UK hypermarkets and superstores open and trading or with planning permission. Locations of multiple retailers and service businesses.
- **Shopping centres and schemes:** details of all large managed shopping schemes together with proposed schemes and those with planning permission. Locations of all major shopping centres including town and city centres with cross references to Goads and Newmans data.

Data and analyses can be supplied with or without associated GIS software.

Gardline Surveys

Gardline's Digital Mapping Division offers a variety of high quality services to a wide range of Digital Mapping and Geographic Information System users.

AUTOMATED MAP DATA CAPTURE All types of maps, charts and drawings efficiently digitised using Gardline's automated flowline

RECORDS CONVERSION Utility and Local Authority foreground records conversion and integration with GIS databases.

DATA PROCESSING & PLOTTING Data structuring, processing and output to a range of devices, media and formats.

CONSULTANCY & TRAINING Practical advice and support services based on a number of years experience using a variety of systems.

Since its formation in 1986, Gardline's Digital Mapping facility has continued to expand and has completed numerous projects on behalf of the Ordnance Survey, Utilities and Oil companies.

Expansion has included ongoing investment in the very latest computer equipment and software from Intergraph, Laser-Scan as well as software applications developed in house.

This combination of advanced equipment and expertise means that we offer the flexibility and knowledge to satisfy the individual needs of our customers.

Geografix Limited

Geografix are involved in a broad range of activities relating to the acquisition, analysis and presentation of geographically referenced data. These include video surveys and the production of street furniture inventories for local authorities, various types of road condition assessment and road geometry surveying, digitizing the assets of utilities and the production of digitized maps, both raster and vector.

Geografix have a rapidly expanding software development facility specialising in land navigation, advanced hydrographic data acquisition and processing systems, mapping and surveying systems based on GPS receivers and a range of software to support activities related to geographically referenced databases.

Glen Computing Limited

In association with our sister company Glen Surveys Limited (Chartered Land Surveyors) we offer an extensive range of data capture and computing services.

DIGITAL MAPPING A fast and economic service for digitising documents of any scale. All work can be certified to NJUG13 standard.

PHOTOGRAMMETRIC MAPPING Employing state of the art, analytical plotters to provide digital and graphical output, from aerial or terrestrial photography.

TERRAIN MODELLING Three dimensional modelling for volumetric analysis, sections, cut/ fill, and contouring.

CAD Using data captured by ourselves, or from other sources for housing layout, drainage, refurbishment and design in general.

GIS/ MIS Data capture and independent consultancy service for analysis of user requirements and supply of systems and spatial data. Typical examples are Grounds Maintenance and housing stocks/ condition surveys for local authorities and Highway Inventory surveys for County Councils.

LAND SURVEYING Any form of land, engineering and building surveys, field verification of aerial surveys, monitoring and inshore hydrographic surveys.

SURVEY DATA PROCESSING Computerised survey processing, plotting, modelling and contouring. Output in graphical or digital format.

DATA CONVERSION For a wide range of input/output to and from a range of magnetic media.

Laser-Scan

Laser-Scan's Mapping Services Department have been operating for over a decade producing plans, topographic, thematic and cadastral maps, air and hydrographic charts. Output can be provided in either hardcopy form or as raster or vector digital data. The service offered includes high quality map conversion services and data transformations.

A typical flowline for a map data capture project undertaken by Laser-Scan's Mapping Services Department would include

- **Project definition and consultancy** –agreement with the client the specification for the work and the requirements and the deadlines for the project.
- **Flowline set-up** –tuning the system to perform the task and staff briefing for any special requirements.
- **Automatic Data Capture** – use of Laser-Scan's advanced data capture systems.
- **Manual Data Capture** –complimenting the automatic data capture with manual additions and data tidying.
- **Further processing** –incorporation of additional processing such as structuring as necessary.
- **Check Plotting** –plot the results supported by data listings for verification.
- **Data Editing** –an interactive cycle of data correction.
- **Quality Assurance** –well established procedures for QA are performed.
- **Final plotting and data dispatch** –final completed files are plotted and the data is despatched to the client in the required format.

Mason Land Surveys Limited

Mason Land Surveys are one of the UK's largest companies involved in digital data capture and conversion with wide experience of different computer systems and client requirements.

The company operates from seven offices nationwide and has over 200 staff many of whom are professionally qualified as engineers, surveyors and computer specialists. Significant and careful investment over the years has resulted in the use of the most modern and efficient systems in the main business units of land and photogrammetric surveys, digitising and scanning, the production and maintenance of graphical and textual databases and project management.

Typical examples of recent projects include:

1. Contracts with the Ordnance Survey for large-scale mapping and digitising.
2. Project management of computerised land records and property terriers for government authorities.
3. Land and aerial survey work for local and central government with output to different computer systems.
4. Database assembly and maintenance.

The offices are equipped with the latest Apricot computers running in-house graphical software together with other branded systems to interface with the systems retained in Dunfermline and our clients systems. CADD systems on which MLS have developed experience and retain in-house include MOSS, Autocad, Intergraph.

National Remote Sensing Centre Limited

NRSCL is the UK's leading supplier of remote sensing imagery. Image data for most parts of the world is acquired under the existing "open skies" policy. Image data is either in digital format, as computer compatible tapes, floppy disks, or in the form of photographic and film products. NRSCL supplies data in pre-processed or processed condition tailored to meet specific customer requirements and application needs.

Image data is available from Landsat, Spot, NOAA, Seasat, Nimbus 7 (Coastal Zone Colour Scanner) and ESA programmes. Archive and quick-look facilities are available of cloud free Landsat and Spot imagery, dating back to 1972 and 1986 respectively. NRSCL will operate the United Kingdom Earth Observation Data Centre which will process, archive and distribute data from ESA's ERS-1 satellite. ERS-1 data products will be available in the latter half of this year. NRSCL will also acquire data from other sources, such as airborne imagery and ground truth data.

The Company offers a comprehensive range of value added and consultancy services. Interactive image processing facilities, housed within a commercially secure environment, are available for those customers wishing to undertake their own processing.

Plowman, Craven and Associates

Plowman, Craven and Associates have been providing surveying and mapping services worldwide for over 26 years. Digital products have been produced for the last 13 years and our dedicated development team is always up-to-date with the latest techniques and equipment.

Of specific interest to GIS users we run 7 interactive graphic workstations for the preparation of digital maps and plans, with associated attribute data. Data input can be direct from field survey instruments, analytical photogrammetric workstations and digitising workstations.

Data translations are achieved by our own software and we have extensive experience with many formats reflecting the variety of clients we serve.

Current projects for GIS users include Property Information for Central/Local Government departments and the creation of databases for space audits of large buildings.

For property departments we present digital floor plans and geographical location along with a database of maintenance and condition survey information for use by Local Authority departments or Chartered Surveyors.

Space audit information is collected with our own data-logging software and transferred directly to a relational database. These databases are used by Architects and Building Surveyors in their analysis of existing building space.

The Robertson Group Plc

The Robertson Group Plc is an integrated group of consultancy companies, involved throughout the world in natural resource exploration, the environment, agriculture, forestry and rural development, and in analytical chemistry, computing and cartography.

Digital cartography, remote sensing and desktop publishing form part of the Petroleum Division, services being provided for other companies within the Group and a complete mapping service is available for companies, governments and aid organisations worldwide. The facilities are used principally in mapping for petroleum and mineral resources, in the assessment of land use applications and environmental management and monitoring.

The Robertson Group Plc are well known for a high standard of conventional cartographic products. The GIS installation reflects the philosophy of using computers to expedite production and enhance product quality. The GIS is linked by Ethernet to our Erdas image processing system which handles raster scan data from both satellite and airborne systems. Software links have been developed to enable data transfer between the GIS, image processing and desktop publishing systems, enabling the Robertson Group to undertake complex work requiring integration of maps, raster images and text.

Satellite Positioning Company Limited

Satellite Positioning Company (SPC) and our American sister Satellite Positioning Corporation are the world's leading suppliers of NAVSTAR Global Positioning Satellite (GPS) based geodetic control and satellite based image orthophoto map products. These products are currently being utilized throughout Europe and North America to "Build the Geographic Foundation", necessary to properly construct and maintain geographic information systems (GIS).

MULTI-BASE™ SURVEY TECHNOLOGY: In 1986 GPS Technology Corporation, the parent company of SPC, tested and developed the Multi-Base technology to minimize the problems of single-offset Differential GPS. Instead of using just one base station, the Multi-Base system uses error measurements from a network of at least three reference stations covering the whole area being surveyed. The instantaneous accuracy obtained using this method has a standard error of better than 1 metre, with reference stations separated by as much as 300 kilometres.

GEODETIC CONTROL: Using Multi-Base technology, SPC can establish a network of Geodetic Control points faster and less expensively than other methods. SPC's Control Points can be either conventional, physical monuments or photo-identifiable locations suitable for use as orthophoto image control.

SATELLITE IMAGE PROCESSING: The French SPOT image collection satellites provide systematic, up-to-date coverage of the Earth's surface. The raw satellite image data, however, contains many distortions which inhibit its use. Errors, which SPC corrects, include: camera tip and tilt angles; the varying distance from the camera to different parts of the image; skew generated by camera motion and Earth rotation; Earth curvature; and parallactic shifts caused by variations in elevation. SPC also sub-samples the data and applies enhancement filters to achieve higher visual resolution of subtle details.

SPC MAPS: For its standard 1:25,000 series of large-scale orthophoto map products, SPC uses a 5 meter sampling interval, coinciding with exact coordinate spacing on a Universal Transverse Mercator (UTM) map projection. These maps can be either rectified to existing maps or, preferably, geodetically tied using SPC's surveyed ground control.

BASE MAPS FOR GIS: SPC Maps, rectified to geodetic ground control, present the solution to the problem of registering a variety of different GIS databases. The SPC Map, as a raster image, can be used in a GIS workstation as a reference underlay, over which all other data can be aligned.

GIS SURVEY PRODUCTS: Using a combination of these technologies, together with SPC's proprietary Attribute Collection System, SPC is also able to offer a flexible range of GIS Field Inventory and Transportation Database services.

Summerside Associates Limited

Summerside Associates currently provide quality GIS data capture and support services to an impressive list of clients including Government agencies, major utility and private sector companies both in the UK and overseas. Particular emphasis is placed on combining the best people, quality training, exceptional technical resources and management expertise into a full service package that puts the interests of the customers first.

Computer Mapping Bureau: Captures background topographic mapping to meet the standards of the Ordnance Survey of Great Britain, together with quality assurance testing to NJUG13 requirements. The bureau particularly specialises in the capture and conversion of 'foreground' mapping into structured digital records. All types of digitising and attribute capture are undertaken using modern Digital workstations, ALTEK digitising tablets and Laser-Scan software products including the semi-automatic VTRAK raster to vector conversion package. Output can be simple map feature data or topologically structured into networks, polygons or surfaces, transferred direct to the Customers system or via National Transfer Format Level 3.

Contract Cartographic Computer Staff: We provide fully trained, experienced computer mapping technicians to work within the client's own premises and in-house digital facilities on a part or full project basis. This service ensures an efficient and cost-effective mapping project solution.

Project Management and Consultancy: We offer project and facilities management services and consultancy for all mapping related projects.

Training: Specialised staff training is provided in digital mapping and GIS applications.

Taywood Data Graphics

Data in whatever form is a key asset of any business. Converting this data currently on paper or film to a digital format is the speciality of Taywood Data Graphics. Maps, engineering drawings, site plans etc., can now be readily transferred by a variety of different techniques.

Maps at 1:1250 and 1:2500 scale are digitized both for the Ordnance Survey and a wide range of utility companies. These are all issued with an NJUG 13 quality assurance test certificate. Maps at other scales can be readily undertaken, and we have just completed the 1:50,000 contours of Scotland for the Ministry of Defence, whilst digitizing the 1:10,000 scale contours for the U.K. is on-going.

Prior to volume data conversion of your records, it is crucial that a pilot project is undertaken. Pilots have been completed for British Telecom, Norweb, and are continuing with three other major Utilities.

Complementing the 40 plus digitizing tables is VTRAK, a semi-automatic vectorizing tool. With this system the power of the computer is harnessed by the intelligence of the operator.

An alternative to vector digitizing is raster scanning which provides an excellent reference or archive base. Documents can be scanned at up to 1,000 dpi, and output to all computer systems that handle raster data.

Terraquest by Sutcliffe and Associates

Nineteen years specialism in land and property information combined with 'hands on' GIS implementation experience since 1988 enables Terraquest to provide unique capabilities to both suppliers and users of Geographic Information Technology. Capabilities encompass a wide range of support services:

- Consultancy
- Systems Development and Integration
- Data Preparation and Capture
- Quality Assurance
- Project Management

These capabilities can make a significant contribution to a variety of GIS projects and are especially applicable to projects with a focus on Land Management or Infrastructure Provision.

Current projects include GIS Implementation for a County Council, Metropolitan District Council and a District Council. Implementation Services being provided range from Initial Consultancy through to Data Entry. Consultancy Services include identification of Business Requirements and advice on planning, particularly with regard to optimum Programme. Terraquest are responsible for data gathering, preparation and entry to the chosen system. Data Entry services range from digitising of base map details at varying scales, to capture of historic departmental data.

Infrastructure related projects include the 100 mile Crossrail Project extending from Reading and Aylesbury through Central London to Shenfield. In this instance a GIS approach is being used to handle land use and property aspects of the project initially related to gaining Royal Assent for the scheme.

Programmable databases

Trading name, phone and contact	Address	Product description	Features for GIS users	Ref.
Auto-trol Technology 021 456 2022 Barry Jones	53-54 Calthorpe Road, Edgbaston, Birmingham B15 1TH	Empress/ Oracle RDBMS, Graphics SQL.	Integration of both RDBMS and graphics.	
Automobile Association 0256 492906 Ralph Robbins	Information Services, Fanum House, Basingstoke, Hampshire RG21 2EA	AA copyright data of GB, Europe & C. London. Structured place names Gazetteers. Roadwatch dynamic road & traffic reports. Restricted road data. AA Milemaster calc. package. European postcodes	GIS, vehicle scheduling & consumer location systems. Tourist & travel information. Route planning & scheduling systems. Databases levelled & feature coded, supplied as vectors or link/ node.	371
Doric Computer Systems 0923 210450 Peter Paisley	Doric House, 23 Woodford Road, Watford, Hertfordshire WD1 1PB	INFO, a complete relational data management system and fourth generation language, provides easy access to information as well as powerful programming ability.	INFO provides easy to use data entry and query facilities along with sophisticated decision support tools. Three alternative report generation formats allow fast production of tabular reports.	
Empress Software UK 0483 861990 Dennis Flavell	Godalming Business Centre, Catteshall Lane, Godalming, Surrey GU7 1XW	EMPRESS: an advanced high performance RDBMS and 4GL with query language, report writer, user defined operators, extended data types and programming interfaces.	Standard product supports special BULK data type suitable for 'object-orientated' coordinate data, SQL and custom query (eg spatial) extensions.	
Greenly's Systems 071 371 9191 Robert Greenly	Riverbank House, Putney Bridge Approach, London SW6 3JD	Land and Estates Terrier for Local Authorities and other public bodies. Records details of land and property holdings including Sales, Leases, Licences, Council House Sales and Appropriations.	Optional Mapping Link to record polygon details of land & related transactions. Area measurement capability. Display of land held by Committees, values, rental etc. Permanent link to text Terrier.	

Trading name, phone and contact	Address	Product description	Features for GIS users	Ref.
Ground Modelling Systems Limited 0908 667799 Steve Gould	Rockingham Drive, Linford Wood, Milton Keynes MK14 6PD	Modular software: database links textual and numeric data to points, strings, polygons.	User specified data fields, integrated software options, GIS plus survey, modelling and design.	
Hoskyns GIS 0962 844188 Julie Hoskyn	Technology House, Victoria Road, Winchester, Hampshire SO23 7DU	IBM relational DBMS DB2 & SQL/DS GEOMANAGER acts as storage & retrieval interface.	Network analysis, feasibility studies, 3D representation, graphic reports.	
IBM United Kingdom Limited 0926 332525 Preet Dhillon	GIS Market Development, PO Box 31, Birmingham Road, Warwick CV34 5JL	IBM GIS products are based on industry standard DB2 or SQL/DS relational database models, using geoManager to ensure data integrity, security & integration with non-GIS applications.	* One repository for attribute, graphic and spatial data * Search by area and object * Viewing on low cost terminals * Corporate information integration	372

AA Information Services

AA Information Services collect, collate and distribute a wide range of time sensitive data. Spatial databases available of GB (digitised at 1:200,000), London Central (digitised at 1:10,000), Europe (digitised at 1:1,000,000). European post codes at district level are also held and available in link and node and vector formats.

Route planning systems, both for UK and Europe contain AA sourced data, including information on banned turns, weight and height restrictions and other restrictions. All UK distances are logged by AA staff. Data can be used in vehicle scheduling, time and mileage calculations and marketing support systems.

The AA GIPSY database holds a huge amount of detail on AA recommended Hotels, Guesthouses and other accommodation both in the UK and Europe, plus information on Places of Interest and leisure facilities i.e. 1,600 UK Golf Courses are held on file.

The AA Roadwatch database is unique, holding data on long term road works and the current traffic problems in congested areas. This can now be merged as graphic overlay on the GB mapbase.

AA Gazetteers are comprehensive and structured interpretations of Britain and Europe in 1991.

AA databases are of modular construction and individual modules or multiples may be licensed. Regional extracts are also possible.

IBM United Kingdom Limited

IBM's Enterprise approach to GIS is based upon the use of industry-standard DB2 or SQL/DS relational database models. GIS functionality is provided by the use of IBM geoManager, which is designed for use by government agencies, utility companies, transportation departments, petroleum companies, telecommunication firms, retail organisations and others seeking more efficient ways of managing their geographically related land and facilities data.

The geoManager product's object-based implementation takes graphic data (what it looks like), descriptive data (what it is), spatial data (where it is) and connectivity data (what it is linked to) and integrates it all within one relational database model. The data is stored as a continuous map base with the ability for a user to specify the portion and type of data needed. Database retrievals can be area or network based. If required retrievals can be modified by user-defined qualification criteria—using layers, object types and attribute data fields.

geoManager is compliant with IBM's Systems Application Architecture (SAA), allowing users to invoke GIS functionality with the same look and feel as all other SAA applications.

The geoManager relational database model offers four primary benefits to GIS users.

- It enables geographic information to be shared between different departments by eliminating the reliance on special-purpose spatial data structures.
- It permits integration of geographic information with other applications within an enterprise—such as work management systems, customer information databases, in-house publishing and corporate decision support applications.
- It delivers a true information system by giving users access to geographic data via low-cost terminals, thus reducing the cost of implementing a GIS.
- It uses industry standard data security, integrity, back-up and recovery features (as provided by DB2 and SQL/DS), augmented by spatial indexing routines designed to enhance GIS performance.

IBM is a registered trademark and geoManager, SAA and DB2 are trademarks of International Business Machines Corporation.

Peripheral equipment

Trading name, phone and contact	Address	Product description	Features for GIS users	Ref.
Auto-trol Technology 021 456 2022 Barry Jones	53-54 Calthorpe Road, Edgbaston, Birmingham B15 1TH	Scanning, plotters and workstations.	Full integration services for hardware and software.	
CalComp Limited 0734 320032 Sue Carter	Vector House, Ruscombe Business Park, Twyford, Berkshire RG10 9NU	High accuracy input from all our CalComp digitisers. High quality and performance from CalComp pen, thermal and electrostatic plotters, high resolution graphics cards and monitors.	Fast colour output from pen, high quality cartography from thermal transfer and wide format high res. electrostatic plotters; dual colour, high throughput direct imaging and thermal plotters.	377
CDD Limited 0494 713769 DJ Pavely	Marlborough, Spurlands End Road, Great Kingshill, High Wycombe, Buckinghamshire HP15 6HY	Calcomp high resolution digitising tables model 95600. Interfaces to plotters and other digitisers.	Digitising accuracy of the highest order for direct use by ARC/INFO or any other Geographical Information System.	
Gammadata Computer Limited 0256 817640 Robert Gilson	Worting House, Worting Road, Basingstoke, Hampshire RG23 8PY	Gammacolor Systems 300 & 400; Gammadata Mac printer.	All 300 dpi thermal transfer colour printers providing fully comprehensive range of connection to applications, networks etc. to provide screen dump, Postscript and plot from files.	
Gimms Limited 031 668 3046 Marlene A. Ferenth	30 Keir Street, Edinburgh EH3 9EU	DIGIT-II digitising package: for precision data capture. Implements full feature coding + topological checking. DIGIT-II can be customised to meet particular requirements.	Wide range of digitisers supported. IBM PC (+100% compatibles) with EGA/VGA graphics and 640kbytes memory are required to run DIGIT-II.	

Trading name, phone and contact	Address	Product description	Features for GIS users	Ref.
Graphtec Precision Image Limited 0635 248192 Mark Plews	9 Langley Business Court, World's End, Beedon, Chieveley, Newbury, Berkshire RG16 8RY	A range of A0/A1 mono and colour, cut sheet, 400dpi electrostatic plotters. On board HPGL, Calcomp, GKS, CGM and raster support with 200 Mips processor.	Unique ability to merge vector drawings and scanned images with the highest achieveable accuracy: better than 0.05%, typically 0.02%.	377
Ground Modelling Systems Limited 0908 667799 Steve Gould	Rockingham Drive, Linford Wood, Milton Keynes MK14 6PD	Turnkey system: PC, graphics screen, plotter, digitiser, printer options, I/O to tape or disc.	Flexible data input and output: survey data, digitiser or digital data. High quality plotted output on economic PC based system.	
Hoskyns GIS 0962 844188 Julie Hoskyn	Technology House, Victoria Road, Winchester, Hampshire SO23 7DU	IBM GFIS peripherals: 618X plotters, 5084 digitisers, 508X displays, PS2 displays. Interfaces to other digitisers and plotters.	GFIS on PS2; existing hardcopy/input devices utilised.	
IBM United Kingdom Limited 0926 332525 Preet Dhillon	GIS Market Development, PO Box 31, Birmingham Road, Warwick CV34 5JL	Full range of IBM workstations (5080, 6090, RS/6000, & PS/2 based), 3270 low-cost terminals, digitisers, plotters, raster screen copy devices and A4 scanners.	IBM provides a range of ergonomically designed input and output equipment, together with links to other industry standard peripherals.	378
Intergraph (UK) Limited 0793 619999 NJ Price	Delta Business Park, Great Western Way, Swindon, Wiltshire SN5 7XP	Intergraph offers a range of hardcopy scanning systems for flexible and rigid materials and with resolutions up to 2000 dpi. Software supports mixed raster and vector formats.	Intergraph provides low cost storage of maps, modification and updating of existing maps, creation of new maps and finally to distribution.	
Laser-Scan 0223 420414 David Boughey	Cambridge Science Park, Milton Road, Cambridge CB4 4FY	Laser-Scan VTRAK scanned data capture systems, are configurations with scanner, workstation(s) and software to provide a unique, efficient system for building optimal GIS.	VTRAK systems provide the best balance between the power of modern workstations and the intelligence of the human being.	379

Trading name, phone and contact	Address	Product description	Features for GIS users	Ref.
Remote Sensing Services Limited 0672 20226 Duncan Milnes	Lychgate House, 24 High Street, Ramsbury, Wiltshire SN8 2PH	A full range from manufacturers who include Calcomp, DBA, Eikonix, Linotronic, Matrix, QCR, Tektronix and ViTec.	Products include film writers and recorders, thermal electrostatic and inkjet printers and pen plotters; scanners and digitizers and attached processors.	
Spectrum Energy & Information Technology 0483 730201 Phil Brown	Spectrum House, 56 Goldsworth Road, Woking, Surrey GU21 1LQ	DIP system that allows input and archival of documents from A4 to A0 in size and storage on optical disk. Images/ documents can be sent/ received via the ISDN network at 64Kb/sec.	Complete DIP solution for scanning, storage and transmittal of documents/ images.	
SSI-Microcad 0672 63000 Gerry Anniss	Fordbrook Business Centre, Marlborough Road, Pewsey, Wiltshire SN9 5NU	Scanners A3-A0. Mutoh pen/pencil plotters. GTCO digitisers: A4-A0. Glaser precision plotters/ Photo-plotters. DARTS—drawing archival retrieval system. Microcad work-stations. Versatec E/static plotters	FMS/AC—Geographic Information System for AutoCAD. CAD overlay esp: Raster Editor. VIEWBASE: Relational Database. V-PACKAGE & CADMATE: raster to vector conversion.	
SUN Microsystems Limited 0276 20444 Jonathan Cooper	Watchmoor Park, Riverside Way, Camberley, Surrey GU15 3YL	SUN SPARCstations and SPARCservers. Full range of powerful UNIX-based desktop workstations and network servers.	SUN's SPARCsystem family is the most popular platform for both GIS users and application developers, offering industry leading price/ performance and flexibility.	
Taywood Data Graphics 081 575 9240 Pat Jarvis	Greenford House, 309 Ruislip Road East, Greenford, Middlesex UB6 9BQ	Microtek and contex low cost scanning systems, together with supply of optical discs, raster editing software for PC systems.	In-house scanning capability ensuring maps do not leave the clients premises.	
TDS-Numonics Limited 0254 676921 Peter Mingins	Lower Philips Road, Blackburn, Lancashire BB1 5TH	TDS-Numonics manufacture a wide range of large format digitizers—opaque and backlit—in sizes from A2 to Super A0 (44″×60″). The range includes the high precision Mantissa series.	TDS-Numonics digitizers provide a cost-effective solution to precision data capture. TDS-Numonics products are used in many Cartographic/ GIS and Surveying applications throughout the world.	

Trading name, phone and contact	Address	Product description	Features for GIS users	Ref.
Wicks and Wilson Limited 0256 842211 Derek Wittig	Morse Road, Basingstoke, Hampshire RG22 6PQ	CF 1000 large document scanner, 1000 dpi resolution, high scanning speed and accuracy especially GIS applications.	1000 dpi resolution, speed and accuracy, dynamic thresholding.	
Xerox Engineering Systems Limited 0635 517200 John Gover	Versatec House, 5 Oxford Road, Newbury, Berkshire RG13 1QD	Electrostatic, Laser and Thermal plotters. Wide format plain paper plan printers.		
Zeh Graphic Systems (Europe) Limited 0306 740105 Paul Popeck	290-294 High Street, Dorking, Surrey RH4 1QT	Sophisticated colour raster plotting software. Designed with special functions, performance and flexibility for GIS and CAD Users. International user base on VAX, UNIX and IBM.	Supports wide range of raster plot devices: electrostatic, thermal transfer and laser film. Combines image and vector data. Performs area colour and pattern fill.	

Calcomp Limited

Calcomp has become a respected name in the computer graphics industry and has been a leader in the supply of graphics systems since 1959. CalComp was the first company to build a commercial graphics device. The product range is broad, enabling CalComp to produce a total solution for all your computer graphics requirements.

CalComp offers very high accuracy, high resolution input via its range of digitizers. These can be used in any number of applications, including drawing, draughting, mapping and desk top publishing.

CalComp are widely recognised as experts in the CAD field and understand the plotting needs of architects, engineers, designers, cartographers etc. We design and build plotters to match your needs, whether it be a pen plotter, a direct imaging plotter, a thermal transfer printer or an electrostatic plotter.

CalComp's colour electrostatic plotters are ideally suited to your plotting needs in geophysical mapping, stress analysis and other applications involving complex multi-layer output.

Also produced by CalComp are high resolution graphic display systems which can be used in a wide variety of applications including mapping (e.g. soil survey research, topography and geodetic survey), architectural design (e.g. landscape design and facilities management) and desk top publishing.

CalComp is your best single source supplier for a full range of computer graphics products.

Graphtec Precision Image Limited

The A0 size 636 model and the A1 size 624 are available for monochrome or colour plotting from £18,000. Precision Image mono plotters can be upgraded to colour in the field. Most working environments are suitable for operation thanks to a low omission toner system which is also self cleaning. Accuracy is improved to 0.05% at 400 dpi, with an advanced twin track drive system. This new generation of electrostatic plotters is engineered with 30% less parts to provide the highest reliability available.

With onboard processing at 20 mips and an 85 Mbyte hard disk to minimise data transfers, an A0 colour plot can be completed in 6 minutes or an A1 monochrome plot in 1 minute. In addition to the best performance in electrostatic plotters, Precision Image customers benefit from sales and support by systems experts, providing total solutions for plotting applications.

IBM United Kingdom Limited

IBM recognises that the peripherals are a very important part of a GIS, as they provide the all important user interface. IBM therefore provides a range of input and output devices, together with links to other industry standard GIS peripherals, e.g. electrostatic plotters, very high resolution digitisers and A0 scanners.

Workstations

A comprehensive range of workstations is provided, allowing the user to select the most appropriate and cost-effective solution. The range consists of:

- IBM RS/6000 based, technical workstations.
- Mainframe attached IBM 5080 and 6090 workstations, together with accessories such as programmable lighted function keyboards and tablets.
- Standalone or mainframe based IBM PS/2 terminals
- Low cost enquiry screens, based on the IBM 3270 family of terminals.

Input and Output Peripherals

The following peripherals are provided, designed for connection to IBM workstations:

- High performance pen-plotters, ranging in size from desk-top, auto-feed A4/A3 size to floor standing, A4-A0 size, roll or cut-sheet feed.
- High resolution digitising tablets ranging in size from 24 × 36 inches to 44 × 60 inches active area.
- A4 flatbed scanners for the capture of appended raster images e.g. technical documents and photographs.
- A4/A3 format, raster screen copy devices.

IBM is a registered trademark of International Business Machines Corporation.

Laser-Scan Limited

VTRAK, from Laser-Scan, is a software package for digitising topographic and other map data in a rapid and cost effective manner. Scanned images of existing paper maps and plans are converted to structured vector form suitably coded with attributes, giving significant productivity gains over other techniques. For certain graphics such as contours VTRAK can run fully automatically. For more complex data an interactive mode is used ensuring that the most efficient use is being made of the system and the operator at all times.

VTRAK is an open systems product which uses X-windows with the MOTIF graphical user interface cutting operator training time and increasing ease of use. The software accepts raster data from a wide variety of scanners and can produce many industry standard output vector formats, providing integration with existing systems. For poor or variable quality source material, greyscale scanned data can be used to produce a result which is just as cost effective as that from high quality originals.

High quality data capture for either cartography or GIS is often an protracted expensive process, absorbing 75% of the total cost of implementing a GIS. Laser-Scan's VTRAK coupled with a complete service including consultancy, project management and full support, will ensure cost reductions, quicker and significant return on investment.

Communications and networks

Trading name, phone and contact	Address	Products or services	Features for GIS users	Ref.
Auto-trol Technology 021 456 2022 Barry Jones	53-54 Calthorpe Road, Edgbaston, Birmingham B15 1TH	Ethernet, TCP/IP, OSI Standards supported.	Full support and integration services.	
Datatrak Limited 0793 722549 Robert Peel	Hargreaves Road, Groundwell Industrial Estate, Swindon, Wiltshire SN2 5AZ	DATATRAK are providers of real time data, in particular vehicle location. Supply PC 'Street Atlas' mapping facility where required.	Real time vehicle location and status, of particular value to maintenance service vehicle in utilities and local government.	
Ground Modelling Systems Limited 0908 667799 Steve Gould	Rockingham Drive, Linford Wood, Milton Keynes MK14 6PD	Data transfer to other systems; proprietary or published data formats. Novell network support.	Compatibility with other sytems; digital OS format input/output. Refer to Ground Modelling Systems for currently supported hardware.	
IBM United Kingdom Limited 0926 332525 Preet Dhillon	GIS Market Development, PO Box 31, Birmingham Road, Warwick CV34 5JL	IBM networking: * LAN/WAN products, architectures * Business network services (VAN)	IBM's network architectures: * Provide an enterprise approach * Enable configuration flexibility * Integrate OSI standards	382
IT Southern Limited 0273 606766 Peter Elkins	Southern House, Lewes Road, Brighton, East Sussex BN1 9PY	IT Southern operates extensive LAN/WANs across South East England for GIS. Networking SUN and ICL. Development of Data Communications software.	Expertise in design and implementation of multi-user communication environments. Mixing centralised and distributed systems.	

Trading name, phone and contact	Address	Products or services	Features for GIS users	Ref.
Remote Sensing Services Limited 0672 20226 Duncan Milnes	Lychgate House, 24 High Street, Ramsbury, Wiltshire SN8 2PH	Networks implemented as appropriate including standard protocols.		

IBM United Kingdom Limited

Maximum GIS benefit is associated with open access to information by users in various departments within an organisation. IBM recognises that this often involves connection of terminals, workstations, PS/2's and host computers across distances that require local or wide area networks (LANs or WANs). These networks often need to work in a multi-vendor environment and IBM can provide practical demonstrations of this capability at a number of "Connectivity Centres" based at IBM branch offices in the UK.

IBM provides a comprehensive set of network architectures and products to fulfil these requirements, based on the IBM Systems Network Architecture (SNA). In addition IBM also has a genuine commitment to the Open Systems Interconnection (OSI) standard and offers a service to help customers understand the varying implementations of OSI across different systems. This understanding is vital if existing and future data systems are to communicate and integrate with the organisation's GIS.

Integration capabilities are further enhanced by the use of DB2 (or SQL/DS) and the geoManager product, together with the published interface format file (IFF). The latter permits external data to be imported by IBM GIS and also enables GIS data to be exported for use by other corporate applications. Third party GIS applications can thus use these interfaces and the communication links to archive or retrieve data from the IBM relational database.

The use of networks can be further enhanced by employing a value-added-network (VAN) of the type provided by IBM's Business Network Services. This can provide a valuable means of both developing network-based applications and of accessing other database information or integrating GIS with other business systems.

IBM is a registered trademark and geoManager and DB2 are trademarks of International Business Machines Corporation.

Education and training

Trading name, phone and contact	Address	Name of course	Type and features	Ref.
Auto-trol Technology 021 456 2022 Barry Jones	53-54 Calthorpe Road, Edgbaston, Birmingham B15 1TH	Graphics, RDBMS management, systems courses provided.	CAD Manager Graphics application SQL and RDBMS Management	
Birkbeck College, London 071 631 6471 Merle Abbott	Department of Geography, Birkbeck College, 7-15 Gresse Street, London W1P 1PA	'Geographical Information Systems'	Suitably qualified candidates may register as non-degree students for a three term course (one evening/week) which covers all aspects of GIS, and is part of the U of L degree in Geography.	
Birkbeck College, London 071 631 6577 Dr Jonathan Raper	Department of Geography, Birkbeck College, 7-15 Gresse Street, London W1P 1PA	'Introducing Geographical Information Systems'	3-day short course held quarterly, designed to introduce new users to the principles and applications of GIS.	
ESRC Regional Research Lab. for Scotland 031 650 2543 /4078 PJ Dowie	57 George Square, Edinburgh EH8 9JU	Short courses in GIS and database management.	Introduction to GIS, GIS methods, relational DBMS design, data management, GIS/ Remote Sensing integration, GIS for resource/ land use planning. Specialist courses on request.	
GMAP, Geographical Modelling & Planning 0532 333450 Dr Martin Clarke	University of Leeds, 3F Springfield House, Hyde Terrace, Leeds LS2 9LN	* Methods and Intelligence for Retail Analysis and Planning * GIS: Financial Services * Introduction to GIS * GIS in the NHS	Short courses on transforming information into intelligence using mathematical modelling and spatial analysis to produce Intelligence Systems for Management and Planning.	

Trading name, phone and contact	Address	Name of course	Type and features	Ref.
Kingston Polytechnic 01 549 1366 Seppe Cassettari	School of Geography, Penrhyn Road, Kingston-upon-Thames, Surrey KT1 2EE	BSc in Geographical Information Systems	3 year full-time course covering all aspects of GIS, digital mapping, remote sensing, photogrammetry and IKBS with applications in the human and physical environment.	
Kingston Polytechnic 01 549 1366 Seppe Cassettari	School of Geography, Penrhyn Road, Kingston-upon-Thames, Surrey KT1 2EE	HND in Geographical Information Systems.	2 year full-time course, from September 1991 covering all aspects of GIS, digital mapping, remote sensing, photogrammetry and IKBS. The course is run in parallel with the BSc in GIS.	
Luton College of Higher Education 0582 34111 Ron Beard	Park Square, Luton, Bedfordshire LU1 3JU	BSc Mapping Science with Geology or Computer Science.	Concerned with data capture, analysis and presentation with emphasis on resources and environment. Software—LAMPS, Gimms, ARC/INFO, Ingres, Oracle, DTP.	
Luton College of Higher Education 0582 34111 Ron Beard	Park Square, Luton, Bedfordshire LU1 3JU	BTEC HND Land Administration (Geographical Techniques).	3 year sandwich course (second in employment), concerned with the collection, analysis and presentation of spatially referenced data. Software—LAMPS, Gimms, ARC/INFO, DTP	
Polytechnic of East London 081 590 7722 /2540 Dr. Brian Whiting	Department of Land Surveying, Longbridge Road, Dagenham, Essex RM8 2AS	BSc(Hons) Geographical and Land Information Management Systems.	3 year full time course covering all aspects of GIS: surveying, photogrammetry, cartography, digital mapping, remote sensing, computer systems, data management & professional practice.	
Polytechnic of East London 081 590 7722 /2540 Patrick Sorrell	Department of Land Surveying, Longbridge Road, Dagenham, Essex RM8 2AS	HND(BTec) in Geographical Information Systems. Also Day & Block release HNC integrated into HND.	3 year sandwich course with one year commercial attachment. Covers all aspects of GIS: surveying, remote sensing, cartography, data handling, digital mapping and environmental issues.	

Trading name, phone and contact	Address	Name of course	Type and features	Ref.
Polytechnic of East London 081 590 7722 /2540 Roger Fisher	Department of Land Surveying, Longbridge Road, Dagenham, Essex RM8 2AS	MSc Cadastre and Land Information Systems. Also Postgraduate Diploma—1 year.	A 16 month full time course which includes the evaluation & design of cadastral systems, land registry, LIS & skills for land management and the application of new technology.	
Remote Sensing Services Limited 0672 20226 Duncan Milnes	Lychgate House, 24 High Street, Ramsbury, Wiltshire SN8 2PH	Bespoke training courses developed according to customer's needs.	Input, processing, GIS etc.	
Silsoe College 0525 60428 /273 Dr Chris Bird	Rural Land Use Department, Silsoe College, Silsoe, Bedfordshire MK45 4DT	P.G. Diploma—Applications of GIS in Land and Water Management / MSc in Applied Remote Sensing / Satellite Remote Sensing for land applications / GIS for land applications	9 month course suitable for recent graduates or experienced planners in need of GIS training. 12 month course covering all aspects of remote sensing and GIS. 4 day short courses.	
Summerside Associates Limited 0480 407240 Mike Quarmby	Atlas House, 17a Huntingdon Street, St Neots, Cambridgeshire PE19 2BU	Comprehensive training in digital mapping and GIS applications for technicians, supervisors and managers. Available to Clients Worldwide.	Work environment training on modern workstations using our in-house bespoke training programmes. (Supported by part-time academic units for suitable candidates.)	
University College London 01 380 7226 Prof. IA Harley	Department of Photogrammetry and Surveying, University College, Gower Street, London WC1E 6BT	MSc courses in Geodetic and Geographic Information Systems, Remote Sensing, Photogrammetry, Surveying and Geodesy. MPhil and PhD work in these fields.	Courses are for one calendar year and comprise lectures, course work and individual project. Some courses give exemption from RICS examinations.	
University of Edinburgh 031 650 2523 BM Gittings	Department of Geography, Drummond Street, Edinburgh EH8 9XP	Diploma and MSc. in GIS. First established in UK with large dedicated laboratory. M.Phil. and Ph.D. research in GIS/ Remote Sensing and other areas at this major GIS research centre.	9/12 month modular courses to suit all backgrounds. Wide range of software in use. Priority rated for NERC and ESRC support. Mature applicants wishing to consolidate skills are welcomed.	

Trading name, phone and contact	Address	Name of course	Type and features	Ref.
University of Leicester 0533 523825 Dr Alan Strachan	Department of Geography, University of Leicester, University Road, Leicester LE1 7RH	Diploma and MSc in GIS.	Diploma 8 months, MSc 12 months. In collaboration with Department of Computing Studies. ESRC α rated. Studentships available. ARC/INFO, ERDAS and Intergraph GIS.	
University of Newcastle upon Tyne 091 232 6447 Dr David Parker	Department of Surveying, University of Newcastle upon Tyne, Newcastle upon Tyne NE1 7RU	Undergraduate degree in Land Surveying Science. GIS Masters degrees by research. Customised short courses for industry.	Specialism in applications based on large scale OS mapping. Facilities include AUTOCAD, SLIMPAC, ARC/INFO, SPANS, SMALLWORLD.	
University of Nottingham 0602 484848 /3381 Dr Roy Haines-Young	Institute of Engineering, Surveying and Space Geodesy, University of Nottingham, Nottingham NG7 2RD	MSc in Geographical Information Systems.	One year full time or two year part time. An introduction to GIS concepts and applications. Includes 5-month project preferably undertaken in conjunction with a sponsor.	
University of Salford GIS Unit 061 745 5478 Dr. James Petch	The University of Salford, Department of Geography, The Crescent, Salford M5 4WT	1 day Awareness Courses / 3 day Hands-on Courses / Tailored courses on GIS topics / Product training: System 9 / Diploma in GIS from 1991/2 / Tailored courses on image processing.	The GIS laboratory provides access to SUN Sparc workstations, PC's, PRIME System 9, SPANS and ARC/INFO software. There is also a GEMS33 satellite image processing system.	387
University of Stirling 0786 67845 David J Gilvear	Department of Environmental Science, Stirling University, Stirling FK9 4LA	Environmental Management.	Full time Diploma/MSc degree which includes Geographic Information Systems.	

University of Salford GIS Unit

Courses

Awareness courses	1 day; general or thematic topics, open or tailor-made;
Hands-on courses	3 day or weekend; open or tailor-made for early or maturing skills;
Tailor-made courses	on specific GIS products or topics;
System 9 Training	UK centre for System 9 Courses for project managers, operators and Analytical Tool Box/ Applications Control Language;
Diploma in GIS	to be introduced in 1991/2. Part time distance learning course;
Satellite Image Processing	Courses to order.

The GIS Unit has laboratories of SUN Sparc 1+ workstations and of IBM PS/2's with other 386 machines. It has PRIME System 9, SPANS, and ARC/INFO software, and a range of A0 and A3 digitisers and ink-jet and laser plotters. In addition it has a GEMS33 image processing system for satellite data processing and video camera digitising. The Unit is experienced in the integration of satellite data into GIS.

The GIS Unit offers training as part of wider contracts for research and consultancy to clients in the UK or overseas.

Its specialist GIS areas are in local government operations and environmental analysis.

PART XIII

AGI membership directory

Sponsor members

Alper Systems Ltd, Milton Road, Cambridge CB4 4FQ

Bull HN Information Systems Ltd, Loudwater, Bucks HP10 9QZ

CMG Information Services, Public Sector Ltd, Tothill Street, London SW1H 9NB

Coopers & Lybrand Deloitte, Plumtree Court, London EC4A 4HT

Department of the Environment, LGS Division, 2 Marsham Street, London SW1 3EB

Digital Equipment Company, Queens House, Reading, Berkshire RG1 3JH

Doric Computer Systems, GIS Division, 23 Woodford Road, Watford WD1 1PB

Geonex UK Ltd, 92-94 Church Road, Mitcham, Surrey CR4 3TD

Hoskyns, GIS Division, 95 Wandsworth Rd, London SW8 2LX

IBM (UK) Ltd, Birmingham Road, Warwick CV34 5JL

ICL (UK) Ltd, 127 Hagley Road, Birmingham B16 8LD

Intergraph (UK) Ltd, Swindon, Wiltshire SN5 7XP

Laser-Scan Ltd, Milton Road, Cambridge CB4 4FY

McDonnell Douglas Info Systems, Guildford Road, Woking, Surrey GU22 7QH

National Rivers Authority, Information Systems, 30-34 Albert Embankment, London SE1 7TL

Ordnance Survey, Maybush, Southampton S09 4DH

Pergamon Press plc, Technical Manager, Pergamon GIS, Headington Hill Hall, Oxford OX3 0BW

Prime Computer (UK) Ltd, 373/399 London Road, Camberley, Surrey GU13 3HR

RICS, Land Surveyors Division, 12 Great George Street, Parliament Square, London SW1P 3AD

Siemens plc, 2 Hanworth Road, Feltham, Middlesex TW13 5BA

Survey International, Chief Surveyor, Flitwick Industrial Estate, off Maulden Road, Flitwick, Bedfordshire MK45 5BY

SysScan (UK) Ltd, SysScan House, Easthampstead Road, Bracknell RG2 1NS

Taywood Data Graphics, 309 Ruislip Road East, Greenford, Middlesex UB6 9BQ

Terraquest, Information Technology, PO Box 9, Union Street, West Bromwich B70 6NB

Unisys Ltd, Government & Public Services, Central Park, New Lane, Leeds LS11 5EB

Corporate members

Abbey National, Resources & Planning, Field Operations, Abbey House, Baker Street, London NW1 6XL

Agencia de Medio Ambiente, Sevicion de Evaluacion de Recursos Naturales, Eritana 2, 41013, Spain

Anglian Water, Regional Systems, Chivers Way, Histon, Cambridge

Ashford Associates Ltd, 72 Harrow Lane, Maidenhead, Berkshire SL6 7PA

Avon County Council, Planning Department, PO Box 46, Middlegate (Floor 11), Whitefriars, Lewins Mead, Bristol BS99 7EV

Baymont Engineering Company, President, 14100 58th Street N Rubin, ICOT Center, Clearwater, Florida 34620, USA

Bedfordshire County Council, Planning Department, County Hall, Cauldwell Street, Bedford MK42 9AP

Berkshire County Council, Shire Hall, Shinfield Park, Reading RG2 9XD

Birkbeck College, Geography Dept, 7-15 Gresse Street, London W1P 1PA

Birmingham City Council, Management & Information Services, 9th Floor, 1 Victoria Square, Birmingham B1 1BD

BKS Surveys Ltd, Computing Department, 47 Ballycairn Road, Coleraine, Northern Ireland BT51 3HZ

Brentwood District Council, Engineering Services, Council Offices, Ingrave Road, Brentwood, Essex CM15 8AY

Bristol Water Company, Operations, Technical Services, PO Box 218, Bridgwater Road, Bristol BS99 7AU

British Aerospace (Military Aircraft) Ltd, Defence Systrems Business Directorate/854/W182M, Warton Aerodrome, Preston, Lancs PR4 1AX

British Cartographic Society, c/o 13 Sheldrake Gardens, Hordle, Lymington, Hampshire SO41 1FJ

British Coal, Information Technology Dept, Coal House, Cannock, Staffordshire WS11 3HZ

British Gas PLC, Production & Supply, Rivermill House, 152 Grosvenor Road, London SW1V 3JL

British Geological Survey, Info & Central Serv Directorate, Keyworth, Nottinghamshire NG12 5GG

British Telecom, UKC/N.NSSD2, 375 West George Street, Glasgow G2 4NN

Broadlands District Council, Information Services, Thorpe Lodge, Yarmouth Road, Thorpe St Andrew, Norwich, Norfolk NR7 ODU

Buckinghamshire County Council, County Engineer's Department, Room 508, County Hall, Aylesbury, Buckinghamshire HP2O 1UY

Burnley Borough Council, Planning & Development Services, PO Box 29, Parker Lane, Burnley, Lancashire BB1 2DT

Byers Engineering Company, 6285 Barfield Road, Atlanta, Georgia 30328, USA

CACl Ltd, Public Services, Regent House, 89 Kingsway, London WC2B 6RM

Cambridgeshire County Council, Property Department, Shire Hall, Castle Hill, Cambridge CB3 OAP

CES Ltd, 5 Tavistock Place, London WC1H 9SS

Chelmsford Borough Council, Technical Services Department, Civic Centre, Chelmsford, Essex CM1 1JE

Cheltenham Borough Council, Architecture & Planning, PO Box 12, Municipal Offices, Parade, Cheltenham, Gloucestershire GL50 1PP

Cheshire County Council, County Planning Officer, Commerce House, Hunter Street, Chester CH1 2QP

City of Bradford Metropolitan Borough Council, Chief Executive's Office, Room 201, City Hall, Bradford BD1 1HY

City of Stoke on Trent, Town Clerk & Chief Exec's Dept, Town Hall, Glebe Street, Stoke on Trent ST4 1HH

Clyde Surveys Ltd, Clyde House, Reform Road, Maidenhead, Berkshire SL6 8BU

Colne Valley Water Company, "GIS Manager, Operations", Blackwell House, Aldenham Road, Watford, Hertfordshire WD2 2EY

Community Communications Ltd, Elsmere House, 12 Elstree Way, Borehamwood, Herts WD6 1NF

Computer Corp of America (Int), 36-38 Market Street, Maidenhead, SL6 8A9

Cornwall County Council, I.T. Officer, County Hall, Truro, Cornwall TR1 3AY

Cumbria County Council, Corporate Information Unit, Corporate Services, The Courts, Carlisle CA3 8NA

Cyngor Sir Powys C.C., County Treasurer, County Hall, Llandrindod Wells, Powys LD1 5LG

Dartford Borough Council, Development Services Officer, Civic Centre, Dartford, Kent DA1 1DR

Data Dictionary Systems Ltd, 85 Deepcut Bridge Road, Deepcut, Camberley, Surrey GU16 6RP

Data General Ltd, Euro GIS Competence Centre, 45 Poplar Road, Solihull, West Midlands B91 3AW

Design Computer Aids Ltd (DeCAL), Business Development Manager, 6 Broughton Street Lane, Edinburgh EH1 3LY

Desktop Engineering Systems, Pembroke House, 5/9 Pembroke Road, Ruislip, Middlesex HA4 8NQ

Devon County Council, Chief Executive's (R&I), County Hall, Topsham Road, Exeter EX2 4QD

Directorate of Military Survey, Elmwood Avenue, Feltham, Middlesex TW13 7AH

Dorset County Council, Transportation & Engineering, County Hall, Dorchester, Dorset DT1 1XT

Dudley Metropolitan District Council, Public Works, Council House, Mary Steven's Park, Stourbridge Dudley, West Midlands DY8 2AA

Dumbarton District Council, Planning & Development Dept, 69 Glasgow Road, Dumbarton G82 1RE

Dumfries & Galloway Regional Council, Physical Planning, Council Offices, English Street, Dumfries DG1 2DD

Durham County Council, Environment Department, County Hall, Durham DH1 5UQ

DWH Associates Ltd, 20 Rother Street, Stratford-upon-Avon, Warwickshire CV37 6NE

Earth Observation Sciences, Branksome Chambers, Branksomewood Road, Fleet, Hampshire GU13 8JS

East Midlands Electricity, Operations & Telecoms Dept, PO Box 4, North PD0, 398 Coppice Road, Arnold, Nottingham NG5 7HX

East Sussex County Council, County Management Services, PO Box 5, County Hall, Lewes, East Sussex BN7 1SF

English Heritage, Fortress House, 23 Saville Row, London W1X 2HE

ESRC, South East Regional Research Lab, Department of Geography, London School of Economics, Houghton Street, London WC2A 2AE

Essex County Council, Planning Department, County Hall, Chelmsford, Essex CM1 1LF

Exploration Consultancy Ltd., Highlands Farm, Grave Road, Henley on Thames RG9 4PS

Gardline Surveys, Data Processing Department, Admiralty Road, Great Yarmouth, Norfolk NR1 3 5AA

Gedling Borough Council, Deputy Planning & Est. Officer, Civic Centre, Arnold Hill, Arnold, Nottingham NG5 6LU

Geobase Consultants Ltd, 28 Church Road, Epsom, Surrey KT17 4DX

Geobuild Projects Ltd, 4 Grange Road, Burley in Wharfdale, Ilkey, West Yorkshire LS29
Geografix Ltd, Hurricane Way, Norwich NR6 6EW

Geographical Association, 343 Fullwood Road, Sheffield S10 3BP

Geomatrix Ltd, Cooper Buildings, Sheffield Science Park, Arundel Street, Sheffield S1 2NS

Gimms Ltd, 30 Keir Street, Edinburgh EH3 9EW

Glasgow District Council, Planning Department, 231 George Street, Glasgow G1 1RX

Glen Computing Services, 309 High Street, Orpington, Kent BR6 0NN

Glenigan Ltd, Marketing Department, 41-47 Seabourne Road, Bournemouth, Dorset BH5 2HU

Gloucestershire County Council, County Planning Officer, Shire Hall, Westgate Street, Gloucester GL1 2TN

Graphical Data Capture Ltd, 262 Regents Park Road, London N3 3HN

Graphics Management Ltd, Director, 9 Braefoot Court, Law, Strathclyde ML8 5HY

HM Land Registry, Plans Practice HQ Division, Lincoln's Inn Fields, London WC2A 3PH

HM Land Registry, Computer Services Division, Drake's Hill Court, Burrington Way, Plymouth PL5 3LP

Haiste Ltd, Computer Applications Dept, Newton House, Newton Road, Leeds LS7 4DN

Halcrow Fox & Associates, 44 Brook Green, Vineyard House, Hammersmith, London W6 7BY

Hampshire County Council, County Planning Department, The Castle, Winchester, Hampshire SO23 8UD

Harrogate Borough Council, IT Division, Dept of Corporate Services, Council Offices, Crescent Gardens, Harrogate HG1 2SG

Hart District Council, Civic Offices, Harlington Way, Fleet, Hampshire GU13 8AE

Havering Borough Council, Chief Executive's Department Town Hall, Main Road, Romford RM1 3BD

Herring, Son & Daw plc, Information Officer, 26/28 Sackville Street, London W1X 2QL

Hertfordshire County Council, Planning & Estates Department, County Hall, Hertford SG13 8DN

Hertsmere District Council, Chief Executives Department, Civic Offices, Elstree Way, Borehamwood, Hertfordshire WD6 1WA

Hillier Parker, 77 Grosvenor Street, London W1A 2BT

Horsham District Council, Directorate of Planning & Law, Park House, North Street, Horsham, West Sussex RH12 1RL

Howard Humphreys, IT Applications, Thorncroft Manor, Dorking Road, Leatherhead, Surrey KT22 8JB

Hunting Technical Services Ltd, Remote Sensing & Info Sciences, Thamesfield House,

Boundary Way, Hemel Hempstead, HP2 7SR

Hydrographic Society, U.K. Branch, c/o Southampton Inst of H.E., Newtown Road, Warsash, Southampton, SO3 9ZL

Inland Revenue, Valuation Office, New Court, Carey Street, London WC2A 2JE

Institute of Hydrology, Catchment Characteristics, Maclean Building, Crowmarsh Gifford, Wallingford, Oxfordshire OX10 8BB

John Bartholomew & Son, Cartographic Director, 12 Duncan Street, Edinburgh EH9 1TA

Kent County Council, Planning Department, Springfield, Maidstone, Kent ME14 2LX

Kerrier District Council, Treasurer's Department, Council Offices, Dolcoath Avenue, Camborne, Cornwall TR14 8RY

Kilmarnock & Loudoun District Council, Ref F256/B, PO Box 13, Civic Centre, Kilmarnock KA1 1BY

Knowlley Metropolitan Borough Council, Planning Department, Municipal Buildings, PO Box 26, Archway Road, Huyton, Merseyside L36 9FB

KPMG Peat Marwick Management Consultants, Development & Infrastructure Practice, 8 Salisbury Square, London EC4Y 8BB

Lancashire County Council, County Planning Department, East Cliff, Preston, Lancashire PR1 3EX

Leominster District Council, Planning Department, Lion House, Broad Street, Leominster, Herefordshire HR6 8BT

Logica Aerospace & Defence Ltd., Dynamics Group, 68 Newman Street, London W1A 4SE

Logica Communications and Electronic Systems Ltd, 64 Newman Street, London W1A 4SE

Logica Industry Limited, Randalls Research Park, Randalls Way, Leatherhead, Surrey KT22 7TW

London Borough of Barking & Dagenham, Borough Treasurer's Dept, Civic Centre, Dagenham, Essex RM10 7BH

London Borough of Barnet, Technical Services Directorate, Engineering services Division, Barnet House, 1255 High Road, Whetstone, London N20 0EJ

London Borough of Hounslow, Director of Engineering, Civic Centre, Lampton Road, Hounslow, Middlesex TH3 4DN

London Borough of Newham, Environment & Planning Dept, Town Hall Annexe, Barking Road, London E6 2RP

London Docklands Development Corporation, Executive Office, Great Eastern Enterprise, Millharbour, London E14 9TJ

London Fire & Civil Defence Authority, 8 Albert Embankment, London SE1 7SD

London Underground Ltd, Information Manager (Eng.) I.T., 2nd Floor East, Ashfield

House, Beaumont Avenue, West Kensington, London W14 9UY

Longman Group UK Ltd, Longman Scientific & Technical, Longman House, Burnt Mill, Harlow, Essex CM20 2JE

Lothian Regional Council, Systems Planning, Computer Services, Computer Centre, Warriston's Close (323 High St.), Edinburgh EH1 1PG

Macaulay Land Use Research Institute, Land Use Division, Craigiebuckler, Aberdeen AB9 2QJ

Malvern Hills District Council Council House, Avenue Road, Malvern, Worcestershire WR14 4AX

Mason Land Surveys Ltd, Dickson Street, Elgin Street Industrial Street, Dunfermline, Fife KY12 7SL

Merseyside Information Service, Suite 301, Royal Liver Building, Pierhead, Liverpool L3 1JH

Mid Glamorgan County Council, Highways & Transportation, County Offices, Greyfriars Road, Cardiff CF1 3LJ

Mid-Suffolk District Council Planning Department, Council Office, High Street, Needham Market, Ipswich IP6 8DL

Midland Regional Research Lab, Bennett Building, University of Leicester, Leicester LE1 7RH

Milton Keynes Borough Council, Forward Planning Department Civic Offices, 1 Saxon Gate East, Central Milton Keynes, Bucks MK9 3HQ

Ministry of Defence, Hydrographic Department, Creechbarrow Road, Taunton, Somerset TA1 2DN

Moss Systems Ltd, Engineering Department, Moss House, North Heath Lane, Horsham, West Sussex RH12 4QE

Museum of London, Dept of Urban Archaeology, 150 London Wall, London EC2Y 5HN

MVA Systematica, 115 Shaftesbury Avenue, London WC2H 8AD

MVM Strategic Consultancy Services Ltd, 13 Great George Street, Bristol, Avon BS1 5RR

NERC, Scientific Services, Polaris House, North Star Avenue, Swindon, Wiltshire SN2 1EU

N Health & Social Services BD, Planning Department, County Hall, 182 Galgorm Road, Ballymena, N Ireland BT42 1QB

N Hertfordshire District Council, IT Manager, Finance Department, Town Lodge, Gernon Road, Letchworth, Hertfordshire SG6 3HN

N Ireland Reg Research Lab, 105 Botanic Avenue, Belfast BT7 1NN

Nat Council for Ed Technology, Unit 6, Sir William Lyons Road, Science Park, University of Warwick, Coventry CV4 7EZ

National Joint Utilities Group, 30 Millbank, London SW1P 4RD

National Rivers Authority, Thames Region, Kings Meadow House, Kings Meadow Road, Reading, Berkshire RG1 8DQ

Nature Conservancy Council, Monitoring & Carto Services, Northminster House, Peterborough, Cambridgeshire PE1 1UA

NERRL, CURDS, The University, Newcastle upon Tyne, NE1 7RU

Nextbase Ltd, Unit 18, Central Trading Estate, Staines, Middlesex TW18 4XE

NI Housing Executive, Housing Centre, 2 Adelaide Streeet, Belfast, Northern Ireland BT2 8PB

NOMIS, Unit 3P, Mountjoy Research Centre, University of Durham, Durham DM1 3SW

Norfolk County Council, Planning & Property, County Hall, Martineau Lane, Norwich NR1 2DH

Northern Electric plc, PO Box 1SE, Carlton House, Newcastle upon Tyne NE99 1SE

Northern Ireland Electricity, Engineering Services, PO Box 2, Danesfort, 120 Malone Road, Belfast, Northern Ireland BT9 5HT

Northumbria Police, Computer Services, Police Headquarters, North Road, Pontleand, Newcastle upon Tyne N62 0BL

Northumbrian Water Group, Information Technology Consultancy, Northumbria House, Regent Centre, Gosforth, Newcastle-upon-Tyne, Tyne & Wear NE3 3PX

Nottinghamshire County Council, Planning & Transportation, Trent Bridge House, Fox Road, West Bridgford NG2 6BJ

NUTIS, Department of Geography, University of Reading, Whiteknights, PO Box 227, Reading RG6 2AB

O.P.C.S., Census Division, Segensworth Road, Titchfield, Fareham, Hampshire PO15 5RR

Overseas Development Administration, Nat Resources & Environment Dept, Eland House, Stag Place, London SW1E 5DH

Oracle Corporation UK Ltd, Broad Quay House, Prince Street, Bristol BS1 4DJ

Ordnance Survey NI, Department of Environment, Colby House, Stranmillis Court, Belfast, Northern Ireland BT9 5BJ

Orkney Islands Council, Planning Department, County Offices, Kirkwall, Orkney KW15 1NY, Oxford City Council, Planning, Clarendon House, 52 Cornmarket Street, Oxford OX1 3HD

Oxford Institute of Retail Management, Templeton College, Kennington, Oxford OX1 5NY

PA Consulting Group, Computer Aided Engineering, Cambridge Laboratory, Melbourn, Royston, Hertfordshire SG8 6DP

P.E. International plc, I T Division, Park House, Wick Road, Egham, Surrey TW20 0HW
Peterborough City Council, Engineering Department, Bridge House, Bridge Street,

Peterborough PE1 1XG

Photogrammetric Data Services, 1 Ham Business Centre, Brighton Road, Shoreham By Sea, Sussex BN43 6PA

Photogrammetric Society, Dept of Photogrammetry & Surveying, University College London, Gower Street, London WC1E 6BT

Pinpoint Analysis, Mercury House, 117 Waterloo Road, London SE1 8UL

Plangraphics Inc, 202 West Main Street, Suite 200, Frankfort, Kentucky 40601, USA

Plowman Craven & Associates, R & D Department, 141 Lower Luton Road, Batford, Harpenden, Hertfordshire AL5 5EQ

Polytechnic of East London, Land Surveying Dept, Longbridge Road, Dagenham, Essex RM8 2AS

Poole Borough Council, Borough Engineer, Municipal Buildings, Civic Centre, Poole BH15 2RY

Property Intelligence Ltd, Ingram House, 13-15 John Adam Street, London WC2N 67D

QC Data Limited, Sales/Marketing, Second Floor, 100 Park Lane, London W1Y 4AR

RAC Motoring Services, Information Services Dept, RAC House, PO Box 100, South Croydon CR2 6XW

Redditch Borough Council, Chief Executive's Office, Town Hall, Alcester Street, Redditch B98 8AH

Registers of Scotland, Meadowbank House, 153 London Road, Edinburgh EH8 7AU

Remote Sensing Society, Univcrsity of Nottingham, Department of Geography, University Park, Nottingham NG7 2RD

Robertson McCarta, Cartography Department, Tyn-y-Coed, Llandudno LL30 1SA

Royal Borough of Kensington & Chelsea, DBSIT, The Town Hall, Hornton Street, London W8 7NX

Royal Borough of Kingston upon Thames, Chief Executives Office, Office Automation Bureau, Guildhall, Kingston upon Thames, Surrey KT11 EU

Scottish Hydro Electric Plc, Commercial Division, Peasiehill Road, Arbroath, Angus DD11 2NJ

SEEBOARD plc, Northern Division, Electric House, Wellesley Road, Croydon, Surrey CR9 2AR

Sefton Metropolitan Borough Council, Planning, Vermont House, 375 Stanley Road, Bootle L49 1RG

Sevenoaks D.C., Central Services Dept, Argyle Road, Sevenoaks, Kent TN13 1HG

Sheffield City Council, Design and Building Services, PO Box 241, City Plaza, 2 Pinfold Street, Sheffield S1 1SB

Shropshire County Council, County Propert & Planning Services Dept, The Shirehall, Abbey Foregate, Shrewsbury SY2 6ND

Soc of Surveying Technicians, Drayton House, 30 Gordon Street, London WC1H 0BH

Soil Survey & Land Research Centre, Computing & Information Systems, Silsoe Campus, Silsoe, Bedford MK45 4DT

South Oxfordshire District Council lnfo Technology Services, PO Box 17, Council Offices, Crowmarsh, Wallingford, Oxford OX10 8NP

South Wales Electricity, Engineering Director, St Mellows, Cardiff CF3 9XW

South West Water Services Ltd, Peninsular House, Rydon Lane, Exeter, Devon

Southern Electric, Information Systems, Bartons Road, Havant, Hampshire PO9 5JB

Southern Water, Guildbourne House, Worthing, Sussex BN11 1LD

SPA Marketing Systems, 1 Warwick Street, Leamington Spa, Warwickshire CV32 5LW

Sterling Surveys Ltd, Grove House, Headley Road, Grayshott, Hindhead, Surrey GU26 6LE

Strathclyde Regional Council, Chief Executive's Department, 20, India Street, Glasgow G2 4PF

Suffolk County Council, Highways Department, St Edmund House, County Hall, Ipswich IP4 1LZ

Summerside Associates, Atlas House, Huntingdon Street, St Neots, Cambs PE19 2BU

Swale Borough Council, Information Technology, Swale House, East Street, Sittingbourne, Kent ME10 3HT

Swansea City Council, Information Technology, The Guild Hall, Swansea, West Glamorgan SA1 4PN

Systems Distribution International Ltd, 36 Lad Lane, Dublin 2, Ireland

Tameside Metropolitan Borough Council, Policy Services Depanment, Council Offices. Room 6.14, Wellington Road, Ashton under Lyme OL6 6DL

Tandridge District Council, Directorate of Planning, Council Offices, Station Road East, Oxted, Surrey RH8 0BT

Tayside Regional Council, Planning Department, Tayside House, Crichton Street, Dundee DD1 3RB

The Advisory Unit for Microtechnology in Education, Endymion Road, Hatfield, Herts AL10 8AU

Thurrock Borough Council, Management Services, Civic Offices, New Road, Grays, Essex RM17 6SL

Touche Ross Management Consultants, Hill House, 1 Little New Street, London EC4A 3TR

Tydac Technologies Ltd, 9th Floor, Colston Centre, Colston Avenue, Bristol BS1 4XE

Tyne & Wear Research & Intelligence Unit, Room 394, Civic Centre, Newcastle upon Tyne NE1 8QN

University College London, Dept of Photogrammetry & Surveying, Gower Street, London WC1E 6BT

University of Edinburgh, Department of Geography, Drummond Street, Edinburgh EH8 9XP

University of Edinburgh, GISA, Department of Geography, Drummond Street, Edinburgh EH8 9XP

University of Newcastle upon Tyne, Surveying Department, Newcastle upon Tyne NE1 7RY

University of Nottingham, Inst of Eng & Space Geodesy, University Park, Nottingham NG7 2RD

University of Reading, Geography Dept, Whiteknights, Reading RG6 2AB

University of Sheffield, Town & Regional Planning Department, Western Bank, Sheffield S10 2TN

University of Southampton, Geodata Institute, Southampton SO9 5NH

Vale Royal Borough Council, Estates Section, Over Hall, Beeston Drive, Winsford, Cheshire CW7 1ES

Versatec Electronics Ltd, Sales Support Dept, 5 Oxford Road, Newbury, Berks RG13 1QD

ViewMAP Spatial Systems Ltd, Clifton Heights, Triangle West, Bristol BS8 1EJ

Wakefield Metropolitan District Council, Town Hall, Wood Street, Wakefield WF1 2HQ

Warwickshire County Council, Planning & Transportation, PO Box 43, Shire Hall, Warwick CV34 4SX

Water Research Centre, Asset Management Department, PO Box 85, Frankland Road, Blagrove, Swindon SN5 8YR

Water Research Centre, Catchment Studies, PO Box 16, Medmenham, Marlow, Buckinghamshire SL7 2HD

Waveney District Council, Computer Centre, Town Hall, High Street, Lowestoft NR32 1HS

Welwyn & Hatfield Borough Council, Planning & Developmenmt Dept, Council Offices, The Campus, Welwyn Garden City, Hertfordshire AL8 6AE

Wild Leitz UK Ltd, Photogrammetry & Systems, Davy Avenue, Knowlhill, Milton Keynes MK5 8LB

Wiltshire County Council, Computer Division, Finance Department, County Hall, Trowbridge, Wiltshire BA14 8JJ

Winchester City Council, IT Section, City Offices, Colebrook Street, Winchester, Hants S023 9LJ

Windsor & Maidenhead Royal Borough Council, Borough Surveyor, Town Hall, St Ives Road, Maidenhead, Berkshire SL6 1RF

Wolverhampton Metropolitan Borough Council, Technical Services, Civic Centre, St Peters Square, Wolverhampton WV1 1RP

Yorkshire Electricity, Energy Supply Division, Scarcroft, Leeds LS14 3HS

Yorkshire Water Services Ltd, Technology & Services Dept, Broadacre House, Vicar Lane, Bradford BD1 5PZ

ZS Associates, Apex Plaza, Forbury Road, Reading, Berks RG1 1ZS

Individual members

Mr V J Abbott, Polytechnic South West, Institute of Marine Studies, Drake Circus, Plymouth PL4 8AA

Mr M P Adam, 49 Histon Road, Cambridge CB4 3JD

Mr N C Adnitt, Hampshire House, Wade Road, Basingstoke, Hampshire RG24 0PL

Mr I Archibald, 10 Charlotte Square, Edinburgh EH2 4DR

Mr GL Ashton, 200 Lichfield Lane, Berry Hill, Mansfield NG18 4RG

Mr K M Atterbury, 4 Marine Drive, Edinburgh EH5 1YB

Mr I D Aufflick, 1 Treasury Place, Melbourne, Victoria 3149, Australia

Dr T C Bailey, Streatham Court, Exeter University, Exeter, Devon EX4 4PU

Mr M Bailey, Taberner House, Park Lane, Croydon CR9 1JT

Mr I D B Ballantyne, Butlers Leap, Rugby, Warwickshire CV21 3RQ

Mr G J Bardill, 200 Lichfield Lane, Mansfield, Nottinghamshire NG18 4RG

Mr R Barr, University of Manchester, Dept of Geography, Manchester M13 9PL

Miss A.K. Bartlett, 16 Campion Road, Abingdon, Oxon OX14 3TQ

Prof P W J Batey, PO Box 147, Liverpool, L69 3BX

Mr R Beard, Park Square, Luton, Bedfordshire LU1 3JU

Prof J Beaumont, University of Stirling, Dept of Business Management, Stirling, Scotland

Mr M Bedford, 24 Repton Court, Repton Close, Basildon, Essex SS13 1LN

Mr M. Bell, W5B Warton, Preston PR4 1AX

Mr F Bennion, 5 Wharf Lane, Solihull, West Midlands B91 2JP

Ms S.J. Benson, Silsoe College, Land & Resource Management, Silsoe, Bedfordshire MK45 4DT

Dr ACH Bird, Land & Resource Management, Silsoe, Bedfordshire MK45 4DT

Mr D Bloom, 47 Old Church Street, Chelsea, London SW3 5ES

Mr C S Bradford, 28 Crichton Street, Dundee DD1 3RQ

Miss D. Brenke, 11 Robinson Court, Lower Earley, Reading RG6 ZYX

Mr A A Britten, Kingfisher House, Goldhay Way, Orton Goldhay, Peterborough PE2 0ZR

Mr A R Bromley, Council Offices, Nottingham Road, Melton Mowbray, Leicestershire LE13 0UL

Mr P.B. Browne, St James's Building, Oxford Street, Manchester M1 6FL

Mr J Buchanan, Rodborough Manor, Bear Hill, Rodborough, Nr Stroud, Gloucestershire GL5 5DH

Dr J C Budd, Northminster House, Northminster, Peterborough, Cambridgeshire PE1 1VA

Dr R.K. Bullard, Duncan House, High Street, London E15 2JB

Mr. M.P. Butterworth, Elettra Avenue, Waterlooville, Hampshire PO7 7XS

MR R Buxton, 14 St Andrews Road, London E13 8QD

Mr D A Byles, West End Road, Ruislip, Middlesex HA4 6NG

Mr N Cabot, Knoll Rise, Orpington, Kent BR6 0XG

Mr J.W. Cadoux-Hudson, Mayfield, Buckland Road, Childswickham, Broadway WR12 7HH

Mr. M. Calvert, RM511, Hydebank, 4 Hospital Road, Belfast BT8 8JP

Mr. A. Carnell, 27 Field Close, Harpenden, Hertfordshire AL5 1EP

Ms AW Carruthers, 30 Keir Street, Edinburgh EH3 9EU

Mr G W Carver, Telford House, Tothill Street, London SW1H 9NB

Dr G A Cassettari, Penrhyn Road, Kingston upon Thames, Surrey KT1 2EE

Mr W B Christian, 4 Boulton Avenue, New Ferry, Wirral L62 1DT

Mr P J Churchward, Suite 27A/34, Concourse House II, 432 Bewsbury Road, Leeds LS11 7DF

Mr P E Clegg, Cliffe House Cottage, High Bradfield, Sheffield S6 6LJ

Mr F G Cole, 197 East Clyde Street, Helensburgh, Dumbartonshire, Scotland GA4 7AJ

Mr P G Coles, Council House, Room75, Earl Street, Coventry CV1 5RQ

Mr A C Collins, Coopers & Lybrand Deloitte, Plumtree Court, London EC4A 4HT

Mr A K Cooper, PO Box 395, 0001 Pretoria, South Africa

Mr A M Coote, Romsey Road, Maybush, Southampton SO9 4DH

Mr R.B. Coote, Vine House, 41 Portsmouth Road, Cobham, Surrey KT11 1JQ

Mr C Copp, Brislington, Bristol, Avon BS4 5PF

Dr J Cordingley, Northampton Health Authority, Community Medicine Department, District Headquarters, Highfield, Cliftonville Road, Northampton NN1 5DN

Mr M Cory, 1A Pine Road, Bournemouth BH9 1LS

Mr M G Coulsen, University College Swansea, Dept of Geography, Natural Science Building, Singleton Park, Swansea SA2 8PP

Ms S Courtney, Fiuroy House, 18-20 Grafton Street, London W1X 4DD

Mr C. Craig, 650 Aztec West, Bristol BS12 4SD

Miss S C Cox, 4 Greenway, Crosby, Liverpool L23 9PX

Mr. W. Craig, 120 Malone Road, Belfast BT9 5HT

Mr D B Cross, 33 Sykes Street, New Hey, Rochdale, Lancashire OL16 4JP

Mr R P E Cunningham, Observatory House, Windsor Road, Slough, Berkshire SL1 2EY

Mr J E Curley, Digital Mapping, Brunel House, St Georges Road, Bristol BS1 5UY

Mr J Curzon, 26 Axminster Walk, Bramhall, Stockport, Cheshire SK7 2QJ

Dr J R Cuthbert, Room 5155, New St Andrews House, Edinburgh EH1 3SX

Prof P F Dale, Dept of Land Surveying, Polytechnic of East London, Longbridge Road, Dagenham, Essex RM8 2AS

Ms J C Danczak, 104 St John Street, London E1M 4EH

Mr J Davie, Scott House, Basing View, Basingstoke RG21 2JG

Mr J H Davies, Horn Hill Road, West Addersbury, Nr Banbury, Oxfordshire OX17 BHN

Mr P G Davies, The Kilnhurst Business Centre, Victoria Street, Kilnhurst, Rotherham S62 5SQ

Mr H Dawoodi, 139 High Street, Longstone, Cambridge CB3 7UN

Mr. J. Day, 116 Barons Court Road, London W14

Mrs K M De Groot, 125 Victoria Road, New Barnet, Hertfordshire EN4 9PE

Mr. C.W. Decker, Terriers House, 201 Amersham Road, High Wycombe, Bucks HP13 5AJ

Mr T.J. Dixon, University of Reading, Whiteknights, Reading RG6 2AW

Mr P J Dowie, Drummond Street, Edinburgh EH8 9XD

Mr D R Drew, 72 Carterhatch Road, Enfield, Middlesex EN3 5LU

Prof. R. Dunn, University of Reading, Department of Geography, Whiteknights, Reading RG6 2AF

Dr N J Eales, Summer Fields, Sketty Road, Swansea SA2 0LH

Mr P A E Eames, Cory House, The Ring, Bracknell, Berkshire RG12 1AX

Mr N D J Edmead, Silver Birches, Gentles Lane, Upper Passfield, Nr Liphook, Hampshire GU30 7RY

Dr C.N. Edmonds, Cranford, Kenilworth Road, Blackdown, Royal Leamington Spa, Warwickshire CV32 6RG

Mr M P Edwards, London School of Economics, Dept of Information Services, Houghton Street, London WC2A 2AE

Col. J P Elder, Elmwood Avenue, Feltham, Middlesex TW13 7AE

Dr C Evans, 40 Priestley Road, Chancellor Court, Surrey Research Park, Guildford, Surrey GU2 5YP

Mr I W Evans, Derlwyn, Lon Y Wern, Tregarth, Bangor, Gwynedd

Mr J. Falconer, Phoenix House, 202 Elder Gate, Milton Keynes MK9 1BE

Mr H M Fingland, 8-9 Bridge Street, Cambridge CB2 1OA

Mr L J Finniear, 67 Forest Road, Loughborough LE11 3NW

Mr M Fisher, PO Box 1732, Riyadh 11441, Saudi Arabia

Mr M.F. Fishwick, 8 Swallowfield Gardens, Theale, Berkshire RG7 5AD

Mr M N Forbes, London Borough of Lambeth, Directorate of Town Planning, Computer Team, 9-15 New Park Road, London SW2 4DU

Mr J Ford, Pegasus House, 375 West George Street, Glasgow G2 4NN

Mr D Forrest, Lyon Terrace, Portsmouth PO1 3HE

Mr T S Fulton, Pegasus House, 375 West George Street, Glasgow G2 4NN

Mr C Garbett, Hayes Lane, Mixenden, Halifax HX2 8UL

Dr C L Garner, Loanstone House, Loanstone, Nr Penicuik, Mid Lothian EH26 8PH

Mr T L Garrett, 49 Lamont Avenue, Bishopton PA7 5LJ

Mr. D.P. Giles, St Ann House, Wellesley Road, Croydon CR9 2UL

Mr D G Goffin, 266 Hatfield Road, St Albans, Hertfordshire AL1 4UN

Mr S J Gould, 86A Florence Road, London N4 4DP

Dr A Greggains, 20 Rutland Park, Sheffield S10 2PB

Mr J Gregory, Attention J Aldridge, c/o American Embassy - Box 12, 24 Grosvenor Square, London W1A 1AE

Dr. P.G. Gretton-Watson, 49 Berners Street, London W1P 4AQ

Mr R.G.W. Griffiths, Hafod Elfyn, Penrhos Road, Bangor, Gwynedd LL57 2LQ

Mr R Grimes, Telford House, Tothill Street, London SW1H 9NB

Mr R W Groom, 5 Rose Avenue, Hazelmere, Buckinghamshire HP15 7PR

Mr G Hamilton, Steers House, Canning Place, Liverpool L1 8JA

Maj. N Hamilton Fletcher, Sronan, Foss by Pitlochry, Perthshire PH16 5NQ

Mr L Hardins, Simonsway, Withenshaw, Manchester M22 5LA

Mr. S. Hawkins, 46 Fortescue Road, Colliers Wood, London SW19 2EB

Mr D J Hendley, 10 Challoners Way, East Molesey, Surrey KT8 0DW

Mr B P F Heran, Park House, 116 Park Street, London W1Y 4NN

Mr FMA Hewlett, Miller's way, 1A Shepherd's Bush Road, London W6 7NA

Mr R M Hirst, Woodcote House, Ashley Road, Epsom, Surrey KT18 5BW

Mr S Hoffman, Fanum House, Basing View, Basingstoke RG21 2EA

Mr N Holmes, County Hall, Atlantic Wharf, Cardiff CF1 5UW

Mr. C. Hookham, Cambridge Laboratory, Melbourn, Royston, Herts SG8 6DP

Mr J R Houghton, No1 The Courtyard, Swan Centre, Fishers Lane, London W4 1RX

Ms K. Howard, Rivers House, 21 Park Square South, Leeds, W Yorks LS1 2QG

Mr E B Howard, Herikebrink 17, 7544ER Enschede, The Netherlands

Mr D P Hughes, 21 Baden Powell Drive, Colchester, Essex CO3 4SL

Mr D C Hughes, 82 Melbourne St, North Adelaide, SA 5006, Australia

Mr. J. Hutchinson, London Square, Cross Lanes, Guildford, Surrey GU1 1UJ

Mr D.W. Hutchinson, Room A4/8 Government Buildings, Leatherhead Road, Chessington KT9 2LU

Mr T J Jeffries-Harris, Rightwell House, Brehon Centre, Peterborough, Cambridgeshire PE3 8DW

Mr P R Jones, Electricity House, PO Box 4, Leatherhead Road, Chessington KT9 2LU

Mr K A Jones, Mucklow Hill, Halesowen B62 8BP

Dr B Kelk, British Geological Survey, Info & Marketing, Keyworth, Notts NG12 5GG

Ms C. Kenyon, 4 Strode Road, Tottenham, London N17 6TZ

Miss S E Kerr, 12-22 Linenhall Street, Belfast, Northern Ireland BT28 BS

Mr C Kirkwood, Meirion House, 6 Guildford Road, Woking, Surrey GU22 7QH

Mr S K Kirwan, Forge House, 66 High Street, Kingston upon Thames, Surrey KT1 1HN

Mr R W Laing, Sapphire Court, 274-276 High Street, Slough SL1 1NB

Mr A G Larner, AGI, c/o RICS, 12 Great George Street, London SW1P 3AD

Mr J D Leatherdale, 10 Bartrams Lane, Hadley Wood, Barnet, Hertfordshire EN4 0EH

Mr. H.M.H. Lee, Silwood Park, Buckhurst Road, Ascot, Berkshire SL5 7QW

Mr G Levett, Consilium House, City Forum, City Road, London EC1V 2NY

Mr A J Lodwick, 17 Clifton Park Road, Caversham, Reading, Berkshire RG4 7PD

Mr I T Logan, 10 Horsebridge Way, Rownhams, Southampton, Hampshire SO1 8AZ

Mr R A Longhorn, 1 Potters Cross, Wootton, Bedfordshire MK43 9JG

Dr. A.A. Lovett, University Plain, Earlham, Norwich, Norfolk NR4 7TJ

Mr R M Lowe, Danesfort, 120 Malone Road, Belfast BT9 5HT

Mr J G Loy, 3 Woodside Place, Glasgow G3 7QF

Mr F F Lyon, 2 Rownhams Close, Rownhams, Southampton, Hampshire SO1 8AF

Mr I Lysenko, PO Box 2310, Amman, Jordan

Mr A S MacDonald, Ordnance Survey, Romsey Road, Maybush, Southampton SO9 4DH

Mr N P Mackay, 8 Blyth Close, Manchester Road, London E14 9DU

Mr K J Maddock, 1 Abdel Salam Aref Street, Orman Giza, Cairo, Egypt

Mr P W Madeley, 24 Repton Court, Repton Close, Basildon, Essex SS13 1LN

Dr D J Maguire, University of Leicester, Dept of Geography, University Road, Leicester LE1 7RH

Mr R P Mahoney, 14 Kings Avenue, Denton, Newhaven, East Sussex BN9 0NA

Miss B Malpass, Civic Offices, Guildhall Square, Portsmouth P01 2BQ

Mr. D. Marsden, Heathwall Pumping Station, 60 Nine Elms Lane, London SW8 5DA

Dr P D Marsh, Walford Manor, Baschurch, Shrewsbury SY4 2HH

Mr S A Matthews, UCLA, Los Angeles, California 90024, USA

Mr H McCracken, 20 India Street, Glasgow G2 4PF

Ms G P Mclntyre, Lovat Bank, Silver Street, Newport Pagnell, Buckinghamshire MK16 0EQ

Mr R A McLaren, 33 Lockharton Avenue, Edinburgh EH14 lAY

Mr F McLeod, Polytechnic of East London, Longbridge Road, Dagenham, Essex RM8 2AS

Mr C.M. McLundie, Granton House, 4 Marine Drive, Edinburgh EH5 1YB

Mrs J M McMorrow, St Helens Road, Ormskirk, Lancashire L39 4QP

Mr D H McPherson, Ministry of Defense (Navy), Hydrographic Department, Taunton TA1 2DN

Mr A Metcalfe, Council Offices, Farnborough Road, Farnborough, Hampshire GU14 6AH

Mr. A.T. Mitchell, 231 Corstorphine Road, Edinburgh EH12 7AT

Mr D W Moat, Council Offices, Wellington Road, Ashton under Lyme, Tameside OL6 6DL

Mr D W Moore, Templar House, 81-87 High Holborn, London WC1V 6NV

Mr R S Morgan, 20 lndia Street, Glasgow G2 4PF

Mr G.J. Morris, 10 Church Way, Yealmpton, Devon PL8 2LA

Mr A Mort, Leicester University, Dept of Geography, Post Graduate Office, Leicester LE1 7RH

Dr. A.M. Murphy, St. Andrews Ho. (SA47), St. Andrews Street, London EC4A 4HT

Mr J.J. Murray, 24 Cambridge Place, Hills Road, Cambridge

Dr T J Nash, Hawthorn House, Edale Road, Hope, Sheffield S30 2RF

Mr P H Neal, 69 Buckland Road, Maidstone, Kent ME16 0SH

Mr D Y K Ng, 9th Floor Empire Centre, 68 Mody Road, Tsim Sha Tsui East, Kowloon, Hong Kong

Mr. J.B. Noon, P.O. Box 525, Hamilton, Bermuda H.M. CX

Dr D P J O'Reilly, 20 Seagoe Industrial Estate, Portadown, Craigavon, N Ireland BT63 5QD

Mr K J O'Shaughnessy, 28 Edgemont Road, Northampton NN3 3PQ

Mr D A O Omoigui, 17 Idita Street, Surulere, Lagos, Nigeria

Ms C.M. Owen, Stokes House, 17-25 College Square East, Belfast BT9 6AT

Mr R F Oxley, Hornsey Town Hall, Crouch End Broadway, London N8 9JJ

Mr N E Pallister, Manley House, Kestrel Way, Sowton, Exeter, Devon EX2 7LQ

Mr C J Parker, Llandudno, Gwynedd LL30 1SA

Mr. T.C. Payne, 126 Pembroke Road, Seven Kings, Ilford, Essex

Mr N J Pearce, 17 Fox Brook, Wootton Bassett, Swindon, Wiltshire SN4 8QD

Ms E J Pearson, Dept of Geography, University of Reading, Whiteknights, PO Box 227, Reading RG6 2AB

Mr T J Perkins, 42 South Oswald Road, Edinburgh EH9 2HH

Mr D. M. Pettit, Admiralty Road, Great Yarmouth, Norfolk NR30 3NG

Ms J. M. Plant, 8 St Bride Street, London EC4A 4DA

Mr J A Polwin, The Civic Centre, Lampton Road, Hounslow, Middlesex TW3 4DN

Mr D. Popovich, 98 Templemere, Norwich NR3 4EG

Ms D M Porter, 5.13 Sudbury House, 15 Newgate Street, London EC1A 7AU

Mr M D Procter, Alston Farm, Alston Lane, Churston, Devon TQ5 0HT

Mr J.N. Rae, 112 Ingram Street, Glasgow G1 1ET

Mr J W Ramster, Pakefield Road, Lowestoft, Suffolk NR33 0HT

Mr. P.F. Raven, Capenhurst, Chester CN1 6ES

Prof D Rhind, 7-15 Gresse Street, London W1P 1PA

Mr C P Richardson, Oxford Polytechnic, Gipsy Lane, Headington, Oxford OX3 9EP

Miss J.E. Rigby, David Horn House, 42 Craigmillar Park, Edinburgh EH16 5PS

Mr R G Robbins, Automobile Association, Fanum House, Basing View, Basingstoke, Hampshire RG21 2EA

Mr A J Roberts, PO Box 3764, Beha, Qatar

Mr T.W. Robinson, Glenmoor, Cheddar Road, Axbridge, Somerset BS26 2BL

Mr G T Robinson, 15 Belsize Park, London NW3 4ES

Mr P G R Robinson, 4 Rosevale Avenue, Newtownards, Co Down, N Ireland BT23 3ND

Mr K W Robinson, 65 School Drive, Newton Longville, Milton Keynes MK17 0DD

Mr. G. Robson, Regional Buildings, Glenurquhart Road, Inverness IV3 5NX

Mr C Roper, Lordship Cottage, Swaffham Bulbeck, Cambridge CB5 0HP

Miss C J Sayer, 91 Hazel Bank Road, Chertsey, Surrey KT16 8PB

Mr. J. Scanlan, PO Box HM 1384, Hamilton HMFX, Bermuda

Mr B W Selby, Telford House, Tothill Street, London SW1H 9NB

Mr P Shand, High Winds, Cassington, Oxford OX8 1DL

Miss W Y W Shiu, 12 Sycamore Gardens, Mitcham

Mr MC Simmons, 22 Cotton Mill Spinney, Cubbington, Leamington Spa, Warwickshire CV32 7XH

Miss E Simmons, Council Offices, Welwyn Garden City, Hertfordshire AL8 6AE

Mr G F Simons, 7 High Holborn, London WC1V 6DS

Mr A N Sinclair, Coopers & Lybrand Deloitte, Plumtree Court, London EC4A 4HT

Mr. L.N. Smith, 9 Davin Crescent, Pill, Nr Bristol, Avon BS20 0LD

Mr M G Smith, Shinfield House, School Green, Shinfield, Reading, Berkshire RG2 9EW

Mr P Southorn, 21 Mowbray Road, Northallerton, N Yorkshire DL6 1QT

Mr. M.J. Stanbridge, Operations Dept, PO Box 1732, Riyadh 11441, Saudi Arabia

Mr M Start, PO Box 13, County Hall, North Humberside HU17 9AB

Mr J D Steel, Jordans, 7O Smallfield Road, Horley, Surrey RH6 9AT

Mr EJ Stephens, 100 The Causeway, Petersfield, Hants GU31 4JS

Mr. R. Stephenson, 22 Sunderland Road, Ealing, London W5 4JY

Mr D.S. Stewart, 23 London Road, Biggleswade, Bedfordshire SG18 8ER

Mr. J.D.S. Stocks, Erw Wen, Fron Park Avenue, Llanfairfechan, Gwynedd LL33 0AS

Mr I Storey, Trawsgoed Mansion, Aberystwyth, Dyfed, SY23 4HT

Mr F. Stuart-Brown, 41 Walton Park, Pannal, Harrogate HG3 1EJ

Ms A Tait, 93 Constance Crescent, Hayes, Kent BR2 7QG

Mr N.C.R. Taylor, Waltham End, London Road, Abridge, Essex RM4 1UX

Mr D R V Thomas, Coopers & Lybrand Deloitte, Plumtree Court, London EC4A 4HT

Mrs M Thomasson, 9 Bramble Walk, Alexandra Road, Lymington, Hampshire SO41 9LW

Gen. C N Thompson, Burgh House, Burgh by Sands, Carlisle CA5 6AN

Mrs J W Thompson, High Cross, Madingley Road, Cambridge CB3 0ET

Mr A C Thompson, 10 Balfleurs Street, Milngavie, Glasgow G62 8HW

Mr H A C Thomson, Tayside House, 28 Crichton Street, Dundee DD1 3RD

Mr S P Todd, Telford House, Tothill Street, London SW1H 2DX

Mr K E A Tolgu, The Oracle Centre, The Ring, Bracknell, Berkshire RG12 1BW

Mr I H Townend, Swindon, Wiltshire, SN4 0QD

Mr S W Trow, 29 John Street, Sunderland SR1 1JT

Mr P Turner, Suite 27A/34, Concourse House II, 432 Dewsbury Road, Leeds LS11 7DF

Mr J Turner, Romsey Road, Maybush, Southampton SO9 4DH

Ms L.E. Twigg, 6 Pains Road, Southsea, Portsmouth PO5 1HE

Mr J R Tyler, 49 Stonegate, York YO1 2AW

Mr D J Unwin, Department of Geography, University of Leicester, University Road, Leicester LE1 7RH

Mr R.N. Vango, 92-94 Church Road, Mitcham, Surrey CR4 3TD

Mr D R J Vaughan, 106 Gloucester Place, London W1H 3DB

Major A J M Vickers, Elmwood Avenue, Feltham, Middlesex TW13 7AE

Mr J S Voller, 33 The Paddock, Busby, Glasgow

Mr B F Wade, 7 Southern Court, South Street, Reading RG1 4QS

Dr P C Walker, St Andrews Hospital, North Side, Yarmouth Road, Norwich NR7 0SS

Dr. P.A. Waters, 2 Prospect House, Prospect Avenue, Kingsdown, Bristol BS2 8EA

Mr T C Waugh, Department of Geography, University of Edinburgh, Drummond Street, Edinburgh EX8 9XP

Dr C D Wallings, 12 Horseshoe Park, Pangborne, Berkshire RG8 7JW

Lt Col P R Wildman, Elmwood Avenue, Feltham, Middlesex TW13 7AF

Mr D A Wilkinson, Jamas Weir Building, Montrose Street, Glasgow G1 1XJ

Mr. J.H. Williams, Bangor, Gwynedd LL57 2UW

Mr. D.I. Wilson, Clumber Park, Worksop, Notts S80 3BE

Mr S M Wise, Claverton Down, Bath BA2 7AY

Mr J W Woodhams, Heathwall Pumping Station, 60 Nine Elms Lane, London SW8 5DA

Ms J L Wright, 1 Charleville Circus, Sydenham, London SE26 6NR

Mr. A.F. Wright, 2 Ridgacre Lane, Quinton, Birmingham B32 1ES

Index

Index to trade directories

Index to advertisers